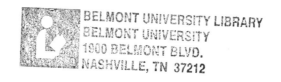

High Performance Loudspeakers

Fifth Edition

High Performance Loudspeakers

Fifth Edition

Martin Colloms
Colloms Electroacoustics, UK

JOHN WILEY & SONS

Chichester · New York · Weinheim · Brisbane · Toronto · Singapore

Copyright © 1997 by   Martin Colloms,
                       Baffins Lane, Chichester,
                       West Sussex PO19 1UD, England

                       National   01243 779777
                       International   (+44) 1243 779777

e-mail (for orders and customer service enquiries): cs-books@wiley.co.uk

Visit our Home Page on http://www.wiley.co.uk
                            or
                    http://www.wiley.com

Other Wiley Editorial Offices

John Wiley & Sons, Inc., 605 Third Avenue,
New York, NY 10158-0012, USA

VCH Verlagsgesellschaft mbH,
Pappelallee 3, D-69469 Weinheim, Germany

Jacaranda Wiley Ltd, 33 Park Road, Milton,
Queensland 4064, Australia

John Wiley & Sons (Canada) Ltd, 22 Worcester Road,
Rexdale, Ontario M9W 1L1, Canada

John Wiley & Sons (Asia) Pte Ltd, 2 Clementi Loop #02-01,
Jin Xing Distripark, Singapore 0512

British Library Cataloguing in Publication Data

A catalogue record for this book is available from the British Library

ISBN   0471 97091 3 PPC
ISBN   0471 97089 1 Pr

Typeset in 10/12pt Times by Dobbie Typesetting Limited, Tavistock, Devon
Printed and bound in Great Britain by Bookcraft (Bath) Ltd
This book is printed on acid-free paper responsibly manufactured from sustainable forestation,
for which at least two trees are planted for each one used for paper production.

To

Marianne and Catherine

Contents

Preface to the First Edition

A high quality loudspeaker is required to reproduce sound with sufficient fidelity to satisfy a critical audience when fed from an accurate electrical signal. It is immaterial whether the listeners are numbered in thousands or comprise only a few individuals: loudspeaker systems can be designed to cater for both situations without compromising the basic standard of performance.

There are thus numerous applications for high quality loudspeakers. For example, broadcast and recording engineers rely heavily on monitor loudspeakers in order to critically analyse the quality of the programme they are producing. Other applications range in scope from the rock festival to the concert and opera hall, and in size from a theatre auditorium to an ordinary living room. Reinforcement loudspeakers are commonly used for sound amplification in live performances today, and while specialized systems are employed for instruments such as an electric guitar, other wider range sounds such as voice and woodwind require high performance speakers with a capability to allow the reproduced level to match that of the accompanying brass or a modern drum kit. Theatres and opera houses often use systems for off-stage sound effects, and most of today's star performers would be unable to reach a large audience without the aid of a microphone and sound reinforcement. Special techniques are, however, required to attain the acoustic outputs necessary to satisfy a large stadium audience, and high efficiency, stacked, horn loaded, directional arrays are commonly employed for this purpose.

The author's aim is to provide an up-to-date analysis and review of high performance loudspeaker techniques. Although it is not intended to be an exhaustive work, reference has been made in the text to original research material including the most important modern work in the field. Precedence is accorded to the moving coil drive unit, as this is by far the most widely used, although some coverage is also given to other viable if less common devices. In addition to the fundamentals — relevant acoustic theory, transducer design, enclosures, acoustic loading, etc. — space is also accorded to developments in electronic crossover design and active speaker systems, as well as to the latest measurement techniques and such controversial questions as linear phase. By using the references supplied, the book can be used as the basis for further research, and as such, not only high fidelity enthusiasts should find it of interest, but also students studying such subjects as electronics, electroacoustics, broadcasting and recording. Even the design engineer and technical author may find it a useful appraisal of current techniques and a convenient source of subject references.

Martin Colloms

Preface to the Fifth Edition

For the fifth edition, my title *High Performance Loudspeakers* has joined technical publisher John Wiley. My initial concern about the transfer was replaced by increasing confidence. The Wiley UK team backed my proposals to substantially expand the text as well as bring the format and layout up to date. Finally, through economies of scale it was planned to significantly reduce the cover price, making the work accessible to a far wider readership.

Many revisions have made the book as up to date as possible, while continuing with that vital critical viewpoint when covering new developments and technologies. Every existing chapter has seen revision and expansion.

Building on the previous editions, the first chapter has been expanded adding an overview of modern design trends and practice.

Almost as this edition was released to the typesetters a new loudspeaker development was announced in London under the NXT brand, patents applied for by New Transducers Ltd. Covering non pistonic, vibrating acoustic panels, there is significant theory to match the wide variety of applications. Press attendance at the launch broke all records with the consensus view that this was an important development in the evolution of the loudspeaker. Accordingly a major section has been included on this technology.

A new chapter appears covering 'Home Theatre Systems' taking account of their special acoustic requirements, Dolby PRO-LOGIC, THX and the more recent AC-3, DTS and MPEG digital discrete, multi channel systems.

The review of computer aided design has been extended, covering both hardware and software systems and including the new generation of low cost audio instrumentation.

Complementing the necessarily academic nature of the theoretical aspects of speaker engineering, there is also a new section which gives much practical advice for real world speaker system design. It has been dubbed 'Hot Tips'.

In 'Systems and Crossovers' new topics include $2\frac{1}{2}$ way system design; external crossovers; D'Appolito types; a distortion analysis of inductors; digital active loudspeakers and low order system design.

There has been a major expansion of the section on sub-woofers, also with relevance to Home Theatre where subs are almost mandatory. Subjective aspects of bass response are explored together with newly expanded sets of boundary matched low frequency alignments.

Speaker placement techniques, multiple driver and port combinations plus adjustable low frequency design are also covered. In 'drivers', there are extensions to include both the metal cone driver and its resonance control.

Design considerations for better dynamic performance are explored, both for overall build and for enclosure construction.

On measurement issues there are more data on absolute phase and the effect of phase on energy decay waterfall displays.

Aspects of running in, quality control and ageing are all considered, together with the effects of tolerances on system performance.

Many new diagrams and illustrations have been included involving an overall 25% expansion for this new edition.

The front cover features the Nautilus speaker, designed by Laurence Dickie and has been reproduced with kind permission of B&W Loudspeaers Ltd.

Many thanks to all of those who have continued to provide constructive criticism and support for *High Performance Loudspeakers*.

Martin Colloms

List of Notation

A_g	Magnet gap area
A_m	Magnet area
a	Piston radius
a_g	Magnet gap radius
B, B_g	Magnetic flux density in the coil and air gap (analogous to current)
c, c_1	Velocity of sound (345 m/s in air), c_1 longitudinal
C_{AB}	Acoustic compliance of enclosure volume
C_{AS}	Acoustic compliance of suspension
C_{AT}	Total acoustic compliance of driver and enclosure
C_{MEC}	Electrical capacitance equivalent of moving mass ($= M_{AC}S_D^2/B^2l^2$)
C_{MS}	Total suspension compliance of driver
E_g	Generator (amplifier) voltage
F	Force
f	Frequency in Hz
f_B	Helmholtz resonance of vented box
f_c	System resonance, driver in closed box
f_o or f_s	Free air resonance of driver
f_3	-3 dB cutoff frequency
g	Gravitational acceleration (981 m/s^2)
$G(S)$	Response function
H	Coercive force, magnetic flux (analogous to voltage)
h	System tuning ratio ($= f_B/f_s$)
I	Current in coil
k	Wave number $= 2\pi/\lambda$
K_B, k_H	Magnetic loss factors
k_n	Reference band efficiency factor
k_p	Power rating constant
l	Length of motor coil wire immersed in magnetic gap field
L_c	Inductance of motor coil
L_{CEB}	Electrical inductance equivalent of box volume compliance
L_{CES}	Electrical inductance equivalent of driver suspension compliance
L_{CET}	Electrical inductance equivalent of system compliance ($= C_{AT}B^2l^2/S_D^2$)
M_c	Mass of motor coil

M_{EC} or (M_{MS})	Acoustic mass of driver diaphragm assembly including air load
M_D	Total driver moving mass (excluding air load)
M_T	Total driver moving mass, $M_D + M_A$
M_{AC}	Acoustic mass of driver and adjacent air load (including baffle effect)
M_{AS}	Acoustic mass of driver and air load
n	Number of turns
P	Sound pressure
P_{ref}	Reference sound level (minimum audible at $1\,kHz = 2 \times 10^{-5}\,N\,m^2$ or $10^{-12}\,W\,m^2$)
Q	Ratio of reactance to resistance (series circuit) or resistance to reactance (parallel circuit) (or alternatively the directivity factor)
Q_B	Driver Q at f_B
Q_T	Driver total Q
Q_E	Electrical Q of driver
Q_M	Mechanical Q of driver
Q_{TC}	Working Q of system of driver and enclosure
Q_{TS}	Driver Q at f_s
R	Resistance (electrical)
R_{EC}	$= B^2 l^2/(R_{AB} + R_{AS})S_D^2$ electrical equivalent of mechanical losses
R_{AB}	Enclosure loss acoustic resistance
R_C	Resistance of motor coil
R_{MA}	Resistive component of air load radiation impedance
R_{ME}	Driver electromagnetic damping factor $(B^2 l^2/R_E)$
R_{MS}	Suspension mechanical resistance
R_t	Reverberation time
r	Distance from source
S_D	Effective projected diaphragm area
s.p.l.	Sound pressure level
U_C	Diaphragm or coil axial velocity
V_{AS}	Equivalent closed air volume of driver compliance, C_{MS}
V_B	Box or enclosure volume
W	Sound power
X_{MA}	Reactive mass component of air load radiation impedance
Z_{MA}	Air load radiation impedance; $R_{MA} + jX_{MA}$
α	Compliance ratio ($= C_{AS}/C_{AB}$ or V_{AS}/V_B)
η_0	System reference efficiency (power in/power out for 2π field) in level range
η_{ref}	Reference power available efficiency
θ	Angle from source axis
μ	Magnetic permeability (analogous to conductivity)
ρ_0	Density of air ($1.2\,kg\,m^{-3}$)
ω	Angular frequency $= 2\pi f$
λ	Wavelength, m

1

General Review

It is now 70 years since the loudspeaker as we know it was first developed, an electrodynamic transducer of respectable loudness, of satisfactory and uniform amplitude versus frequency, reliable in use and with the potential for economic manufacture. That device is the familiar moving coil loudspeaker, whose principle is so effective that its key elements have remained essentially unchanged to this day.

To build one, take an affordable magnet and add a simple arrangement of magnetically permeable 'soft' iron to help concentrate much of the available magnetic flux into the narrow radial gap formed on a cylindrical pole. A small light coil or solenoid is wound onto thin card or similar low mass former, and suspended freely in the magnetic gap, allowing axial motion of half a centimetre or so. Following Maxwell's electromagnetic equations, an axial force is generated on the coil when a current flows through it. This force is the product of B, the magnetic field strength, l the length of the wire immersed in that flux field and I, the current flowing. The force relationship is fundamentally linear and, ignoring minor effects at high amplitudes of motion, where geometry of the coil and flux field may affect performance, there is no perceptible distortion. In fact there is no lower resolution limit for a moving coil transducer. An infinitely small electrical input will produce an equivalent and infinitely small sound output. Another excellent feature of the moving coil transducer, generally taken for granted, is that despite its operation as a moving mechanical device, it is essentially noiseless. It does not grate, or scrape or whirr.

Apply a sub audible 5 Hz sine wave current and you can see the coil move, but silently. It is these fundamental strengths that make the moving coil principle so effective, and so justly popular. Over 99% of all loudspeakers ever made are moving coil. The principle may be used over a very wide range, from low power speech reproducers of just 2.5 octave bandwidth and a modest 75 dB of sound pressure output, built on a frame just 40 mm in diameter, up to low frequency monsters of 60 cm diameter, capable of generating 20 Hz sound waves at body shattering 110 dB pressure levels. Used alone, the moving coil itself generates almost zero sound output as the radiated sound level is proportional to the area of air load driven by the transducer element, and for the coil alone that is merely a thin ring element.

To couple the moving element more effectively to the air load a rigid, light diaphragm is attached to the coil. Typically, larger diaphragms have their own flexible surround suspension, coupled to a skeletal support frame or chassis, thereby aiding centration of the moving system.

Paper is a suitably strong and lightweight material for a diaphragm. As a flat sheet it is stiff in tension but very weak in bending. However, curl it up to form a cone and this structure exhibits an extraordinary axial stiffness for its mass, a marvellous means of coupling a large area of air load to the moving coil. The latter is bonded to the cone apex.

Acting as an impedance transformer the cone matches the lower acoustical impedance of the air load to the driving force impedance of the coil, maximizing the energy transfer in the path from electrical input to mechanical force, leading to radiated sound pressure.

In specialized smaller drivers optimized for high frequencies, the cone may be replaced by a light dome formed of paper, moulded sheet plastic, resin doped fabric or metal foil. In sizes down to 19 mm effective radiating diameter, the frequency response may thus be extended to beyond audibility, up to 40 kHz. By apportioning the audible frequency range, appropriate combinations of moving coil driver sizes may cover a frequency range of 10 Hz to 40 kHz; a ratio of no less than 4000 to 1 in acoustic wavelength, 34 m to just 8.5 mm.

Loudspeaker systems with such a wide range are actually made today for costly high fidelity installations, the near 12 octave span achieved with typically four frequency-dedicated moving coil drivers. Such systems can cost as much as a luxury car, yet the humblest moving coil speaker element for speech use only may cost tens of cents in reasonable trade order quantities.

When the diaphragm of a moving coil driver is appropriately horn loaded, the horn further improves the matching efficiency between air load and transducer. It is then possible to stretch the conversion efficiency to almost 50% compared with the typical 1% efficiency of a high fidelity speaker. With horn designs a fairly easily obtained 40 electrical watts will result in a seriously loud 20 acoustic watts, sufficient to effectively address large audiences at realistic volume levels.

The moving coil driver has proved to be remarkably durable with many examples operating for 50 years and longer. Like the wheel, alternatives have been proposed, but it still reigns supreme.

It seems that new inventions appear almost monthly in the loudspeaker field, many claimed to supplant the moving coil. However, no serious rival has as yet emerged to challenge it, and it remains pre-eminent in terms of efficiency, economy, wide performance range and application.

While this review concentrates on the moving coil applied to loudspeakers, the principle is also widely used in precision actuators such as the high-speed focus and fine tracking mechanisms for laser optical heads, compact disc and optical data discs. It is also used for the most popular form of microphone and, not least, for almost all headphones and earpieces, as well as for many related communication systems.

The world-wide acceptance and growth of the high fidelity market and the high standards achieved in recording and broadcast studios has given great impetus to high performance loudspeaker design in the last 30 years. The loudspeaker, however, has been the most argued-over component in the entire high fidelity chain, with every aspect of its design execution subject to lengthy and involved discussion. Although acoustic engineers like to deal in facts, much to their dismay, fashion plays a considerable part and loudspeakers are no exception to this rule. Occasionally an 'unbalanced' system will find public favour, such a model often being claimed to have a 'new sound', derived from a different bass loading principle, or a new transducer and dispersion method. Unfortunately, other important aspects of its performance are often neglected by the designer in his efforts to incorporate this 'special' effect.

In the professional field, users are also inevitably conditioned by past experience and are often suspicious of any change, even for the better. Only those whose judgement is free of prejudice and who have frequent contact with live programme sources can reliably discuss reproduced sound quality.

Recently there has been an encouraging development, in that a degree of rationalization of performance standards has occurred on both the domestic and professional fronts. Designers are beginning to agree on a common standard of performance based on factors such as a natural frequency balance, uniformity of response and low distortion and colouration. This common ground has developed in spite of dissimilarities of design approach and philosophy, and it implies that sufficient objective and subjective data concerning speaker performance is at last becoming freely available.

Such a situation presents a dramatic reversal of the state of affairs prevailing some 35 years ago. A marked divergence of opinion then existed over subjective sound quality — indeed, this was so extreme that the products of the major manufacturers could be identified by a specific 'in-house' sound, which pervaded all their designs. A typical domestic 'hi fi' speaker system then comprised a 250 or 300 mm chassis diameter bass unit, with a light paper cone incorporating a 33 or 50 mm voice coil on a paper former. A separate paper-cone tweeter covered the treble range and was often concentrically mounted on the bass unit frame. The drivers were rear mounted on the inside face of the front panel, the enclosure was likely to have a typical volume of between 50 and 100 litres, and probably employ reflex loading.

It is interesting to examine the 'ideal performance' the contemporary speaker designer then aimed at achieving, even though the typical speaker, outlined above, in fact fell far short of this standard (Table 1.1).

The ideal specification was limited by the level of achievement currently attained by designers (Table 1.2 and Figure 1.1). Relative to the typical commercial system of the time, the ideal efficiency is placed at 100 dB for 1 W input at 1 m, which is 6 dB more sensitive than the typical specification. Presumably this difference reflects the relatively low power output of contemporary amplifiers as 10–20 W models were commonplace. Only a mild improvement in response flatness or bandwidth was then thought possible;

Table 1.1. Idealized loudspeaker system specification, circa 1965

Efficiency	100 dB at 1 m for 1 W at 100 Hz
Frequency response	100–10 000 Hz, ±4 dB; 35 Hz at −10 dB; 15 kHz at −10 dB
Polar response	100–10 000 Hz, less than 6 dB down at a 60° arc limit
Distortion	Less than 11% at 35 Hz, level unspecified
	Less than 2% at 100 Hz, level unspecified
	Less than 2% above 100 Hz
Cabinet volume	50 litres

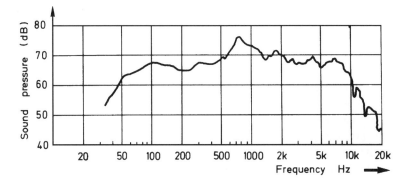

Figure 1.1.   Typical response curve of two-way domestic system in Table 1.2

Table 1.2.   Typical specification of domestic two-way system, circa 1965

Efficiency	93 dB at 1 m for 1 W at 100 Hz
Frequency response	100 Hz to 10 000 Hz, ±6 dB
35 Hz limit	At −17 dB
15 kHz limit	At −12 dB
Polar response	Less than 6 dB down over a 60° arc, 100 Hz–5 kHz
Distortion	Above 200 Hz not quoted; at 100 Hz 4%, at 35 Hz, 10%
Power rating	25 W programme
Cabinet volume	75 litres

the typical speaker gave a 35 Hz point at 17 dB down, and a 15 kHz point 12 dB down, which contrasts with the −10 dB limit of the ideal system.

While it is important to view the specification in Table 1.2 in its proper context, i.e. as an example of typical contemporary commercial practice, it is surprising to discover that the basic technology and theory essential to good loudspeaker design was well known to advanced specialists in the field. Furthermore such work was well documented in many papers, periodicals and books; for example, although designers were aware of colouration effects they appear to have done little about them, despite the fine research that had been conducted almost 20 years earlier by Shorter at the BBC, concerning delayed resonances. Much of the currently accepted loudspeaker technology and principles were rarely applied, and the overall approach to design was a rather haphazard exercise. However, some companies were researching highly advanced designs and a few were even in production, albeit in limited quantities. In 1967, K.E.F. Electronics (UK) released a costly experimental system incorporating a highly developed mid-band transducer. Covering a 250 Hz to 4 kHz range, this latter driver employed a 65 mm hemispherical dome formed in a rigid polystyrene/neoprene polymer, and was fitted with a double suspension and loaded by an 0.8 m pipe filled with long-fibre wool, for absorption. The use of an aluminium voice coil former resulted in a high power handling capacity (Figure 1.2).

At this time very few mid-range domes were available, the other well established example being that employed in the classic American design, the Acoustic Research

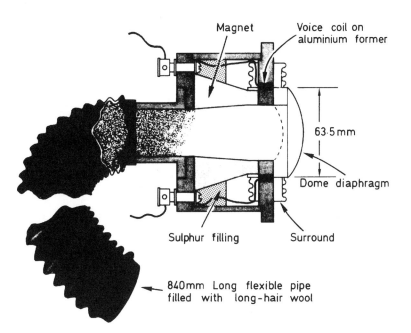

Magnet   Voice coil on
          aluminium former

63·5 mm

Dome diaphragm

Sulphur filling   Surround

840 mm Long flexible pipe
filled with long-hair wool

Figure 1.2. KEF's mid-range loudspeaker with absorbent load

AR3. One loudspeaker system which survived the passage of many years is the Quad full-range electrostatic loudspeaker. Accepting that moderate power handling and low practical efficiency are its specific limitations, its performance continues to bear favourable comparison with many current designs.

The conservative atmosphere then pervading the consumer market in the mid-sixties may be judged from the following example. At that time the best systems were relatively large (50–100 litres) and when a new high performance model of compact dimensions became available, it was viewed with considerable suspicion. The Spendor BC1 (40 litres) sounded quite different to the systems currently available, and in fact was far nearer the live source in character than its contemporaries. It represented a skilled balance of the important factors responsible for realistic sound quality, and yet it took almost a decade for this system to become widely accepted.

By the mid-1960s, the BBC's work on a new generation of monitoring loudspeakers incorporating bextrene cones was well advanced. It proved to be of great significance as it was clear that a major improvement in loudspeaker quality had been achieved. The high standards set by these designs acted as a stimulus to the industry, and through attempts to attain this standard at a commercial level, many new developments and designs have appeared, some strongly related to the BBC originals.

The performance of today's typical high quality domestic systems would have been unbelievable in 1965, for they exceed the majority of requirements of the 1965 ideal specification by a handsome margin (Figure 1.3, Table 1.3). This particular example is a bass reflex design, employing a plastic coned 160 mm diameter bass mid-range unit in conjunction with a 25 mm diameter dome tweeter. System sensitivity/efficiency is lower than the mid-1960s target by some 12 dB.

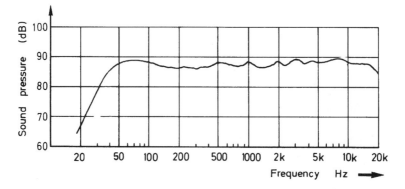

Figure 1.3. Typical response curve of good quality two-way domestic system

Table 1.3. Typical specification of domestic two-way loudspeaker system, circa 1984

Sensitivity	88 dB W at 1 m, 1 W input
Frequency response	50 Hz–18 kHz, ±3 dB; 40 Hz, −6 dB
Polar response	Within ±3 dB of axial curve over ±30° lateral arc, 50 Hz–15 kHz
Distortion (90 dB)	Less than 0.5% at 100 Hz–10 kHz; less than 3% at 35 Hz
Power rating	100 W peak programme
Volume (internal)	25 litres (25×10^{-3} m^3)
Drivers	160 mm diameter bass-midrange
	25 mm diameter dome tweeter
Crossover frequency	3 kHz

This is the inevitable outcome of the modern system's wide bandwidth in a compact enclosure, and its attainment of vastly lower subjective colouration. The narrower amplitude response limits is also important; simultaneously they contain a far greater response range, serving to illustrate the considerable improvement in uniformity and consistency of response. The standard achieved for distortion and polar response are both vastly improved, and the power rating of 100 W programme (see Table 1.4) is 6 dB higher than the typical equivalent for 1965. This is necessitated by the reduced efficiency of the system as well as the considerably higher power ratings of modern amplifiers. In the light of this current level of attainment and contemporary technology, Table 1.4 gives target specifications for a range of high quality loudspeaker systems.

The recent improvement in quality is not confined only to the high performance end of the market: in fact all loudspeaker systems have advanced similarly if not equally over the same period. For example, many of the causes of colouration in both cabinets and drive units have been identified and can now be adequately suppressed. Further key factors concern a refined understanding of diaphragm behaviour, with the successful application of synthetic materials to drive-unit manufacture. A sufficient variety of well designed drivers are now available, which cover specific sections of the audio spectrum over a range of different power levels and allow the designer considerable latitude when determining the size and cost factors for a given system. A key unit in the success of a number of the original BBC and subsequent design derivatives is the Celestion high

Table 1.4. Proposed loudspeaker specification

Axial pressure response	60 Hz–15 kHz, ± 2 dB (sine)
	100 Hz–10 kHz, ± 1 dB (octave averaged)
	Response below 60 Hz tailored to boundary conditions
Off-axis response	$\pm 10°$ vertical, within 2 dB of axial output
	$\pm 30°$ lateral, within 4 dB of axial output
Harmonic distortion (90 dB)	100 Hz–20 kHz, $<0.3\%$
	Below 100 Hz, $<2\%$
Harmonic distortion (96 dB)	100 Hz–20 kHz, $<0.5\%$
	Below 100 Hz, $<6\%$
Sensitivity (2.83 V)	Greater than 88 dB/W at 1 m
Power rating	$\geqslant 100$ W peak programme
Impedance	8 Ω nominal, $6 < Z < 20\,\Omega$, phase angle $<30°$,
	100 Hz–20 kHz
Maximum sound pressure	Domestic >105 dB, unweighted at 1 m
(application)	Monitoring >115 dB, unweighted at 1 m
	Stage amplification 120–130 dB, unweighted at 1 m
Size (internal volume)	Domestic 25–50 l
	Monitoring 50–150 l
	Stage 100–200 l or horn loaded

frequency unit, the HF1300 series. Designed around 1957, it is suitable for medium power applications and was first used by GEC with their aluminium cone drivers. Some 35 years later it was still a popular unit and was employed in several high performance systems. It was primarily based on a pressure unit for a high frequency horn, and was later modified for use as a direct radiator, notably by the addition of the phase correcting front plate. It was pistonic over the working range.

However, there still remains a major problem for drive unit manufacturers, namely suitable cone materials. Bextrene had proved highly successful for the manufacture of vacuum formed cones and had gained wide acceptance among the major UK drive unit/speaker manufacturers, although it was almost a chance discovery, as the material was originally designed for use in the production of low-cost moulded packaging. However, in the more critical loudspeaker application, experienced drive-unit manufacturers had discovered that Bextrene's acoustical and mechanical properties can show variations from batch to batch. The chemical industry is not particularly interested in solving these problems as the requirements of the loudspeaker industry are small compared with total sales. It is thus essential to carefully quantify the mechanical properties of the material to be used and to continue to do so for each batch ordered. Despite these technical difficulties, with careful design and manufacture, plastic cones can be superior to pulp/paper composition types in colouration, uniformity of response and sample-to-sample consistency. In recent years varieties of polypropylene have largely displaced Bextrene and, like its predecessor, characteristic sounds remain which are very difficult to avoid entirely. This and the pursuit of higher sensitivity has led to experimentation with many composites and in some cases a return to treated pulp and aluminium diaphragms.

Concerning frequency response, the fundamental analysis of loudspeakers at low frequencies conducted by Thiele, and first presented in 1961, has been recognized for its

Figure 1.4. Exploded view of a modern multi-way loudspeaker system. Two 200 mm framed LF units drive the bandpass enclosure. Narrow fronted, the system has good uniformity in the lateral plane. Heavy damping is used on the panels of the front module while the rest of the enclosure is extensively braced (courtesy KEF Electronics: R104 II)

true worth in later years, and the subsequent research on the subject by Small has also proved of great value to designers. Papers by these and other authors provide a remarkably complete theoretical analysis of the one area of loudspeaker design where the results are highly predictable. (A summary of this work is given in Chapter 4.) Armed with such theory there is no reason why any loudspeaker designer worthy of note should fail to produce a loudspeaker with a less than optimal low frequency characteristic (see Figure 1.4).

Refined electronic crossover techniques are responsible for further improvements to the modern generation of active loudspeakers. Although the idea is not new, the early active filters were clumsy to execute with valve amplification and found little favour. In recent years the development of active-filter theory and inexpensive operational amplifier circuit modules, together with the low cost of transistor power amplifier units, has given renewed impetus and several active designs have been produced, including some for domestic use. If their performance advantage is seen in full measure, then one would expect active 'electronic' speakers, especially those with a 'digital' content, to assume increasing importance in high quality applications.

1.1 DEVELOPMENTS IN SYSTEM DESIGN

Given the extreme difficulty in attempting to cover the whole audible range with a single drive unit, a high quality speaker will of necessity comprise a system composed of an enclosure, several optimized drivers (frequency range specified), and a crossover network — a passive (non-powered) set of filters which direct the correct input frequency range to the appropriate drive units.

System design is the process of creating a speaker that meets the target specification, both technically and subjectively. The enclosure must provide the right non-resonant support, internal and external acoustic loadings as well as the required style, appearance and finish.

Drivers must be chosen or custom designed to meet this system specification, not just for fundamental aspects such as sensitivity, power capacity and bandwidth, but also with regard to the unique cone characteristics and how the resulting natural acoustic 'signature' is weighted and valued in the final sound. In addition, driver size has a significant influence on frequency range and output power and also controls the directive properties according to frequency. Ideally, there should not be too great a difference in effective acoustic size at the crossover point between adjacent drivers, otherwise a step may occur in the off-axis frequency response and in the related power response through the crossover transition.

Some of the variables involved in system design are extraordinarily subtle and prove a source of continuing frustration for inexperienced designers. For example, long known but often overlooked is the surprising sensitivity of overall sound quality to small changes in high frequency level relative to the mid band. The upper crossover between mid and treble is usually placed in the 2.5–3.5 kHz region.

The correct, critically natural timbre for the human voice, violin, acoustic guitar and the like can only be achieved when the high frequency energy is within ± 0.5 dB of the ideal target. This is rather smaller than the tolerances available for both measurement and for driver production.

If the treble range is set too 'dull', then the speaker system can sound too warm, veiled and muffled, lacking a sense of both air and atmosphere. Set too bright, and the result may be a sharper sound, perhaps attractive on percussive sounds but adding a 'nasal' effect to voices as well as imparting a hardened colouration and a sibilant emphasis. Vocals may also sound too close. Violin acquires a steely harshness and may dominate the instrumental grouping. In stereo reproduction the sense of depth in the image illusion is generally impaired when the treble is set too bright.

Unfortunately, a touch of treble brightness helps counter a lack of definition and clarity in the mid range and many designers resort to this damaging short cut to superficially better performance.

Over the past decade we have seen a shift away from a relatively inflexible textbook approach to system design, one where frequency ranges are neatly compartmentalized, to a point where designers now have a far greater awareness of the broad interaction which may be judged via the subjective interface.

Assessing sound quality is an important discipline, and despite a critical awareness of the many technical factors affecting the sound, it is virtually impossible to separate or filter them out sufficiently to give precise definitions for the many associations between the objective and subjective factors.

For example, two views may be obtained for the overall frequency balance of a given speaker, and in this context the word 'balance' is of particular relevance. One critic describes it as 'bass light', while the next as 'treble bright'. The measuring microphone has no trouble in making the correct identification but it cannot take into account human perception which seeks a balance. In this latter context both critics are in fact correct. On one programme excess treble may well be heard as a lack of bass weight or balance while on another the treble error may be recognized directly.

For the designer there are some interesting options. He or she could readjust the treble but might find the resulting sound less satisfactory, perhaps owing to deficiencies in the mid range. Alternatively they could look to improve the low frequency performance and thus help to redress the overall balance.

In some designs this is surprisingly easy and may involve no more than a percentage reduction in the amount of acoustically absorbent stuffing in the enclosure, and/or a reduction in the length of the reflex port duct, if such an acoustic low frequency equalizer is fitted.

Increasing awareness of the global scope of design parameters allows today's designers to take a less dogmatic view of system design and to exploit more subtle blending and balancing methods for driver output. Aware of the need for subjectively accurate timbral balancing in the face of insufficiently accurate measured frequency responses, designers continue to use measurement as a development tool, but nonetheless rely on critical listening to help blend the response curves to their intention.

To this end, drive units are now designed to operate more smoothly over wider bandwidths. Designers are taking advantage of this and are reducing the complexity of their crossover networks. A decade ago manufacturers proudly boasted of the high complexity of their passive networks, highly toleranced and fully compensated for input impedance as well as for driver acoustic variation. It is now felt that such complexity runs counter to perceived naturalness, and to the ability of a good recording to communicate the composer's musical message to the listener via the loudspeaker.

Thus, those 40 and 50 element crossover filters are gradually being supplanted by much simpler arrangements offering a more direct link between amplifier and drive unit. In one exceptional example, a high quality three-way speaker system, aided by naturally well-tailored intrinsic driver responses, was completed with only three elements in the crossover network. Ten years ago the design would have used typically 10–12 elements without, and 30 elements with, impedance compensation.

1.2 PHYSICAL AND ACOUSTIC PERFORMANCE

With regard to physical and acoustic properties, loudspeaker designers are still busy trying to improve the uniformity of frequency response, not just at one, more or less arbitrary axial point, but over a range of angles and distances. Their aim is to generate a neutral energy balance over a forward directed solid angle encompassing the listening area. Target beam shapes are ±10 or ±15 degrees in the vertical plane, and ±25 or ±35 degrees in the horizontal plane. There is a continuing requirement to reduce enclosure size to improve acceptability in the domestic environment, especially in view of the multi-speaker surround sound applications and home cinema developments. This reduction is in conflict with the quest for greater low frequency extension and uniformity, one of the major factors that distinguishes real hi fi from mid fi. There is also a trend towards genuinely higher efficiency leading to higher maximum sound levels, also in conflict with smaller enclosure sizes. The market expects speaker systems to operate at ever increasing loudness without commensurate increase in input power and, preferably, without adverse consequences, such as a compromised electrical loading on the amplifier.

1.3   SENSITIVITY AND IMPEDANCE

One of the anomalies in specification concerns sensitivity. Objectively, sensitivity is accepted as a measured sound pressure level at a 1 m distance where the input voltage is 2.83 V rms, corresponding to 1 W into a standard 8 Ω resistance. By implication there is an association with efficiency in its pure sense. However, very few loudspeakers have a uniform 8 Ω loading over their working frequency range. Even the standard allows for a range between 6.4 and 10 Ω; such a variation is, by implication, associated with reactive regions where the load impedance passes through resistive, inductive and capacitative values. In practice, loudspeaker systems exhibit impedance peaks well beyond 10 Ω, often to 50 Ω, but these are considered to be harmless since they do not prejudice the nominal sensitivity value. Frequency regions where the impedance falls significantly below the 8 Ω mean are prejudicial. First, designers may deliberately choose to ignore the standard and work to a lower impedance, thereby stealing greater current from the source amplifier (in practice, this is a voltage source of negligible output impedance and thus is capable, within limits, of providing greater current on demand). Greater current provides higher sound levels and thus a superior voltage sensitivity.

Unfortunately there is a penalty to pay here. Higher currents lead ultimately to non-linearity in the magnetic components of the driver. It is possible to generalize loudspeaker distortion as strongly dependent on input current and not on the more obvious parameter, sound level. In addition, cables and amplifiers are subjected to higher stress; indeed, the complex nature of the electrical input impedance of loudspeakers may evoke premature current limiting or protection in the driving amplifier.

Thus good voltage sensitivities, uncompromised by significant regions of low impedance, are to be encouraged.

1.4   ENCLOSURES

Advances in enclosure design have been numerous. Undeniably the trend is towards a heavier, more rigid construction with a double purpose: (a) to control and minimize spurious resonances in the enclosure panels and structure, and (b) to provide an inertial platform against which the moving coil drivers may reference themselves. If their foundation — the termination for their chassis/structure — is not solid or of sufficient mass, then the motion of the moving system will carry reaction errors. It is surprising how subtle those errors can be and still remain audible. For example, the tightness of the fixing bolts attaching a driver frame to an enclosure panel is a significant factor, affecting clarity, colouration and the subjective naturalness on dynamics, i.e. the loudness contrasts heard with live sound. The difference between 'just tight' and correctly torqued may only be a quarter of a rotation in a wood or wood composite panel, yet the resulting change is often audible and significant.

MDF board has generally eclipsed older plywood and chipboard panels for enclosure construction, while complex internal bracing, effective in several planes, is common-place. Bracing is intended to subdivide the panels into smaller unequal areas, thus helping to disperse the natural acoustic resonant signature of the panel. The critical importance of this aspect can only be appreciated with the understanding that even in the case of costly speaker designs, much of the false tonal 'colour' in the sound of a speaker system is still a result of the enclosure and not the drive units or the crossover.

Treatments may also be applied to enclosure walls such as layering with tough phenolic laminates or with steel plates. Fibrous bitumen loaded pads offer a high mechanical resistance, effective in damping higher frequency modes.

More recently, catalytic polymer resins have become available, with useful properties for making loudspeaker enclosures. Heavily mineral loaded, the polymer mix endows the easily cast material with stiffness, mass and resonance damping. The results are encouraging and good examples show a welcome absence of woody panel sounds, hitherto a generally accepted component of loudspeaker performance.

Enclosures are now keyed to the floor on which they sit via hardened steel spikes, with sufficiently narrow points to pierce the usual carpet (though not your best Persian!) and thus engage the floorboards beneath. Surprising improvements in overall system definition and stereo focus result from the improved stability of the enclosure location. For tiled floors, thin felt pads are optimal; any untoward elastic coupling results in audible and measurable secondary resonances between enclosure and floor. For highly critical use, some installations at ground level have apertures cut in the floor, with the speakers mounted on brick piers supplied with their own foundations.

The external appearance of enclosures is also changing. Reduced diffraction is important; edges are bevelled or rounded, and overall shape may include tapered surfaces to enhance the smooth wavefront of the acoustic output and reduce secondary stray or parasitic sources such as re-radiation from sharp edges or corners. These can impair stereo image focus and add audible roughness to the treble range.

A slant or angle to the front panel may help to compensate for the differential time delay between the multiple drive units at the listener position, thus improving phase response and acoustic integration through the crossover regions.

1.5   DRIVE UNITS

The widespread use of metal diaphragms is the most obvious development in high fidelity loudspeakers. While paper or pulp cones are still popular and widely used at all quality levels, the goal of a resonance-free, pure piston performance for the cone still fascinates designers.

Distortions arise when a typical polymer or pulp cone material flexes in its naturally resonant regions. This is because these materials do not have the properties of linear springs. In these materials, often chosen for their good internal damping, deflection is not directly proportional to force. Several higher order terms are required for the force equation, these quantifying the non-linearity. Non-linearity implies the generation of false sounds together with some shift in energy from the fundamental to the harmonic. Changes in perceived timbre may result, together with a masking effect for other lower level sounds in the harmonic masking range. Thus, distortion from mid-range sounds can mask quieter fundamental signals in the treble.

Certainly there are other sources of distortion, but these can be satisfactorily dealt with through, for example, improvements in magnet and coil design.

The adoption of formed sheet metal for the diaphragm, usually a light alloy, typically based on aluminium or magnesium, provides such a high stiffness that the natural resonance modes (typically 700–1.5 kHz for a conventional cone) are pushed up to the 5–7 kHz region usually beyond the crossover point for a multi-way speaker design.

When a metal cone does 'break up' and enter partial resonance, it does so with greater vigour because of the much lower mechanical losses compared with polymer or pulp constructions. What may be an amplitude 'bump' or 'glitch' of 1–3 dB for a high quality plastic cone will now appear as a severe resonance peak, 8–12 dB high (Figure 1.5).

In return for a transparent, distortion-free linear performance in the range below resonance, the designer must suppress resonant peak if it is not to interfere with the performance of the usual high frequency drive unit, married to it via the system crossover network. It is thus customary to fit a special filter to trap electrical input at the main cone resonance.

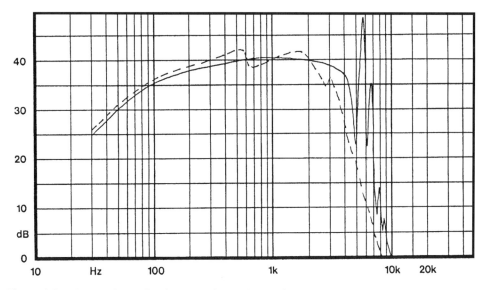

Figure 1.5. Comparison of polymer and metal cone frequency responses in a low diffraction enclosure (- - - - polymer ──── metal)

While metal cones are often considered a recent development, and are increasing in popularity, light alloy cone drivers were in fact developed at the audio division of GEC as long ago as the late 1950s, and also by Jordan in the 1960s. Smaller pressure drive units for horn loaded public address systems have also used metal foil diaphragms for many decades.

Other developments include modern, high tensile strength, low mass fibres used in moving coil driver cones, either as a woven formed matrix reinforced with a catalytic bonding resin, or as a reinforcement to an existing diaphragm structure. Early trials with glass fibre have more recently given way to Kevlar and carbon fibre forms.

Another goal for the drive-unit designer has been realized in recent years. Conventionally, polymer cones were made by hot forming a sheet of thermoplastic — Bextrene, vinyl or polypropylene. This technique tended to thin the regions of greater stretch — the apex — and left the cone rim near the original thickness. This is precisely the opposite of what is required, namely a strong stiff driving point at the apex, the point of attachment to the moving coil, with a lighter, thinner more easily driven region leading out to the edge.

Only recent advances in moulding precision, and the development of a free flowing, mineral reinforced, hard setting grade of polypropylene have allowed the development of close tolerance injection moulded cones. These have a near ideal mass and stiffness distribution. An additional bonus has been the successful addition of the surround suspension, simultaneously co-moulded in the same operation. The result is higher performance polymer cone assemblies of greater consistency and significantly lower cost (Figure 1.6).

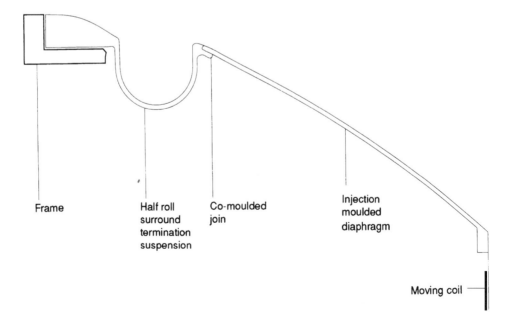

Figure 1.6.   Section through injection co-moulded polymer cone and surround

1.6   THE ROOM

While domestic listening rooms have not really altered, our understanding of the way that this limited acoustic space is used has improved.

Designers are now aware of the effects of local boundaries near to a positioned loudspeaker system; for example, the destructive interference in the lower mid range resulting from out-of-phase reflected acoustic images. Then there is the augmentation at still lower frequencies, where the reflection is now in phase, thanks to the larger wavelength providing good coupling.

The relative contribution of the floor, rear wall and side wall reflections can be accounted for in a practical way. Stable stereo images are aided by good left–right symmetry in placement, including the precise angling and positioning of the loudspeakers in the room.

Cognisant of an 'acoustic centre' for a given speaker system, the most uniform low frequency drive to the room will be obtained if the distances from that 'centre' to the three nearest boundaries are displaced in an inharmonic relationship, e.g. the golden ratio for optimum listening room dimensions. By this means adverse standing wave modes resulting in errors in room/speaker frequency response of up to $\pm 8\,$dB may be subdued to a satisfactory $\pm\,3\,$dB (weighted in third octave bands) (Figure 1.7).

Designers now work to such criteria helping to achieve more uniform frequency response in real rooms and not the artificial textbook design conditions of 2π or 4π steradian anechoic spaces.

The past decade has seen a consolidation of speaker design and technology. Radical new inventions in the field of sound reproduction are rare. Such consolidation is

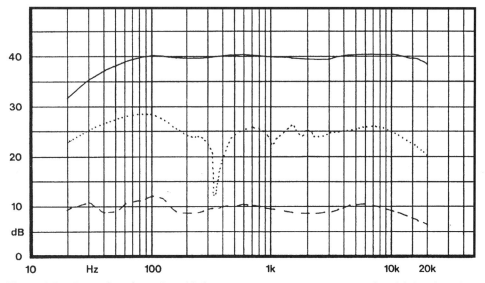

Figure 1.7.   Room interface. One-third octave measurements, average of multiple microphone positions. Loudspeaker positions: ———— in free field (anechoic) ·········· in corner (worst case) - - - - in optimum location (0.3:0.7) (Curves are vertically separated for clarity)

primarily directed at the most enduring and effective sound transducer principle, i.e. the moving coil, which continues to satisfy requirements in a wide range of applications, from the least to the most expensive. Even the acoustic panel loudspeaker (see Section 2.5) which offers an interesting mix of properties and may be driven by several transducer types has so far seen its best results with moving coil based exciters.

As to the future, we can only hope that through the application of both extant and future research material, loudspeaker designers will continue to support the common standard of performance that is beginning to emerge today. No car would find acceptance if it failed to meet basic requirements of handling, braking efficiency, acceleration and comfort, yet the existence of such standards has not prevented the automobile industry from producing a wide range of models of diverse style and size. Similarly, there is no reason why an interesting range of speakers should not continue to be available, while aiming to meet or exceed a common standard of bandwidth, response uniformity, colouration and distortion.

BIBLIOGRAPHY

Borwick, J. (Ed.), *Loudspeakers and Headphone Handbook*, Butterworth (1988) (2nd edn 1994)

Briggs, G., *Loudspeakers*, 5th edn, Wharfedale Wireless Works, Idle, Yorkshire (1958)

Briggs, G., *More About Loudspeakers*, Wharfedale Wireless Works, Idle, Yorkshire (1963)

Cohen, A. B., *Hi Fi Loudspeakers and Enclosures* (1975)

Colloms, M., 'Developments in loudspeaker system design', *Acoustics Bulletin*, **20**, No. 6 (Nov.–Dec. 1995) Institute of Acoustics

Hiraga, J., *Les Haut-Parleurs*, Editions Fréquences (1981)

Jordan, E. J., *Loudspeakers*, Focal Press, London (1963)

Kelly, S., 'Transducer Drive Mechanism', *Loudspeakers and Headphone Handbook*, (Ed. Borwick, J.), Butterworth, 1988

Rice, C. W. and Kellogg, E. W., 'Notes on the development of a new type of hornless loudspeaker', *J.A.I.E.E.*, reprinted *J.A.E.S.*, **30**, No. 7/8 (July 1982)

Tremaine, H. M., *Audio Cyclopedia*, 2nd edn, Howard Sams, New York (1974)

Note: A considerable number of the *J.A.E.S.* papers cited in this book have been published in three volumes by the Audio Engineering Society. These are: R. E. Cooke (Ed.), *Loudspeakers*, Vol. 1 (1953–1977), Vol. 2 (1978–1983) and Vol. 3 (1984–1994).

2

Theoretical Aspects of Diaphragm Radiators

This chapter contains a brief review of those factors that control the behaviour of loudspeaker drivers. (Acoustic loading, particularly at low frequencies, is dealt with in Chapter 4.) Numerous technical works are currently available which cover this field, but they require a good working knowledge of acoustics, mathematics and physics. Although such a background is essential to a thorough understanding of the subject, it is often sufficient for a loudspeaker engineer to be aware of the basic theory rather than delving too deeply into acoustic principles.

If the air is regarded as a fluid medium, then an object in vibration generates vibrating pressure waves which radiate into the medium like ripples generated by a stone dropped into a calm pond. With an infinite pond or an unbounded fluid the ripples continue expanding on a circular wavefront, eventually decaying to infinity. The simplest approach to the properties of sound waves in air is to imagine a pulsating point source in an unbounded air medium, i.e. uncomplicated by reflecting boundaries. An anaechoic or echo-less chamber is a good approximation and is extensively used in acoustic and loudspeaker research and engineering.

Sound is a small pressure change travelling through the air at an essentially constant speed (344 m/s at 20 °C and standard atmospheric pressure).

A property common to waves is the relationship between frequency, the wavelength λ and its velocity c, where

$$f = c/\lambda$$

For example, in air, the wavelength at 1 kHz is

$$\frac{344 \times 10^2}{1 \times 10^3} = 34.4 \, \text{cm}, \, 13.54 \, \text{in}$$

Decibels or dB

The agreed minimum audible sound pressure level or s.p.l. is defined as $2 \times 10^{-5} \, \text{N/m}^2$ (Pa). Note that $0 \, \text{dB s.p.l.} = 0.0002 \, \text{dyne/cm}^2$ (0.00002 pascal (Pa)), $120 \, \text{dB} = 20 \, \text{dyne/cm}^2$.

Human hearing is approximately logarithmic and the reference sound pressure level is set to zero on a logarithmic measure called the bel, B, which is defined for power ratios. Thus a given acoustic power W is defined as

$$W = \log_{10}\left(\frac{P_1}{P_{\text{ref}}}\right) = \log_{10}\left(\frac{P_1}{2 \times 10^{-5}}\right)\text{B}$$

where P_1 is the given sound pressure.

In practice, the bel is too large a unit of loudness and the decibel or dB, one-tenth of the size, has been adopted for convenience. In electronics it is customary to deal with voltages expressed in dB, and the acoustic equivalent is sound pressure rather than acoustic power. Acoustic power is proportional to the square of pressure, i.e. $W \propto P^2$, just as electrical power in watts is proportional to the square of voltage. In converting the above equation to express sound pressure level, s.p.l. in dB, a figure of 20 appears outside the logarithm, a factor of 2 for the square of the pressure and a factor of 10 for the conversion to decibels. Thus

$$\text{s.p.l.} = 20\log\left(\frac{P_1}{2 \times 10^{-5}}\right)\text{dB}$$

Given the equivalence of voltage and pressure, within normal limits a 10 dB increase in input voltage level to a loudspeaker will cause a corresponding increase of 10 dB in the sound pressure level.

For example, the standard test input for a loudspeaker is 1 W, generally referenced to an 8 Ω load, corresponding to an input of $V_1 = 2.83$ V. Suppose an s.p.l. increase of 10 dB is required, and this requires a new input of $V_2 = 8.94$ V. Expressed in dB this voltage increase is

$$20\log_{10}\left(\frac{V_1}{V_2}\right)\text{dB} = 20\log_{10}\left(\frac{8.94}{2.83}\right)$$
$$= 9.99\,\text{dB}$$

Alternatively, this electrical power increase is also a linear factor of 10, i.e. $P_1 = 1$ W to $P_2 = 10$ W, and expressed in dB we have,

$$P = 10\log_{10}\left(\frac{p_1}{p_2}\right) = 10\log_{10}(10)\,\text{dB}$$
$$= 10\,\text{dB}$$

Thus determining maximum sound pressure levels in dB from a speaker with a given amplifier is a simple matter. If the amplifier power is given in dB referred to 8 Ω/W, 0 dB/W (2.83 V) then that figure may be added directly to the loudspeaker sensitivity, in this example, 88 dB/W. A 10 W amplifier is 10 dB/W, a 40 W amplifier is 16 dB/W, i.e. $[10\log_{10}(40/1)]$ and with the example speaker the 40 W amplifier will deliver $88 + 16 = 104$ dB, s.p.l. of undistorted maximum level.

2.1 DIAPHRAGM OR PISTON

Except for ionic and flame type transducers, all drive units possess some sort of vibrating diaphragm, although the way in which this is energized may vary considerably. The diaphragm may be one of a variety of shapes, a cone, a flat surface or piston, a dome, etc. Its method of suspension or support can also vary from none at all, to a separate half-roll of multi-roll structure.

Over the frequency range where the diaphragm moves as a whole, without breakup, it can be equated to a rigid piston. Furthermore, the shape is relatively unimportant as it has little effect on the mid-frequency radiation properties which are the concern of this chapter. (The shape or profile is dictated by considerations of material strength, among other factors, and strongly influences the location of the breakup frequencies.) The behaviour of a circular, rigid piston is well defined, and is essentially derived from the fundamental work of Lord Kelvin, the Victorian physicist.

Several situations are pertinent, and are illustrated below.

Derivation of Sound Pressure at a Given Point due to Vibrating Piston

The analysis involves the division of the piston surface into small elements, vibrating in phase. The sound pressure at a given point, distance r from the piston, at any angle θ is formed by summing the contributions from the individual components through mathematical integration. For example the resulting equation for sound pressure from Figure 2.1(a) is

$$P_{\text{peak}} = \frac{j\rho_0 U_0 \pi a^2 f}{\lambda} \left\{ \frac{2J_1(ka\sin\theta)}{ka\sin\theta} \right\} e^{jw(t-r/c)} \qquad (2.1)$$

where $\rho_0 =$ the density of the gas, i.e. that of air in kg/m^3; $c =$ speed of sound in metres per second, i.e. 345 m/s at 20°C; $J_1 =$ a Bessel function, where

$$J_1(x) = \frac{x}{2} - \frac{x^3}{2^2.4} + \frac{x^5}{2^2.4^2.6} - \frac{x^7}{2^2.4^2.6^2.8} \cdots$$

$a =$ piston radius (metres); $\lambda =$ wavelength of sound in air (metres); $k = \omega/c$ or $= 2\pi/\lambda$, the wave number; $U_0 =$ piston velocity in metres per second (peak).

Directivity

The ratio of sound pressure P_θ at an angle θ degrees off-axis to the on-axis pressure P_0 at the same distance is

$$\frac{P_\theta}{P_0} = 1 - \frac{1}{2}\left(\frac{\pi a\sin\theta}{\lambda}\right)^2 \quad \text{or} \quad 1 - \frac{1}{8}(ka\sin\theta)^2$$

$$\text{where } ka = \frac{\text{piston circumference}}{\text{emitted wavelength}}$$

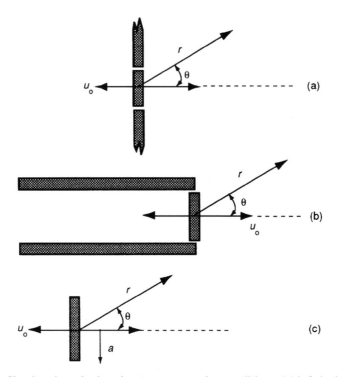

Figure 2.1.   Circular piston in three important mounting conditions: (a) infinite baffle, (b) on face of long cylinder, (c) in free space

DI, the directivity index is defined as

$$DI = 10 \log_{10} \left(\frac{\text{intensity in the } \theta \text{ direction}}{\text{intensity from omnidirectional equal power source}} \right)$$

$$= 10 \log_{10} \left(I_0 \bigg/ \frac{W}{4\pi r^2} \right) = \frac{I_\theta}{I_0}$$

(following the inverse square law for sound power W, where $r =$ distance from source). Intensity is proportional to the square of sound pressure, hence

$$DI = 10 \log_{10} \left(\frac{P_\theta}{P_0} \right)^2 = 20 \log_{10} \left(\frac{P_\theta}{P_0} \right)$$

Taking Figure 2.1(a) at low frequencies, a vibrating piston of radius a, which is small by comparison with the emitted wavelength, has a radiating pattern essentially that of a hemisphere (Figure 2.2(a)) with half the total power radiated on each side of the baffle. Hence the directivity index is equal to 3 dB as the baffle is 'infinite', and only the frontal radiation is available. This case is true in principle for any driver, mounted in a large baffle, from a cone bass to a dome treble unit.

   Two practical points are of interest here. When a driver is mounted in a sealed box which approximates to an 'infinite baffle', the front baffle size is necessarily finite. Below the frequency at which the baffle dimensions approximate to a sound

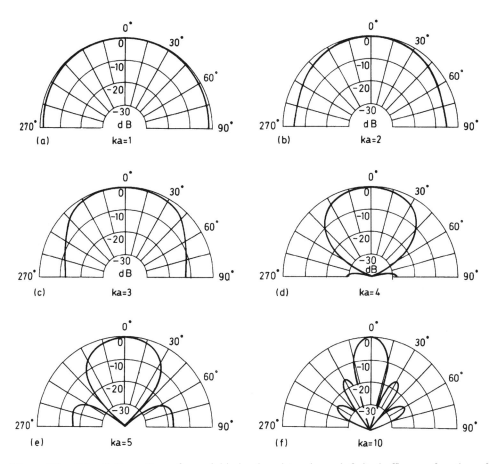

Figure 2.2. Directivity patterns for a rigid circular piston in an infinite baffle as a function of $ka = 2\pi a/\lambda$, where a is the radius of the piston, e.g. a 110 mm diameter piston at 1, 2, 3, 4, 5 and 10 kHz (after Beranek [1])

wavelength, the radiation begins to revert to an omnidirectional or spherical pattern. With increasing wavelength the sound energy begins to diffract around the box (Figure 2.3(d)).

Other factors can also influence directivity. Departure from piston operation is usual at higher frequencies, resulting in areas of the diaphragm vibrating out of phase with the main drive, with consequent amplitude and directivity irregularities. A designer may also employ short horn or plate structures to control the directivity over a specific range.

The velocity of sound in the diaphragm may be another complication with larger units when used at higher frequencies. With typical materials sound wave velocity is variable, a function of frequency. In the upper range the diaphragm may no longer be viewed as a uni-phase element.

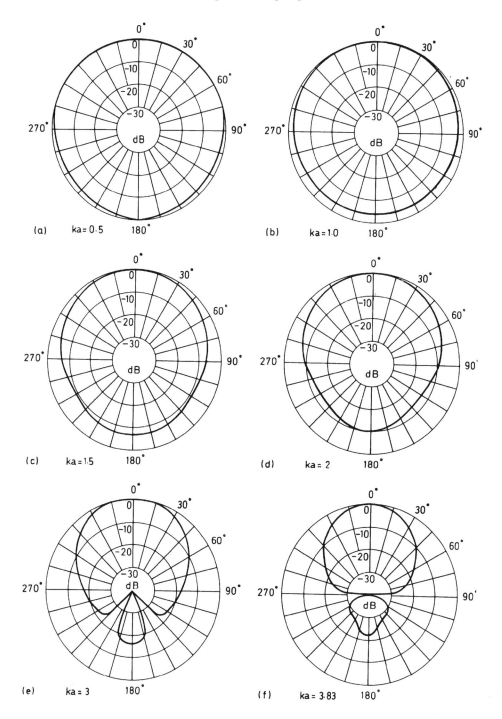

Figure 2.3. Directivity patterns for a rigid circular piston in the end of a long tube as a function of $ka = 2\pi a/\lambda$, where a is the radius of the piston (after Beranek [1])

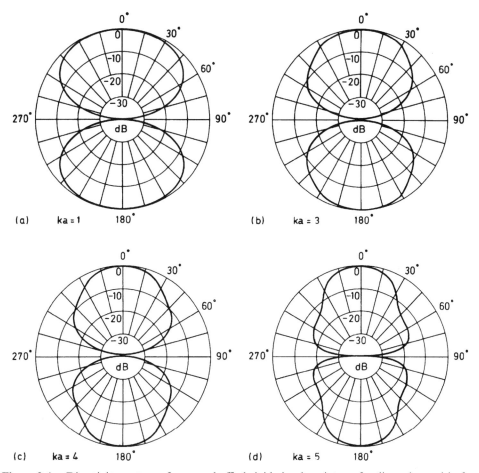

Figure 2.4. Directivity patterns for an unbaffled rigid circular piston of radius *a* located in free space at an angle *θ* a large distance *r* from the point of measurement. For *ka* < 1, the directivity pattern is the same as that for the doublet

Figure 2.1(b) will apply to baffled drivers where the box is deep and the front panel area barely greater than the driver diaphragm; for example, a column enclosure or a compact system with a large LF driver.* The directivity is obtained by analysing the combined effects of the piston area and the diffraction that occurs at the enclosure edges. As with the other two cases, the directivity diagram in this instance is only valid for the far field where the distance *r* is large compared with the piston radius (Figure 2.3).

Finally we come to the unbaffled piston, Figure 2.1(c), which although less common than the first two cases, is nonetheless important (Figure 2.4). Many open backed electrostatic or magnetic drive thin-film diaphragms belong to this dipole group.

---

*Or a miniaturized dome tweeter on a short cylindrical mounting.

Control of Directivity

For any piston radiator, natural directivity is dependent on the frequency of the sound radiated. The audible frequency range spans ten octaves, with an accompanying range of wavelength from 17.2 m to 17.2 mm.* From directivity considerations alone, such a range is beyond the compass of a single driver, irrespective of the practical problems of design and assembly.

For this reason a high quality system will contain several drivers of graded size, which allow a uniform directivity to be maintained over the frequency range. In general one driver should be crossed over to a smaller unit when the diameter of the former is equivalent to one wavelength.

Where narrow directional properties are required, for example to project a sound field to specific regions of a large audience, a converse design may be used. Arrays can be built where the radiating properties of sources which are large compared with the emitted sound wavelength are deliberately exploited and the crossover points are then chosen to maintain the effect over the required bandwidth. Horn systems are naturally more directional than simple radiators and are widely used in this application, especially with sub-divisions or 'multi-cells' which maintain the required radiation pattern to higher frequencies.

A narrow single diaphragm or array constitutes a line source where the radiation remains broad in the transverse direction though narrows in the plane of the line. Line sources are sometimes used domestically, usually where the subject is located in the near field where the vertical beamwidth is still adequate.

In practice directivity will also be controlled by the geometry of the surface or enclosure on which a diaphragm is mounted. For example, Figure 7.16 shows the effect of enclosure form on frequency response. Directivity differences are also associated with these variations.

2.2 PISTON RADIATING ACOUSTIC POWER

So far we have considered directivity, which is a function of frequency, driver size and baffle type, while the sound pressure developed at a distance by a diaphragm is also dependent on other factors such as the air loading. The following brief look at diaphragm radiation mainly concerns the axial response, and the directivity effects so far discussed may be combined with these to give the overall picture.

Equations have been established for the mechanical impedance of the air load presented to a piston, those reproduced here applying to an infinite flat baffle [1].

Z_{MA} (mechanical impedance of air load)

$$= R_{MA} \text{ (resistive or real part)} + jX_{MA}\text{(reactive or imaginary part)}$$

or

$$Z_{MA} = \pi a^2 \rho_0 c \left[1 - \frac{J_1(2ka)}{ka}\right] + \frac{j\rho_0 c}{2k^2} K_1(2ka) \qquad (2.2)$$

*20 Hz to 20 kHz.

where Z_{MA} is in Ns/m or mechanical ohms, and R_{MA} = mechanical resistance in ohms. J_1 and K_1 are the respective appropriate Bessel functions which may be represented by the series

$$J_1(x) = \frac{x}{2} - \frac{x^3}{2^4.4} + \frac{x^5}{2^2.4^2.6} - \frac{x^7}{2^2.4^2.6^2.8} \cdots$$

$$K_1(x) = 2\left(\frac{x^3}{3} - \frac{x^5}{3^2.5} + \frac{x^7}{3^2.5^2.7} \cdots\right)$$

where $x = 2ka$.

For most work the acoustic impedance equation (2.2) may be greatly simplified by utilizing only the early terms of the Bessel series, and this form will be utilized for later sections of this chapter [2]. At low frequencies where $ka \ll 2$, the first two terms of the $J_1(x)$ and $K_1(x)$ series are sufficient which leaves

$$Z_{MA(LF)} \simeq \tfrac{1}{2}\rho_0 ck^2 \pi a^4 + j\tfrac{8}{3}\rho_0 cka^3 \qquad (2.3)$$

At high frequencies where $ka \gg 2$, only the first terms are required, so

$$Z_{MA(HF)} \simeq \rho c \pi a^2 + j(2/k)\rho ca \qquad (2.4)$$

Inspection of the acoustic impedance curves (Figures 2.5 and 2.6) shows that at low frequencies, where ka is considerably less than 2 (1 is the approximate threshold) the mass or reactive component is dominant, whereas at higher frequencies, where ka is much greater than 2 (3 being the approximate threshold) the total acoustic impedance is primarily resistive at 420 Ω (S.I.) and is relatively constant.

It is worth examining the influence of these effects on the radiated power. In order to do this, our massless, resonance-free piston must be energized sinusoidally, by a force F. The air load on both faces must be driven, despite the fact that the rear radiation is unused in an infinite baffle. The piston velocity U is given by the following equation, where

$$U = \frac{F}{2Z_{MA}} \text{ (rms)}$$

If the driving force is assumed constant with frequency, the radiated power equals $U^2 \times R_{MA}$, and the radiated power usefully available is thus

$$W = \frac{F^2}{Z_{MA}^2} \cdot R_{MA} \text{ (from one side)} \quad \left(W_{PK} = \frac{U^2}{2} \cdot R_{MA}\right)$$

If the above approximations concerning the dominance of X_{MA} at low frequencies and R_{MA} at high frequencies are employed, then separate relationships may be obtained for the radiated power under the two conditions. These are as follows:

$$W_{LF} = \frac{F^2}{4X_{LF}} \cdot R_{LF} \quad \text{since } R_{LF} \ll X_{LF}$$

$$W_{HF} = \frac{F^2}{4R_{HF}^2} \cdot R_{HF} \quad \text{since } R_{HF} \gg X_{HF}$$

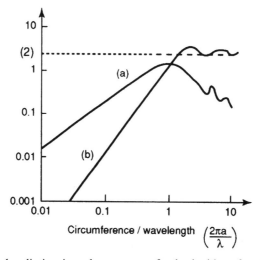

Figure 2.5. Mass and radiation impedance curves for both sides of a circular piston with an infinite baffle where (a) is the normalized curve due to air mass ($X_{\mathrm{MA}}/\pi a^2 \rho_0 c$) and (b) is the normalized resistance curve ($R_{\mathrm{MA}} = \pi a^2 \rho_0 c$)

So

$$W_{\mathrm{HF}} = \frac{F^2}{4R_{\mathrm{HF}}}$$

Over both these regions the sound pressure is level (this includes real cases such as the magnetic and electrostatic film driven designs) but a step in power occurs between the two. This is obvious if the ratio of the two power relations is considered.

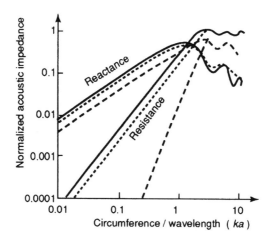

Figure 2.6. Comparison of X_{MA} and R_{MA} for —— infinite baffle, · · · · · on cylinder face, - - - - in free air. Loading on one face of piston

$$\frac{W_{\text{LF}}}{W_{\text{HF}}} = \frac{R_{\text{HF}}R_{\text{LF}}}{X_{\text{LF}}^2} = \frac{9\pi^2}{128} = \frac{1}{1.44} \quad \text{or} \quad 1.6\,\text{dB}$$

While diffraction effects blur the transition, this power relation implies a 3 dB change in sound pressure.

At higher frequencies the piston will become more directional where ka is greater than 3, and will cause the available power to increasingly concentrate on the forward axis with rising frequency (Figure 2.2(d)), and in consequence the axial pressure will also increase. At lower frequencies, where ka is less than 1, the radiation is virtually hemispherical and the axial pressure will be approximately equal to that off-axis.

A small digression is essential at this point in order to introduce the equivalent circuit of a moving coil motor system.

2.3   A HYPOTHETICAL MOVING-COIL DIRECT RADIATOR

Acoustic-electronic Analogues

Powerful analysis techniques exist for electronic circuits and may be applied to complex acoustic and mechanical problems using analogues. For example, mass can be equated to inductance, mechanical compliance to capacitance and friction to electrical resistance. In the SI unit system direct conversion between acoustic, mechanical and electrical quantities is possible. It is worth pursuing this concept as it can considerably aid understanding. An analysis based on analogues assumes that the mechanical-acoustic quantities behave as single simple entities and do not have complex or wavelike properties. A diaphragm is assumed to be a pure piston with no breakup or bending and the air volume in a box acts as a simple linear spring with no standing wave or other resonance within the enclosure (see Figure 2.7 and Table 2.1).

Invoking Newton's second law of motion for any simple lumped element involving stiffness (the inverse of compliance), frictional resistance and mass,

force = mass × acceleration
force = resistance × velocity
force = displacement × stiffness
or force = displacement/compliance

Remembering that when integrated, velocity U becomes displacement and when differentiated (with respect to time) becomes acceleration, the force equations may be rewritten in terms of U giving

$F = j\omega M_{\text{m}}U$ (sine wave acceleration)
$F = R_{\text{m}}U$
$F = U/j\omega C_{\text{m}}$

Re-arranging for U, we have

$U = f/j\omega M_{\text{m}}$
$U = f/R_{\text{m}}$
$U = f j\omega C_{\text{m}}$

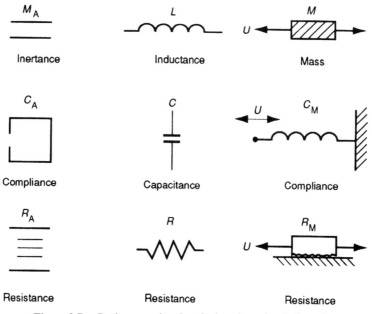

Figure 2.7. Basic acoustic, electrical and mechanical analogues

Similar relations for the alternating current I in an electrical circuit can be written, I, the current equivalent to U, the velocity, in the above relations.

$$I = V/j\omega L$$
$$I = V/R \qquad \qquad \text{(from Ohm's law)}$$
$$I = Vj\omega C$$

In Table 2.1 and Figure 2.7, electrical, mechanical and acoustic analogues are compared. For example, acoustic mass is represented by a volume of air trapped in a section of tube, compliance by an air spring contained in a specific volume, acoustic resistance by a partial obstruction of fine holes, slits or porous fabric.

Table 2.1.

Electrical			Mechanical			Acoustic		
e.m.f.	Volt	V	Force	Newton	F_M	Pressure	P_M/m^2	P_a
Current	Amp	I	Velocity	m/s	U_M	Volume current	P_A/R_A	U_A
Resistance	Ohm	R_E	Resistance	F_M/U	R_M	Resistance	$\dfrac{8nl}{\pi a^4}$	R_A
Inductance	Henry	H	Mass	kg	M_M	Inertance	$\rho_0 V$	IM_A
Capacitance	Farad	F	Compliance	D/F_M	C_M	Compliance	$V/\rho_0 c^2$	C_A

s = second, m = metre, n = viscosity of air = 1.86×10^{-5} Newton/m^2, ρ_0 = density of air = 1.18 kg/m^3, c = velocity of sound = 344 m/s, V = volume in m^3, a = radius of tube, and l = length of tube.

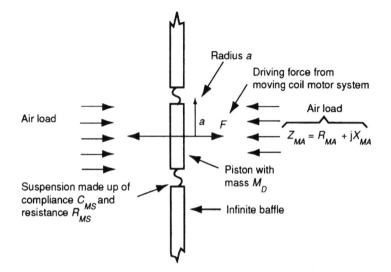

Figure 2.8. (a) Moving coil energized piston in infinite baffle (showing mass, stiffness and resistance)

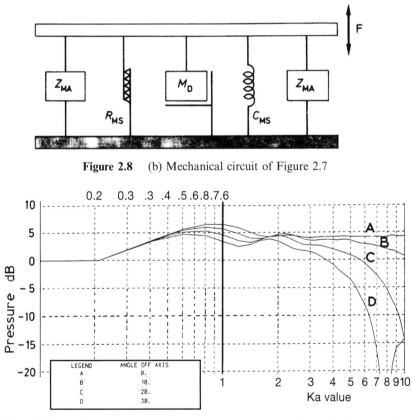

Figure 2.8 (b) Mechanical circuit of Figure 2.7

Figure 2.8. (c) Theoretical on- and off-axis responses of a circular piston with Ka value. The piston is set at an effective 250 mm diameter, offset mounted in a 0.6×0.8 m baffle (after Porter and Geddes)

So far the discussion has been confined to an imaginary piston with no mass or resonances whatever. In practice, however, a drive unit will possess a diaphragm and accompanying moving system of mass $= M_D$, a suspension or flexible support of compliance C_{MS}, and a mechanical resistance R_{MS} (Figure 2.8). We shall ignore for the moment the other components of the motor section — magnet, voice coil, etc.

While Figure 2.8 shows a mechanical diagram, this system may be represented by electrical components where mass becomes capacitance, and inductance represents compliance. The resulting mechanico-electrical circuit is represented by Figure 2.9.

The worth of the electrical equivalent lies in the ease with which standard electrical circuit theory may be applied to analyse what would otherwise be a difficult acoustical problem. We shall now include the moving coil components comprising the coil d.c. resistance R_c, inductance L_c and mass M_c (the latter is included in the total moving mass M_D) which are incorporated in the equivalent circuit, resulting in Figure 2.10; this also includes the 'source' or generator components, resistance R_g and voltage E (see Figure 3.22).

Figure 2.10 includes a transformation step between the electrical and the mechanical parts (albeit in electrical component form). This transformation derives from the motor principle whereby an electric current flowing in a coil immersed in a radial magnetic field produces a mechanical force in the coil along its axis.

The factors are flux density B, the coil wire length l actually immersed in the magnet flux, and the current in the coil, I. For a unit current, $F = Bl$, and hence the transformer may be represented by Figure 2.11.

Transformer theory gives the relation

$$\frac{I}{F} \propto \left(\frac{n_1}{n_2}\right)^2$$

hence

$$\frac{I}{F} \propto (Bl)^2$$

The electrical quantities may be transferred via the 'motor transformer' to the mechanical side utilizing the $(Bl)^2$ constant where their combined effects may be assessed. This complete mechanically based electric circuit, shown in Figure 2.12, which in mechanical terms is referred to as a mobility arrangement, may be rearranged into the equivalent impedance form of Figure 2.13, for ease of analysis. In this latter form the mechanical circuit can be subjected to rigorous mathematical analysis but for the purposes of this chapter it is sufficient to simplify its treatment by examining convenient divisions.

In the circuit of Figure 2.13, at the junction between the electrical and mechanical sections, we have the voice coil velocity U_c, which is directly related to the driver excursion, and thus controls the output power and hence the sound pressure level.

The circuit of Figure 2.13 provides the foundation for the analysis. For example, at medium to low frequencies L_c is negligible, leaving the simplified circuit as shown in Figure 2.14. This represents a simple LC resonant system with air loading Z_{MA} and damping $R_e + R_c$, which possesses the universal resonance curve for U_c, the coil velocity. If we assume for the moment that the directivity of this hypothetical driver is constant

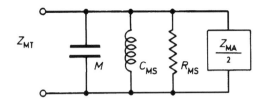

Figure 2.9. An electrical impedance equivalent circuit of Figure 2.8

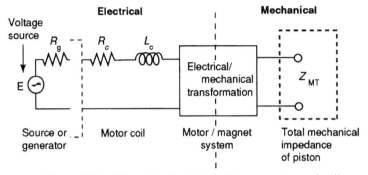

Figure 2.10. Electrical circuit including generator and coil

over the frequency range, then the sound pressure and coil velocity will follow the same variation as that for a simple resonance circuit (a plot of current against normalized frequency for a series RLC circuit).

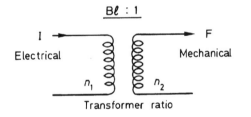

Figure 2.11. The 'motor transformer'

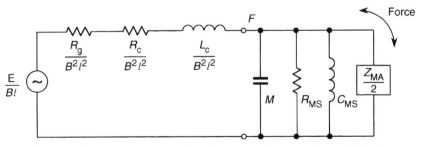

Figure 2.12. Electrical parts referred via 'transformer' to mechanical side (represented in mobility form)

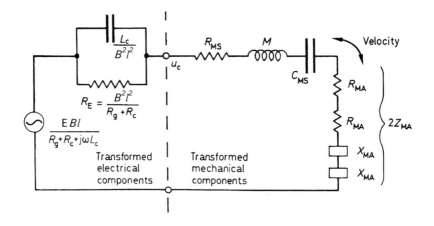

Figure 2.13. Transformation of mobility circuit of Figure 2.12 to the impedance or velocity form. In this circuit, F, the driving force, is transformed into U_c, the piston or coil velocity, which is equivalent to a voltage or pressure in a conventional electric circuit

In decibel form, this curve gives slopes on either side of the peak tending to 6 dB/ octave. When the velocity is translated into sound pressure, or acoustic power, and plotted against frequency, the additional linear frequency term tilts the curve, so that below resonance, the roll off approaches 12 dB/octave (6 + 6 dB), and above resonance is actually flat ($-6 + 6$ dB $= 0$ dB). (Alternatively, this follows from the acoustic power relation, $W = U_c^2 R_{MA(r.m.s.)}$, i.e. $W \propto f U_c^2$.)

Region A of Figure 2.18

Figure 2.15 is a simplification of Figure 2.14 which is valid at very low frequencies. At very low frequencies C_{MS} is dominant, and the other components, such as the air load, are negligible. The output is proportional to the fourth power of frequency, which implies a fall of 12 dB/octave with reducing frequency in the range below the fundamental resonance ($W \propto f^2 . f^2 C_{MS}$).

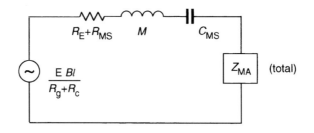

Figure 2.14. Simplified circuit of Figure 2.13 at medium to low frequencies (regions A and B of Figure 2.18)

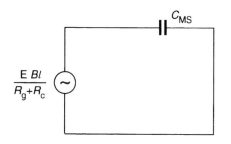

Figure 2.15. Simplified circuit of Figure 2.14 at very low frequencies (region A of Figure 2.18)

Region B of Figure 2.18

Figure 2.16 shows the components of Figure 2.14 which are relevant at resonance, a condition where the reactive parts cancel leaving resistive effects only.

At resonance L_c is still negligible; L and C disappear as the velocity/power is wholly resistive and X_{MA} is zero. The output at resonance is thus dependent on the total resistance, including that of the generator or source R_g. This figure shows that as R_g is increased the losses are reduced; that is, Q is inversely proportional to R_g. This is the reverse of what might be expected, and the source resistance can be seen to be an important damping element in the fundamental resonance of the system.

The coil resistance is the other major factor as, in practice, the suspension losses are usually low. If R_g is large or B^2l^2 small, Q can exceed unity and thus develop a peak in the sound output. This will be covered in greater detail in the chapters on LF loading (Chapters 3 and 4).

Region C of Figure 2.18

Above resonance (Figure 2.17), at medium frequencies, the radiation is still essentially hemispherical so that the directivity is constant, and the sound pressure and power output are directly related. It is possible for a driver to have a level output in this range, as the frequency squared term in the radiation resistance R_{MA} is balanced by the

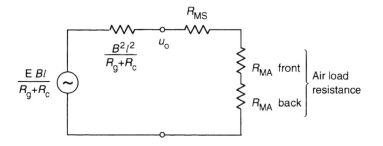

Figure 2.16. Simplified circuit of Figure 2.14 at resonance f_o, reactive components disappear (region B)

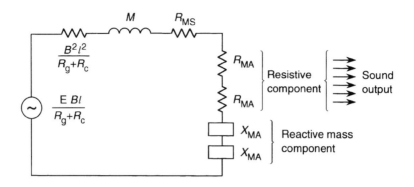

Figure 2.17. Simplified circuit of Figure 2.13 above resonance, medium frequencies (region C)

frequency squared term in the mass reactance (piston and air mass load). This condition is known as 'mass controlled', resulting in a constant output over this range. It assumes that the resistive components R_{MS}, etc. are small; if not, the resonance may be overdamped, 'rounding the corner' and tilting the flat region to some degree. (This is represented by the dotted area in Figure 2.18.) Excessive Bl will also reduce the span of this region through overdamping.

Region D of Figure 2.18

At higher frequencies, a second resonance is theoretically possible. Inspection of Figure 2.14 shows that the coil inductance can become significant at higher frequencies (as

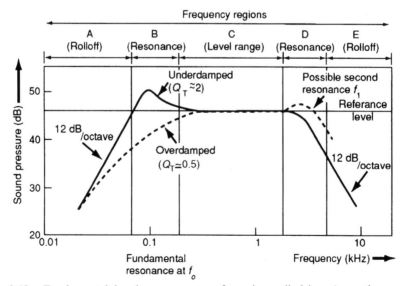

Figure 2.18. Fundamental bandpass response of moving coil driver (over the range where directivity may be assumed constant, sound pressure and power may be shown by the same curve)

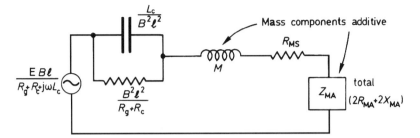

Figure 2.19. High-frequency equivalent circuit of Figure 2.14 (resonances involving summed masses and L_c/B^2l^2) (region D)

represented by the capacitor in our impedance equivalent circuit) and it may resonate with the mass component remaining in the loop, the network given in Figure 2.19.

At resonance only the resistive components apply, as seen in Figure 2.20. The resonant frequency is given by

$$f_1 = \frac{\omega L_c B^2 l^2}{\omega^2 L_c^2 + (R_g + R_c)^2} = \omega M + 2X_{MA}$$

Interestingly, if $(R_g + R_c)^2$ is large compared with $L_c^2 \omega^2$, then the parallel combination of L_c/B^2l^2 with $B^2l^2/(R_g + R_c)$ produces a negative inductance, $-B^2l^2 L_c/(R_g + R_c)^2$, and no resonance lift will occur.

Region E of Figure 2.18

At very high frequencies, above second resonance, we reach the upper range where the acoustic resistance is virtually constant and is thus independent of frequency, while the acoustic mass contribution has fallen to negligible proportions (see Figure 2.5).

The simplified circuit in Figure 2.21 is now dominated by only two components: the constant radiation resistance and the mass reactance of the moving piston itself (including motor coil, etc.). The normal 6 dB/octave rolloff for this single-pole filter is

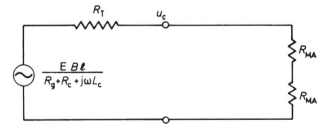

Figure 2.20. Simplified circuit of Figure 2.19 at resonance (region D). Total resistance

$$R_T = \frac{(R_g + R_c)B^2l^2}{(R_g + R_c)^2 + \omega^2 L_c^2}$$

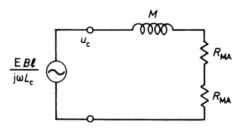

Figure 2.21. Simplified circuit at very high frequencies, i.e. above second resonance (region E). The coil inductance, L_c is dominant and R_{MA} is constant at $\rho_0 c \pi a^2$

modified by the linear frequency term, producing a final rolloff (of 12 dB/octave) after second resonance.

Inspection of the pressure response (Figure 2.18) shows that the most uniformly wide characteristic is obtained if the fundamental resonance is critically damped and the compliance is high, giving a low resonant frequency. This is the theory underlying the widespread use of small, high compliance bass/mid drivers for domestic systems.

2.4   DEVELOPMENT OF THE AXIAL SOUND PRESSURE

The preceding analysis assumed a constant hemispherical directivity and pure piston operation. It is now necessary to apply further factors such as directivity variation and diaphragm breakup in order to develop a theoretical basis for the axial sound pressure as produced by a practical drive unit.

Initially we will consider the case of an infinite baffle-mounted driver radiating into a half-space (hemisphere or 2π steradians). This condition is approximated by a wall-mounted, flush-fitted box system or, more simply, a wall-mounted driver. The sound pressure will follow the power curve (Figure 2.18) until the diaphragm diameter is comparable with the wavelength of sound in air.

If we begin at region C, the level power range in Figure 2.18, both power and axial pressure response are uniform at low frequencies. With increasing frequency a point is reached where the diaphragm begins to concentrate sound pressure on axis, and the sound pressure will in consequence rise with increasing frequency. Reference to the section on directivity shows that when ka approaches unity, i.e. the piston circumference approximates to one wavelength, the piston becomes increasingly directional. The axial gain, which may be expressed as Q, the directivity factor, approaches 6 dB/octave.

It is happens that the hazy division between regions C and D of the power curve (Figure 2.18) lies between ka values of 1 and 2. If the second resonance is overdamped, then the directivity gain may offer some compensation, which will then result in a response extension with certain diaphragms. The 12 dB rolloff beyond second resonance will now tend to a 6 dB/octave slope, due to the narrowing directivity. This is by no means coincidental, as the equations for directivity are derived from the same acoustic effects as the radiation impedance.

For example, consider a 200 mm diameter driver chassis with a well behaved cone. If we take $ka = 2$ as the upper limit of the level region, the transition point from hemispherical to forward radiation is given by

$$f_t \simeq \frac{C}{\pi a} \simeq \frac{345}{\pi \times 0.09} \simeq 1.2\,\text{kHz}$$

where $9\,\text{cm} = $ nominal piston radius. The sound pressure is likely to be reasonably maintained for an octave beyond $1.2\,\text{kHz}$, above which the sound output should fall by $6\,\text{dB}$ per octave.

In practice, most cone diaphragms enter the first major breakup mode or departure from true piston behaviour at values of ka between 1 and 2, with the poorer examples producing audible breakup effects at half that frequency. When a cone is in breakup, the theory described above is of little value, as the air load impedance and moving mass will be indeterminate, according to the specific breakup mode present at each frequency. This behaviour will be discussed in more detail in the later section on cones (Section 3.4).

Reference Efficiency

The level region for a moving coil driver provides an important reference for the calculation of power output and efficiency [1]. It also gives a starting point for a system designer to match to the power level of drivers covering other sections of the range. The reference power available efficiency is given by

$$\eta_{\text{ref}} = \frac{800 R_g B^2 l^2 \rho_0 \pi^2 a^4}{2c(R_g + R_c)^2 (M_D + M_A)^2}\ \% \tag{2.5}$$

or in more typical form using $S_D^2 = \pi^2 a^4$ (where $S_D = $ surface area)

$$\eta_{\text{ref}} = \frac{\rho_0}{2\pi c} \cdot \frac{B^2 l^2 S_D^2}{(M_D + M_A)^2} \cdot \frac{800 R_g}{(R_g + R_c)}\ \% \tag{2.6}$$

Equation (2.6) applies to the total output from both sides of the cone where the unusual condition of electrical power matching from a significant generator resistance R_g is included. In practice R_g tends to zero with low output impedance amplifiers and the above equation then makes little sense.

Taking an efficiency criterion based on power input versus power out, the reference efficiency is

$$\eta_0 = \frac{\rho_0}{2\pi c} \cdot \frac{B^2 l^2 S_D^2}{(M_D + M_A)^2} \cdot \frac{200}{R_c}\ \% \tag{2.7}$$

Comparing equation (2.7) with (2.6) it may be seen that if the optimum power transfer condition is allowed, i.e. $R_g = R_c$, then the two agree. For one side, as is usual for a box enclosure, the value in equation (2.7) should be halved.

It can be shown that

$$\eta_0 = 100 \cdot \frac{4\pi^2}{c^3} \cdot \frac{f_0^3 V_{AS}}{Q_E}\quad \text{(Small [3])} \tag{2.8}$$

where f_0 is the driver free air resonance, V_{AS} the air volume equivalent compliance, and Q_E the driver electrical Q value. With substitution for the constants for example, $\eta_0 = 9.6 \times 10^{-5} \dfrac{f_0^3 V_{AS}}{Q_E}$ % $= 0.39\%$

for the driver described in Figure 3.38, (with $Q_E = 0.5$). This is a low value and illustrates the inefficiency of such small 'long-throw' low-colouration drivers.

In equation (2.7) we see that the total mass term is squared. For the above driver size

the air mass, $M_A = \dfrac{8\rho_0 a^3}{3} = \dfrac{8 \times 1.18 \times 0.086^3}{3}$ kg $= 2$ g per side

which implies that the cone mass is 16 g, quite high for a 200 mm frame unit. If a light pulp cone were substituted and the compliance adjusted to maintain the resonance frequency, the cone mass might be reduced to 8 g giving a total of 12 g. This would result in a power output improvement of nearly three times, giving an $\eta_0 = 1.1\%$.

Sensitivity and Efficiency

It should not be forgotten that the above change in moving mass will have other consequences, such as a relative increase in electrical damping at the fundamental resonance and thus a major change in the low frequency response. In the reference frequency region where the sound radiation from a system is omnidirectional, or hemispherical in the case of an infinite baffle, the sound pressure level (s.p.l.) at a given distance is proportional to the square root of the radiated total acoustic power. Assuming a nominally fixed electrical input resistance which is dominated by R_c the electrical power input is proportional to the square of the voltage applied. This is the basis for equation (2.7) for η_0.

In practice the concept of efficiency is not much used in loudspeaker design and measurement, except on a theoretical basis. Accepting that the source resistance is low for both cables and amplifier it is usual to specify the sensitivity of a loudspeaker for a given voltage input, usually 2.83 V, which produces 1 W into a nominal 8 Ω standard load. Thus a specification for the example unit would read, 'sensitivity $= 84$ dB/W at 1 m'.

While the concept of efficiency has some relevance in the case of a single drive unit operating in the reference piston range it becomes meaningless in the context of a complex multi-unit loudspeaker system. Factors such as the variable and reactive input impedance and variable directivity exhibited by such systems over their working frequency ranges makes a power in/power out relation unusable.

A method which can be used, though probably of limited significance, consists of driving a speaker system in a reverberant chamber with random noise excitation. The power input is assessed by monitoring the full bandwidth integrated value of the voltage and current input, and the acoustic output from the integrated sound power assessed from measurement of the reverberant field.

For a half space, 2π, then s.p.l. per 1 W per 1 m $= 112.2 + 10 \log (\eta_0)$ dB. The loudness, i.e. sound pressure, at a point is proportional to the square of the following parameters, flux density in the gap, B, wire length, l, immersed in that gap field, and to the radiating area of the diaphragm, S_D.

Again for a hypothetical driver, the loudness is inversely proportional to the square of the moving mass, M, but is simply inversely proportional to the coil impedance, R_E.

If all other parameters are held constant, then consider halving the coil resistance (impedance) from 8 Ω to 4 Ω.

On the basis of the efficiency equation the change in power is calculated in logarithms for the s.p.l. difference change:

$$dB = 10 \log \eta_8 - 10 \log \eta_4$$
$$= 3 \, dB$$

The 4 Ω coil draws twice the current from the source and uses twice the power, in agreement with the above.

For a 100% efficient unit, 1 W input at 1 m will provide 112 dB s.p.l. (assuming spherical radiation and the power from one side only).

To calculate the sound level for speakers of lower efficiency η_0

$$s.p.l. = 112 - 10 \log 10 \, [1/\eta_0]$$

where the efficiency is specified in %.

For example, a typical good quality loudspeaker of 1% efficiency will have an output of $(112 - 20)$ dB, which equals 92 dB. For 0.5% efficiency, the output falls a further 3 dB to 89 dB, an average value for medium size box loudspeakers generally augmented by reflex loading at low frequencies.

2.5 NON-PISTONIC RADIATORS AND THE DISTRIBUTED MODE ACOUSTIC PANEL

Resonant behaviour in plates and other structures has been studied for some centuries, for example by Chladni in Italy in the late 18th century and more recently by Lord Rayleigh in the 19th century. Chladni showed the fascinating resonant modes possible in a glass or metal plate, fixed only at its centre, exited by a violin bow, and damped at nodal points by touch, the modes made visible via a dusting of sand (Figure 2.22). Almost all objects have significant acoustic properties in that they can be set into vibration and in consequence radiate sound energy. In general these resonances are highly selective for each object, their shape and combination of materials defining a unique resonant structure and thus a related series of modal

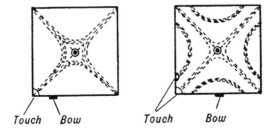

Figure 2.22.   Two examples of modal resonance patterns in a free plate. (After Chladni 1787)

frequencies. These constitute an acoustic signature, usually so specific that when struck or resonated, common objects may be identified by their acoustic signature alone, for example, a glass bottle, a door, a table top, a window or an aluminium saucepan. While these are rather obvious examples it confirms that the ear/brain processor is highly analytical even in the face of short term, high complex acoustic signals.

Reduction of noise and the control of its transmission in buildings is a major research activity. The distribution of resonances in structures is studied in depth with the help of advanced mathematics allied to computer generated models, especially using FEA (Finite Element Analysis) and BEM (Boundary Element Modelling) methods.

In particular, research into noise suppression in aircraft structures by Heron [4] at the DRA, resulted in a better understanding of the complex modal behaviour of stiff, light panels, which when tested unexpectedly turned out to have greater than anticipated acoustic transparency. This inherent, associated sound radiation property was modelled further by V Labs of Huntington on behalf of NXT, this further analysis extending the mathematical understanding from the two or so upper range octaves originally pertinent, to encompass almost the whole audio spectrum.

Even at the earlier stage of research a key result was established, namely that if mechanical losses were held low, a light very stiff panel possessed a radiation efficiency closely approaching that of a pure piston of the same area, in fact up to 98% with practical material science.

At this stage the directional properties of the acoustic panel were poorly understood but it was known that they would differ fundamentally from a piston.

Resonances for a Free Rectangular Plate

While the following represents a considerable approximation, for a basic understanding of panel behaviour it is instructive to consider the resonant modes for the two orthogonal axes of a simple rectangular plate.

Each axis has a series of resonant modes commencing with the fundamental, these defined by mass versus stiffness for the plate in that direction. By mass we mean mass per unit of surface area, while the effective stiffness depends on the bending equation relating length, thickness and Young's modulus.

In a stretched membrane, for example seen in a magnetic or electrostatic surface driven loudspeaker, the wave speed is constant, the medium is non dispersive and obeys the wave equation with the general solution:

$$\Psi[x,t] = f[k.x + c.t] + g[k.x - c.t]$$

where f and g are any functions whose wave shapes are preserved during wave propagation.

For a distributed mode acoustic panel the general wave equation is not obeyed. The wave speed varies with frequency and is generally higher than in air. In such a panel the wave speed is given by

$$v(\omega) = \sqrt{\omega}.[B/\mu]^{1/4}$$

Note the fourth order dependency of velocity on the B/μ ratio, and the dispersive nature of the medium, here waveshape is not preserved during propagation.

In essence the distributed mode design requires that at least two resonance series interleave in an optimized manner. The third significant contender in this simplified treatment is the diagonal axis, obviously differing from the orthogonal modes, and providing further opportunity for resonance distribution. Those familiar with plate dynamics will quickly note the presence of the higher order elliptical and related degrees of freedom, these at first sight not obvious for a rectangular plate.

Solving the wave equations for a plate does not give the entire answer. In theory such bending waves propagating in an unbounded medium do not generate acoustic power at infinity, the differing phases of the local bending regions sum to zero in the far field. In practice the application of finite boundaries, arising from choice of panel size, confirms that useful acoustic power is available after all.

Loudspeaker Theory Turned Upside Down

Whereas in piston based loudspeakers, resonance is intentionally subdued, moderated and in general avoided due to the largely unpredictable changes in behaviour which occur in resonance or 'breakup' as it is termed, for the distributed mode loudspeaker natural diaphragm resonances are positively encouraged.

In piston loudspeakers, the resonances are effectively levelled down until they have little effect on uniformity of amplitude response and on energy the decay rate. Conversely for the distributed mode panel, a maximized amplitude and density of resonance is the objective. Here the resonance peaks are levelled up to a new reference line.

Coincidence and the Bending Wave Panel

For any chosen panel axis there will be a defined relationship of acoustic wave speed from the relevant bending stiffness. Depending on the panel construction, over the frequency span the wave speed may cover a wide range, above, across or held below the velocity of sound in air, the latter essentially constant with frequency. For a given individual axis or degree of freedom in a particular panel; a correspondence is likely for panel and air velocity at a particular frequency. Here the two are aligned. The axis directed set of panel bending waves launches a coherent, precisely directed acoustic wave.

For a theoretical single mode panel this is clearly an undesirable result unless a directed output, almost tangential to the surface, is the design intention. However the complex distributed resonance behaviour of a well specified distributed mode panel means that any measuring point or axis will be almost saturated with a fine structure of contributions from many points on the flexing panel.

Nevertheless consideration of coincidence* does give an indication that the sound radiation from an acoustic panel will not be the usual, axially-directed, concentrated beam. Just as the modal resonance operation of the panel is counter intuitive, so also is its radiation property.

*See also section 7.3.

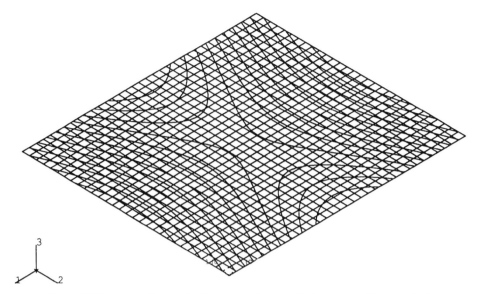

Figure 2.23.   Acoustic Panel; Contours of equal displacement. Eigenmode 8

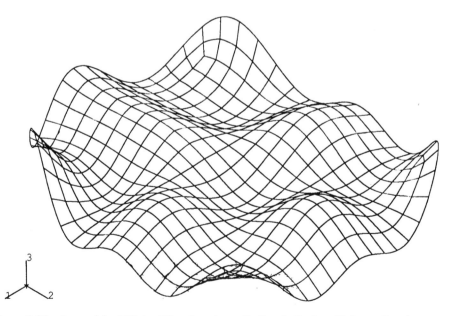

Figure 2.24.   Imaged by F.E.A., Vibrating Acoustic Panel: Surface Deformation (exaggerated for clarity)

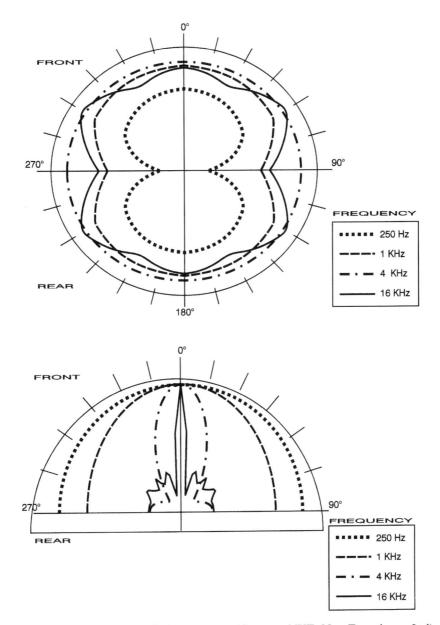

Figure 2.25. Comparison of radiation patterns. (Courtesy NXT, New Transducers Ltd)

Directivity of a Bending Wave Panel

Calculation of the directivity of a well distributed resonator is very complex. Acoustic modelling using FEA software is the most effective approach. This, and published data (Figure 2.25) based on actual measurement shows an unusually wide polar pattern, averaging 120 degrees of beam width. There is a mild null when observed exactly on the panel edge. A low degree of dipolar correlation between the radiation from the front and rear surfaces is also indicated, somewhat confirming the claim for largely bi-polar radiation [5].

When compared with a semi pistonic cone radiator the wide directivity of the distributed mode panel speaker is more or less independent of frequency, a valuable property for sound distribution.

Natural Diffusion

The defined complexity of bending waves for a distributed mode panel results in a large area sound source, one which may be regarded as essentially diffuse over its entire surface. This has important implications with regard to the effect of room boundaries and also for the relationship of loudness with distance.

Upon auditioning an energized distributed mode panel it is not possible to identify one particular source or region of excitation; the whole area appears to be evenly illuminated over the working frequency range. This would normally imply plane wave radiation and despite the diffuse nature of the output it does have some of the properties of a plane radiator.

Intensity Fall With Distance

The contributory factors of bi-directionality; essentially bi-polar radiation; large source area, and a high degree of diffusion result in a departure from the usual square law relationship for sound pressure reduction with distance normally operative for loudspeakers.

In free space a small or near point source radiates spherically. The intensity at a distance 'r' is inversely proportional to 'r squared'.

However, for the distributed mode panel at moderate distances, e.g. 1 to 5 metres, the intensity fall approximates to a linear function of distance.

Comparing the two cases, where the sound pressure level for a normal source is assessed at four times the distance the sound pressure is reduced by a factor of 16 or 12 dB. For a large diffuse source this reduction is a lower factor of 4 or 6 dB.

In a typical room environment, the bi-polar wide angle directivity of the acoustic panels results in a still greater differential for the reverberant power compared with a cone type speaker. Subjectively, surprisingly little diminution in volume is observed in the far field for the panel radiator.

Excitation of Room Modes

Almost perfectly representing a diffuse source, an energized distributed mode panel cannot generate specular reflections from local room boundaries. Consequently the usual comb filter interaction resulting from direct and reflected wave paths is absent.

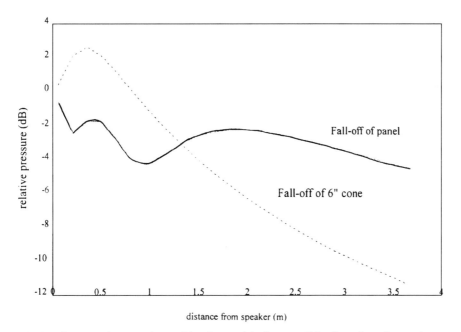

distance from speaker (m)

Figure 2.26. Computed comparison of loudness with distance: Distributed mode panel versus a 6″ 150 mm piston. (Courtesy of NXT, New Transducers Ltd)

In a sense the acoustic coupling to the room is softened, resulting in more uniform coverage. Judged subjectively high clarity and intelligibility is maintained under room loaded conditions. (See Figure 2.27)

Mechanical Acoustic Impedance

Solving the complex 'multi dimensional' wave equation delivers a remarkable result: from the start of the specified frequency range of good modal density, the mechanical impedance Z_m is found to be entirely resistive, this given by

$$Z_m = 8.\sqrt{B.\mu}$$

where μ = mass per unit area
B = bending stiffness

Almost all energy in the panel is radiated as sound. We can therefore say that the acoustic output of such a panel is independent of frequency. Moreover, provided that the panel material properties are appropriate, there is no acoustic limit to the upper frequency point.

If fitted with an electrodynamic transducer of negligible inductance the electrical input impedance is also almost entirely resistive.

There is no motional impedance effect; the measured value is essentially the coil resistance itself since the practical electroacoustic efficiency of the complete loudspeaker

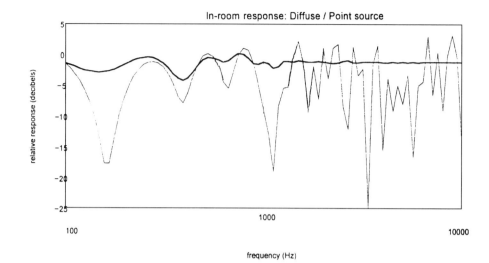

Figure 2.27   Computed Effect of Room Boundary Reflections: —— for a conventional point source: —— for a simulated NXT diffuse source. (Courtesy NXT, New Transducers Ltd)

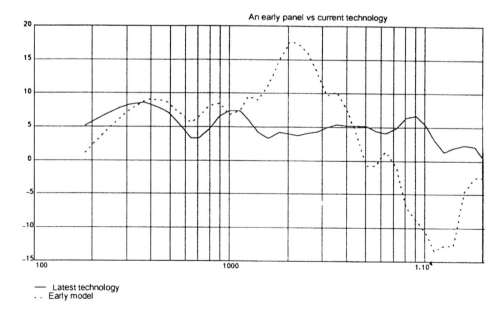

Figure 2.28.   Amplitude/frequency response for an early acoustic panel and for later development. Measurement uses spatial averaging. (Courtesy of NXT, New Transducer Ltd)

is similar to normal speakers, 0.5 to 2%. The influence of the acoustic impedance is small when viewed at the input terminals.

Other Factors

A number of design factors for the acoustic panel are subject to commercial and patent protection. Nevertheless it may be surmised from basic bending theory that the practical performance of a distributed mode radiator will depend on the following, noting that the literature available describes the use of light, stiff forms implying the use of selected composites:

- Mass per unit area.
- Bending stiffness.
- Geometry including complex forms.
- Mechanical loss including frequency dependency if any.
- Area.
- Driving region or regions.
- Support and or termination.
- Baffled or unbaffled installation.
- Type of exciter technology, velocity or displacement.
- The shear modulus of the core (if a composite).

Sensitivity/Efficiency

Engineered with normal transducer systems of typical force factor, a well designed distributed mode panel speaker provides comparable efficiency to a moving coil cone type.

Power Handling and Distortion

Typically the panel operates well within the elastic limit for bending and hence is highly linear. Distortion is dominated by residual non linearity, eddy currents and pole piece distortion for the magnet system. Advanced magnet design will reward the engineer with reduced distortion levels potentially approaching that of the push–pull electrostatic.

   Power handling is essentially a thermal matter. As with conventional designs, the use of conductive structures local to, and in thermal contact with the motor coil, provides augmented heat dissipation.

Distributed Mode Panel Applications

Passive

While the theory has described an acoustic radiating panel, it does not become a loudspeaker until an appropriate transducer is fitted. For the panel itself, there are applications in architectural acoustics and mechanico-acoustic musical instruments. New understanding and control may well be brought to these and related applications.

Active, driven

Operating in bending mode over a typical surface area say $0.2\,\mathrm{m}^2$ the amplitude of motion for a panel is very small, typically $100\,\mu\mathrm{m}$ or less, generally invisible to the naked eye.

Short throw transducers, closer in design to vibration exciters, are a logical choice and these, even in miniaturized form will give good $B.l$ figures and hence worthwhile conversion efficiency. It is possible to incorporate such a transducer within the panel itself.

A constant velocity transducer, e.g. an electrodynamic type, will match the resistive mechanical impedance of the panel, resulting in a naturally uniform amplitude/frequency response.

Operating as a loudspeaker the following applications have been cited for the distributed mode acoustic panel.

Sound distribution

The wide directivity, the lack of undue intensity increase for proximate listeners and large area coverage are positive factors here. NXT have adopted the name SonTile (tm) for their ceiling type speaker.

Laptop computers

The demand for an improvement on the poor quality, very limited bandwidth loudspeakers presently fitted to portable computers is self evident now that they are CD and movie capable. Box type speakers are inappropriate on grounds of space and weight.

An acoustic panel speaker may be 20- to 30 grams in weight and just a few millimetres thick. One suggestion is for thin, slide out panels released from the computer lid. Path lengths to the operator are consistent with good, nearfield stereo, augmented by the partial baffle action of the screen itself.

Home theatre

Two distinct specializations have already emerged for the distributed mode panel.

Figure 2.29. Multi media computer with distributed mode loudspeaker panels

a) Acoustic Projection Screen

An active acoustic panel employed for centre channel or 'dialogue' use has the advantage of easy placement due to a flat thin form, and the diffuse, wide frequency range radiation property. For the observer, the sound source is beneficially unlocalized, increasing the sense of realism.

In addition, for an audience the sound level is more constant over the listening region, improving coverage.

An important bonus arises when a movie or video image is projected onto the appropriately dimensioned and reflectively finished surface of an acoustic panel speaker. (NXT 'SoundVu' (tm). Observers report a strong synaesthetic response whereby the virtual sound source is perceived to closely match and track the perceptually dominant visual image. Both lateral localisation and virtual acoustic depth qualities have been noted (London Press Show, 27th September 1996).

b) Creation of Acoustic Soundfield e.g. rear channel

Reduced proximity effect, the advantageously weak source localization, plus wide directivity are all design factors for reproducers in creating an even soundfield for ambience signals. The radiation properties of the acoustic panel, distributed mode loudspeaker, are well suited to this application.

c) Stereo — two channel

At first sight the diffused, output from an acoustic panel would indicate a loss of virtual image performance when used for stereo reproduction.

Early demonstration models have given a good impression of height and width. In addition the sense of spaciousness was favourable, while the lack of sharp images reflected from the room boundaries and the good energy spread emanating from the bi-polar radiation pattern were also evident.

Observers reported a positive balance of opinion considering the greater uniformity of the two channel coverage for multiple listeners, this offset by the reduced level of specific image focus for a centrally placed listener, compared with conventional line or near point source reproducers.

In manageable sizes it is expected that many designs of good quality, free mounted acoustic panel loudspeakers will be augmented by conventional low frequency drivers, i.e. local sub woofers for the 20 Hz to 100 Hz range, or perhaps 150 Hz for the smaller examples. Output below 100 Hz is the province of larger panel, e.g. greater than $0.4\,m^2$ or when partially baffled, for example in a ceiling array.

Other applications for this interesting new technology include audio-visual panels for exhibitions, advertising and display. Several grades of panel are unimpaired by decoration, e.g. paint, paper and thin veneers, aiding integration for style and architectural purposes.

REFERENCES

[1]   Beranek, L., *Acoustics*, McGraw-Hill, London (1954)

[2]   Jordan, E. J., *Loudspeakers*, Focal Press, London (1963)

[3]   Small, R. H., 'Direct radiator loudspeaker system analysis', *J. Audio Engng Soc.*, **20**, No. 5 (1972)

[4]   Heron K. H., *Acoustic Radiation From Honeycomb Sandwich Plates*. 5th European Rotorcraft and Powered Lift Aircraft Forum. September 4–7 1979 Amsterdam. Paper 65

[5]   NXT 'White Paper' New Transducers Ltd, Huntingdon, 27th September 1996

[6]   International Patent WO 92/03024, PCT/GB91/01262. Heron K. H. Ministry of Defence (UK). Panel Form Loudspeakers'…"exited above the fundamental and above coincidence".

BIBLIOGRAPHY

Bauer, B. B., 'Equivalent circuit analysis of mechano-acoustic structures', *J. Audio Engng Soc.*, **24**, No. 8, 643–655 (1976)

Borwick, J. (Ed.), *Loudspeaker and Headphone Handbook*, Focal Press, (2nd edn 1994)

Gayford, M. L., *Acoustical Techniques and Transducers*, Macdonald & Evans, London (1961)

Geddes, E., Porter, J. and Tang, Y., 'A boundary-element approach to finite-element radiation problems', *J. Audio Engng Soc.*, **35**, No. 4 (1987)

Kelly, S., 'Loudspeaker enclosure survey', *Wireless World*, 552–8, November (1972)

Olsen, H. F., *Modern Sound Reproduction*, Van Nostrand, New York (1972)

Porter, J. and Geddes, E., 'Loudspeaker cabinets', *J. Audio Engng Soc.*, **37**, No. 11 (1989)

Rettinger, M., *Practical Electroacoustics*, Thames and Hudson, London (1955)

Walker, P. J., 'Wide-range electrostatic loudspeakers', *Wireless World*, May, June and August (1955)

Walker, P. J., 'New developments in electrostatic loudspeakers', *Audio Engng Soc. 63rd Convention Preprint* 1472 (D-10) (1979). Also *J. Audio Engng Soc.*, **28**, No. 11 (1980)

3

Transducers, Diaphragms and Technology

So far we have examined a hypothetical diaphragm — a flat infinitely rigid piston. This is a reasonable approximation for high performance drivers, from the very lowest frequencies to the reference level range. For example, a good quality 300 mm chassis size driver should still approximate to a piston up to 500 Hz; a 200 mm cone driver to 1 kHz and a 25 mm rigid dome unit to 10 kHz, or possibly beyond. While such operation is generally valid with regard to the uniformity of sound pressure response, colouration due to suppressed resonances within this working range may still be audible.

The factors responsible for departures from pure piston performance may be divided into two interdependent categories; first, those due to geometry, i.e. the diaphragm shape, and secondly, those resulting from the mechanical properties of the chosen diaphragm material. Indeed perfect piston operation is not necessarily desirable, a view that apparently is in conflict with some modern designers absorbed in the quest for ultra-rigid diaphragm constructions. Consider the abrupt change in directional properties of a system where the bass unit is a large true piston and the treble, the usual small dome. The resulting step in the power response will undoubtedly upset the uniformity of the reverberant field, especially if the axial response is equalized to flatness.

3.1 DOME RADIATORS

Phase Loss

Even if a dome is made from our hypothetical, perfect material it will have a series of dips or nulls in its pressure response, at frequencies where the path difference between the apex and the rim is a multiple of a half-wavelength of that frequency in air. A phase difference exists near these nulls which causes the axial output to fall gently towards the null. This phenomenon is termed 'phase loss' [1,2] (Figures 3.1 and 3.2).

The frequency f_1 at which the dip occurs is dependent on h, the dome height. A shallow profile will place f_1 high.

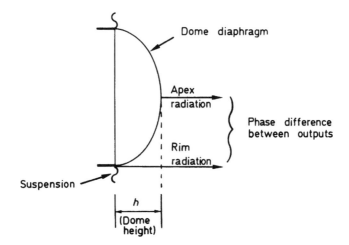

Figure 3.1. Dome geometry and phase loss (first dip when $\lambda = 2h$)

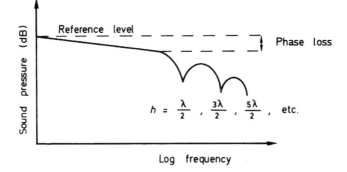

Figure 3.2. Typical dome response dominated by phase loss

For example, in a typical 25 mm doped fabric dome, h is commonly around 7.5 mm.*

$$f = \frac{c}{\lambda} = \frac{345\,\text{m/s}}{7.5 \times 10^{-3}\,\text{m}} = 46\,\text{kHz}$$

and the first dip occurs at $\lambda/2$, i.e. at 23 kHz.

A smaller 19 mm diameter thermoplastic diaphragm, with h typically 3.75 mm, will possess a first null at 46 kHz, well above audibility. However, with some of the larger 34 mm dome HF units, where h may be 10 mm or so, this axial null may occur as low as 17 kHz and a gentle but progressive phase loss droop may be significant from 5 kHz upwards, assuming that no breakup occurs. Phase loss also applies to inverted domes, i.e. cones, but the breakup modes in these generally larger diaphragms usually begins at such low frequencies that the phase loss is unimportant.

Small inverted domes are best described as concave as opposed to the conventional convex type. Their theoretical performance differs in important respects [3]. On axis, the

*5 mm is typical for a metal dome.

concave diaphragm frequency response begins to rise at high frequencies due to the resonance effect of the frontal cavity formed by its shape, at around the frequency where $ka = 1.3$. Flat piston and concave diaphragms have a very similar efficiency and directivity while the convex dome has a reduced absolute radiating efficiency but, in compensation, is less directive except at the highest frequencies (see Figure 3.3(a) and (b)).

Breakup Modes

Even with dome units that are generally driven by a motor coil of the same diameter

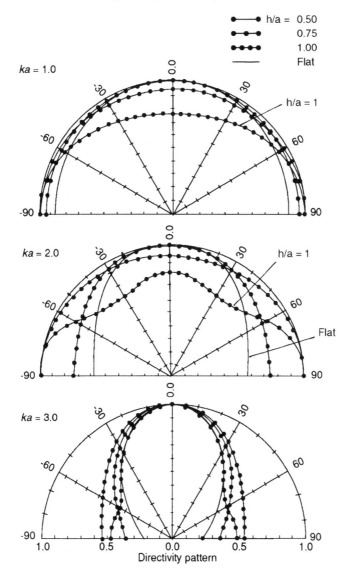

Figure 3.3. (a) Directivity patterns of a convex dome and a flat piston (after Suzuki and Tichy [3]). Note differences between shapes

attached to the rim, the first breakup resonance usually occurs near or just before the first phase null, and the total output will then depend on the sum of the two effects.

The dome profile is chosen with due consideration for the strength, density and damping of the material used. Clearly a flat piston of given thickness is far less rigid and will have a much lower resonant frequency than a dome or cone structure of the same material. For this reason the dome shape may also be exploited to control resonances when they do occur; the contour may be selected so as to place the first resonance as high as possible, always remembering that any undamped resonances are undesirable. The degree of internal damping offered by the diaphragm is a further factor.

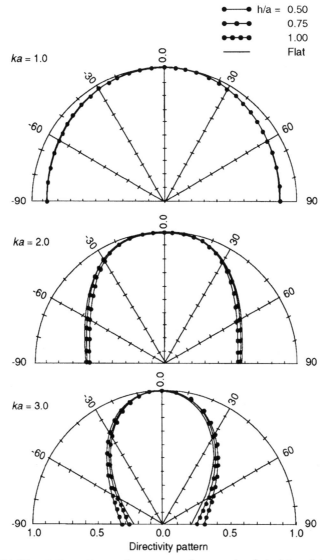

Figure 3.3. (b) Directivity patterns of a concave dome and a flat piston (h/a = height/radius), note close correspondence (after Suzuki and Tichy [3])

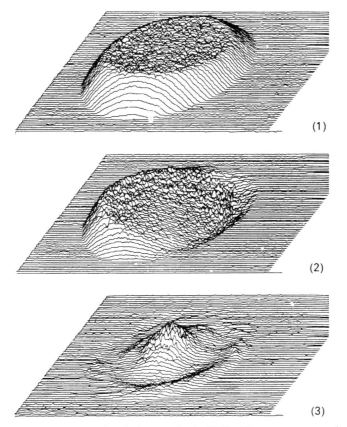

Figure 3.3. (c) (1) 1 in (25 mm) soft dome unit, $f = 30$ Hz (slopes are surround, central area pure piston); (2) $f = 1.2$ kHz, near to fundamental resonance (rocking mode); (3) $f = 9.5$ kHz, axisymmetric model, first central breakup (after Bank and Hathaway [19])

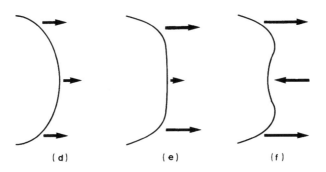

Figure 3.3. (d)–(f) Dome behaviour leading to first breakup: (d) piston, (e) first bending and (f) breakup, centre out of phase with rim

The first breakup resonance is an arbitrary name for the frequency at which the apex area of the diaphragm moves in anti-phase with the rim (Figure 3.3(d), (e) and (f)). The output at resonance depends on the damping and may, if suitably coincident with the phase null, provide some compensation for the latter towards the upper end of the frequency range. The first breakup or resonant frequency f_h can be described as

$$ f_h = 0.523 \frac{T}{Q} \sqrt{\frac{E}{\rho}} $$

where T = thickness, a = piston radius, E = Young's modulus and ρ = density. E/ρ is clearly a crucial factor, suggesting a reduction of ρ and maximization of E (see Chapter 5, Table 5.2).

3.2   VELOCITY OF SOUND IN A DIAPHRAGM

The velocity of sound in a structure is not so much dependent on its density and the elasticity of the material used, as on the actual structure or shape into which it is formed. At low frequencies, where piston operation holds, the sound velocity is

$$ U_s = \sqrt{\frac{E}{\rho(1 - \gamma^2)}} $$

where γ is Poisson's ratio and U_s is the velocity in sheet form. In a cone U_s is high, i.e. three or four times that in air. With increasing frequency the velocity reduces, following an almost hyperbolic law. Breakup modes result in sharp irregular departures from this curve, but a cone with well controlled breakup modes may follow it quite closely. For example, a 200 mm frame size, pulp-paper cone possesses a sound velocity at 750 Hz (approximately its first breakup mode) which is about equal to that in air. With a further increase in frequency, the velocity approaches the value for a flat sheet of the same diaphragm material of infinite size, in this case about 100 m/s or less than one-third of the sound velocity in air.

The concept of sound velocity in diaphragms is not clear cut and may be regarded more as an alternative view of the greater rigidity provided by a cone as opposed to a flat sheet of a given material. It may be argued that regardless of shape the transmission velocity in the surface plane of the medium remains constant and that the apparent velocity of the total forward radiation emitted by the diaphragm is the important factor. Its phase and amplitude, and hence power response, are controlled by geometry and by the distribution of nodal circles and related breakup phenomena.

Hence, we have established another factor, namely a source of 'phase loss' caused by the time difference between radiation from the edges and centre of a diaphragm due to propagation velocity. Strictly speaking, this is another way of looking at the breakup or loss of the rigidity phenomenon. Taking the example of the pulp cone mentioned above, a dip in response will occur when the delay generates a half-wavelength difference in the radiation between the centre and rim; the propagation time lag due to the cone height must also be considered (Figure 3.4).

It is interesting to speculate on the effect of this propagation delay. For pure piston operation the apex or centre radiation lags the rim output for a distant observer, this being equivalent to the dome phase loss discussed. The effect of this propagation delay

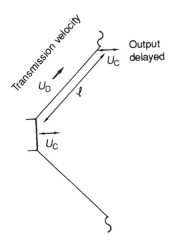

Figure 3.4. Velocity of sound in cones. (Assuming cone is shallow and hence depth negligible, the transit time $t = U_D/l$, where U_D = velocity in diaphragm)

is a retardation of the rim energy, thus bringing it closer to the phase of the centre energy. By suitable control of geometry and other constants, it is possible to balance these two effects to some degree. Inspection shows that this same balance is also possible with a dome over a limited frequency range. In practice both cone velocity and geometric phase loss effects must be considered of secondary importance by comparison with the dominant breakup modes.

3.3 COMPENSATION OF DOME CHARACTERISTICS

In most dome radiators the three effects described above work in unison. There is a propagation delay in the diaphragm which produces a phase lag between the apex and rim outputs, while 'phase loss' due to the height results in a phase lead, and finally the first breakup mode gives a more or less well damped resonance.

With suitable choice of motor design and careful choice of diaphragm thickness, material and contour, it is possible to obtain a usefully wide and uniform characteristic.

3.4 CONE BEHAVIOUR

With dome diaphragms the working frequency range rarely ·extends significantly beyond the first resonance mode described. However, the larger cone-type diaphragms are more susceptible to undesirable breakup or departure from true piston operation at relatively low frequencies, and of necessity they are often used well into their breakup range, even in high performance systems. The best known example is the extensive use of a type of thermoplastic coned, 200 mm and 170 mm frame, bass–mid range unit, covering the range 50 Hz to 3 kHz in compact enclosures of 20–50 litres volume. This is particularly prevalent among UK system designers. It would be no exaggeration to place the first breakup mode in this example as low as 700 Hz, and yet with careful

design the colouration can remain low. The equalized axial response may hold to within
±1.5 dB limits right up to the 3 kHz crossover point.

A dome diaphragm is normally driven at the rim or perimeter and hence the path
length is correspondingly short. In contrast, cones are driven from a relatively small
apex area, the cone diameter generally five or ten times larger. In this case the path
length is appreciable and is responsible for the earlier breakup.

3.5 CONE PARAMETERS

The variables involved in the design of a cone are as follows (Figure 3.5):

1. Material thickness and density, which may be non-uniform: some cones are
    intentionally thicker at the apex and thinner towards the edge.

2. Stiffness (or Young's modulus) of the material used.

3. Cone angle and profile: whether straight sided or flared.

4. Absolute dimensions of the cone: its height, rim and apex diameter.

5. Rigidity and mass of the attached motor coil.

6. Suspension and mechanical properties of the surround.

An elementary model of cone behaviour may be established from vibration theory
but this bears little relevance to real cones, which are often highly variable and hence do
not correspond well with theory. This failure is partly due to the complexity of the
analysis involved but is mainly attributable to the anisotropy of the cone material. This

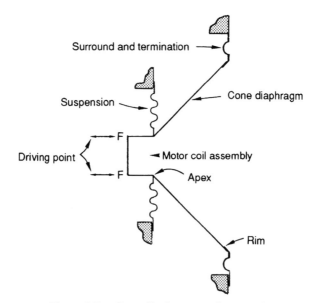

Figure 3.5. Cone diaphragm and suspension

is commonly an impregnated felted paper, whose mechanical properties can vary up to 100% in a random manner over the surface area. However, the use of vacuum formed plastics for cones and recent improvements in paper technology have resulted in rather closer correspondence with the theoretical predictions.

While it is possible to make some analysis of free cones a practical diaphragm will have its behaviour strongly modified by the effects of the surround, compliance, resistivity and, to a lesser extent, mass. In addition the motor coil mass must also be included — a highly significant factor at higher frequencies. A transmission line model may be employed for energy propagation in a diaphragm and although of some interest one has to make strong assumptions that all the breakups are numerous, well distributed and well damped.

Transmission-line Analogy

A cone possesses finite stiffness, distributed mass and some mechanical damping. Assuming for the moment that the damping is high with negligible 'free' resonances, the area of the cone vibrating will tend to contract towards the driving point with increasing frequency. Thus the outer areas become increasingly decoupled from the centre due to the elasticity and mass of the material and the cone may be likened to a lossy transmission-line or low-pass filter (Figure 3.6).

This aspect of behaviour may be exploited by suitable cone design. One effect of a reducing cone area is to maintain an adequate directivity to higher frequencies than would be expected from this size of driver, while the reducing mass results in augmented efficiency as the working cone mass approaches the mass of the motor coil.

Termination

Sound energy applied to the cone apex by a motor coil propagates towards the rim. It is undiminished at low frequencies but will be attenuated at higher frequencies due to the low-pass transmission filter effect described. If poorly terminated, as in the extreme case

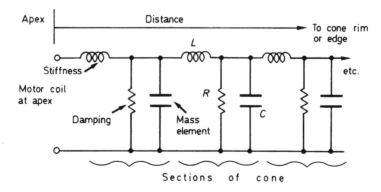

Figure 3.6. Transmission line analogue of loudspeaker cone (direction of reducing cutoff-frequency from left to right, i.e. low-pass filter action)

when the edge is directly clamped to the frame (a short circuit), or alternatively left free (an open circuit), the incident energy will be strongly reflected back towards the apex. This will not only interfere with the prime outward travelling energy, thus setting up standing waves, but it may also excite more complex vibrational modes in the cone.

For this reason, the surround must present a well matched resistive, or lossy mechanical termination to the cone rim. It should efficiently absorb the incident energy at higher frequencies while not introducing excessive mechanical loss at low frequencies where the cone excursion may be considerable. The surround profile should also be dimensioned so that it has no significant self-resonances (see Chapter 5).

The Perfectly Homogeneous Cone

In theory, a homogeneous or uniform cone should possess only one group of nodal resonances, i.e. these occurring in the axis of excitation (see Figure 3.7(a), (b)). This behaviour is illustrated by the aluminium-foil/polystyrene-foam core 'sandwich' cones, which are extremely uniform or isotropic over their surface area.

Normal Cones

While the basic nodal or concentric resonances certainly occur in normal cones, they may be heavily disguised by other modes which can be excited as a result of diaphragm non-uniformity. The concentric modes are present at high frequencies, while lower frequency irregularities may be produced by 'bell mode' or radial vibrations (Figure 3.8). Since their greatest amplitude is at the rim, both the rim contour and surround are critical factors in the control of these breakup effects. The 'bell' modes may be properly terminated, but the mechanical impedance match differs from that required to terminate the longitudinal or concentric mode energy.

With certain examples, notably flared thermoplastic diaphragms [4], a suitable high loss surround, e.g. plasticized pvc, may be capable of successfully dealing with both modes. The surround's bulk properties in sheet form terminate the concentric modes while an extra flat section of surround profile controls the radial modes. With the pvc materials finding less favour on chemical safety grounds, grades of nitrile rubber have also proved to be a good substitute, together with pvc/rubber mixtures (Figure 3.9).

Subharmonics and Rocking

The mechanical non-linearities in a vibrating cone can give rise to subharmonics of the fundamental excitation, that is at one-half, one-third frequency, etc. although only the first mode is generally of significance. Subharmonics usually occur in thin straight sided cones. Cone angle and the other associated constants are all influential factors.

Subharmonics are more common in planar diaphragms, such as leaf or ribbon elements, and may severely restrict the usable range at lower frequencies. In this category may be included one more mode, namely rocking, or as Japanese engineers call it, rolling. Here the diaphragm literally fails to move axially due to poor balance or suspension stiffness asymmetry (see Figure 3.3(c,2)).

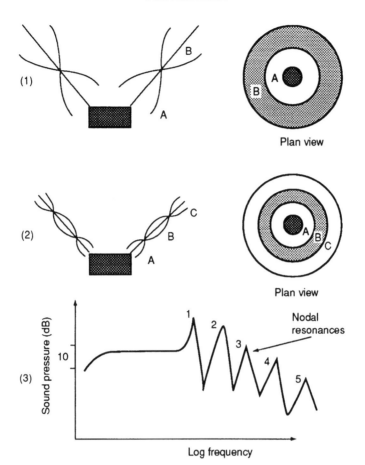

Figure 3.7. (a) Nodal resonances (concentric); (1) first nodal resonance, (2) second nodal resonance, (3) effect of nodal resonances on response

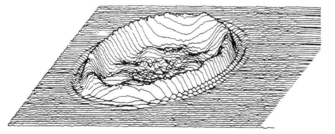

Figure 3.7. (b) 8 in (200 mm) curved paper cone, with surround, $f = 2.9$ kHz, bending resonance (nodal). (Celestion laser scan)

Cone Material Damping

The cone resonance modes may also be controlled by using a material that has a high internal damping. Although many cones are described as being made of paper, this material is more correctly specified as a dense felt, impregnated with a suitable stiffening

---

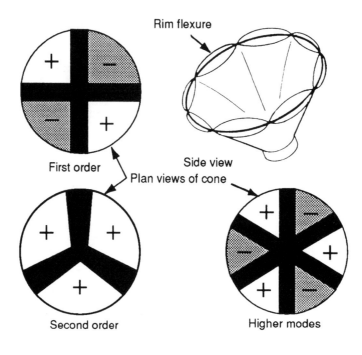

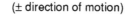

(± direction of motion)

Figure 3.8. (a) Radial or circumferential 'bell' modes

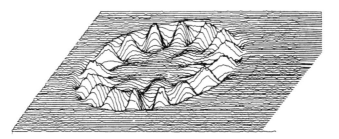

Figure 3.8. (b) 8 in (200 mm) straight-sided paper cone, with surround, $f = 2$ kHz, 16th order bell mode. (Celestion laser scan)

agent. The loss factor may be varied at will, but it should be noted that the high frequency response extension will be strongly related to the amount of damping. With thermoplastic cones the polymer may be selected for its high damping coefficient (e.g. Bextrene, a polystyrene/neoprene mixture, or polypropylene with good self-damping) and may have an additional surface treatment or coating such as polyvinyl acetate.

Experience has shown that it is better to get the cone behaviour right and employ matched surrounds for good mode termination rather than rely on internal damping. The latter can only ameliorate diaphragm shape faults, not solve them. The very best examples of cone driver employ every favourable combination of construction and material to achieve their high performance.

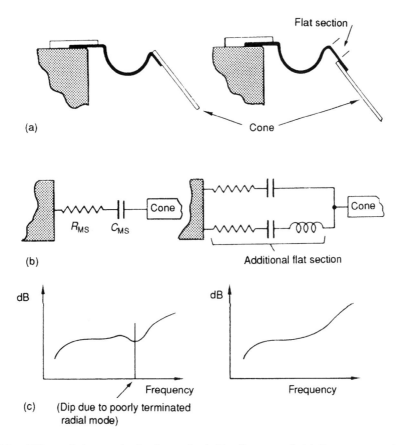

Figure 3.9. Effects of changes in the form of a half-roll surround. (a) Form: conventional and
with flat section added, (b) equivalent circuit, (c) sound pressure response

3.6 CONE SHAPE

It has been demonstrated that cone shape is of great importance, and this fact cannot be
emphasized too strongly. Even where a basic shape has been determined in advance, if
the highest standards are demanded, as many as 10 or more cone profiles may be
experimented with before the optimum is finally obtained. Individual cone shapes will
now be examined in greater detail.

Straight Sided or True Cones

The extreme case of a wide angle cone (i.e. 180°) is a flat disc. If light and thin, it will
possess very little rigidity, and hence will enter breakup at quite low frequencies. This
may be regarded as a special case which is seldom encountered. (See resistance
controlled moving coil driver in Chapter 5, Section 5.6).

  Moving coil drivers with rigid flat diaphragms have become popular with some
designers on the grounds that linear phase is an important objective. With flat

diaphragms, in theory at least the same plane can be easily aligned for all types — bass, mid and treble drivers.

Special structures are required to produce the vastly increased stiffness needed to place the first bending mode well above the required operating range for each planar driver. Foil skinned aluminium honeycomb structures have been used with special care taken over the driving points to minimize resonance excitation. This implies placing the driving points at the first resonance node or minimum and has required very large diameter motor coils or multiple coil driver assemblies [5].

As the cone angle is reduced from 180° the axial stiffness will increase, with a consequent rise in wave velocity in the cone. The first resonance then occurs at a higher frequency and both circumferential and concentric modes appear (Figure 3.10). With further angle reduction the increased axial rigidity continues to defer the first appearance of resonance and the concentric modes increase in strength lifting the high frequency response, often with considerable irregularity. Somewhere in this range, with a given material, there may be an optimum angle where the smoothest extension of response is obtained without serious lower frequency resonances.

Curved Profiles

One obvious defect of a straight sided cone, namely its strong circumferential mode resonance, may be effectively suppressed by the adoption of a flared or curved profile. This confers greater rigidity with respect to these bending resonances.

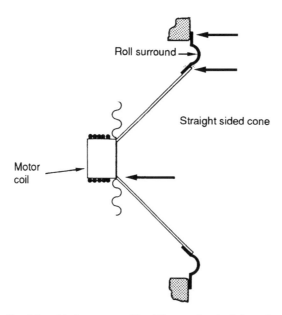

Figure 3.10. Straight sided cone profile. The mechanical impedance changes at the points arrowed, due to both material and rapid contour changes

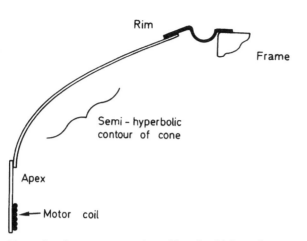

Figure 3.11. Example of cone surround profile of a high performance 305 mm driver (after Harwood [4])

A further defect concerns the discontinuities or corners presented by a straight sided cone at the two junctions; motor coil and apex, and rim and surround. A flared cone may be shaped to provide a smooth contour from motor coil to rim surround, which considerably reduces the severity of the standing wave reflections at these boundaries. A flared cone will give less output over the upper resonant frequency range than will a straight sided cone, and hence will not be as 'loud', but its overall range exhibits greater uniformity and hence it is more suitable for high performance purposes. However, the flare must not be too shallow, otherwise the cone as a whole may prove insufficiently rigid to act as a piston under high-power, high excursion conditions at low frequencies. Additionally an excessively open flare may allow the outer area to flex in anti-phase to the main output at mid frequencies.

Flares close to a hyperbolic law (Figure 3.11) have been successfully manufactured in many materials ranging from doped paper-felt to thermoplastics, and also in metals, notably aluminium and titanium.

While for theoretical purposes the motor coil or driving point is assumed to be rigid, in practice this is not so. Depending upon the apex geometry, adhesive stiffness, whether the neck or apex is open or closed, the behaviour of the whole diaphragm may be affected [6]. Usually the apex is open, but recently designed thermoplastic cones have been produced with the apex 'blank' section left intact as a stiffener. Alternatively, a rigid plug may be fitted or a stiffening ring employed. Coil diameter is also a crucial factor here and results with a 25 mm motor coil may not be reproducible with a 33 mm coil for the same size of diaphragm.

Analysis of a Straight Sided Cone

The speaker research division of Philips in Holland developed a method of analysing straight sided cones [7]. The typical form illustrated in Figure 3.12 is subject to bending (circumferential) and longitudinal (concentric/nodal) resonance modes. The arc of the bending wavelengths is typically much shorter than those equivalent frequencies in air, and hence has less effect on the output than the longitudinal resonance wavelengths

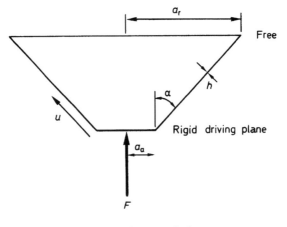

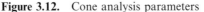

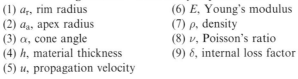

Figure 3.12. Cone analysis parameters

(1) a_r, rim radius   (6) E, Young's modulus
(2) a_a, apex radius   (7) ρ, density
(3) α, cone angle   (8) ν, Poisson's ratio
(4) h, material thickness   (9) δ, internal loss factor
(5) u, propagation velocity

whose dimensions are generally much greater than in air. The latter thus produce phase cancellation effects with consequent irregularities in the axial pressure response and the polar distribution.

In the analysis, the cone is described by four geometrical and four material parameters. The apex is rigidly driven by an axial sinusoidal force whose magnitude is independent of frequency. The radiation impedance is assumed to be negligible (this is valid for most 'hi-fi' drivers with efficiencies of 0.5%–2.0%), and the outer edge is assumed 'free'. Internal losses are also taken into account.

The breakup behaviour is represented by 12 simultaneous differential equations which were numerically solved over a wide frequency range. In the example quoted a paper-felt cone with an included angle of 100° and an outer diameter of 160 mm, was measured and its sound pressure and breakup patterns calculated.

Figures 3.13 and 3.14 illustrate the theoretical and measured curves. The discrepancy at low frequencies between the measured and calculated output is due to the additional contribution from the surround, a factor neglected in the theoretical approach. Likewise, the additional measured output around 10 kHz is attributed to the finite rigidity of the motor coil as opposed to the infinitely rigid structure assumed by the theory, and clearly the latter is quite realistic. The analysis also indicates that the transmission line analogue (Figure 3.6) is oversimplified.

Finite element analysis [8] has proved to be a powerful though complex technique for the analysis of vibrating structures, both for the transducer diaphragms and also their enclosures. Simplified analysis is also possible (see Figure 3.15, Shindo *et al.* [6]). Using a similar design of driver, an interesting comparison of conical, convex and concave diaphragms has been made [6]. Here the practical effects of surround compliance and motor coil mass have been accounted for in the analysis.

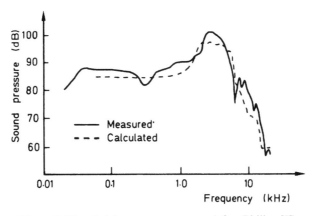

Figure 3.13. Axial pressure response (after Philips [7])

The behaviour may also be modelled using a quite simple equivalent circuit (Figure 3.16). In the figure example values of $M_c = 6.4 \times 10^{-3}$ kg, $M_m = 2.3 \times 10^{-3}$ kg. $C_s = 1.6 \times 10^{-2}$ n N^{-1}, the first eigenfrequency above free air resonance (at 23 Hz) is at 2 kHz approximately. Higher order eigenfrequencies or resonances follow with increasing frequency.

3.7 MOTOR SYSTEMS

Up to now this chapter has mainly been concerned with pistons and diaphragms driven by hypothetical motors, i.e. convertors of electrical power into mechanical force. However, the elements of the moving coil motor were introduced in Section 2.3 dealing with the analysis of a direct radiator cone driver and these will now be examined in greater detail.

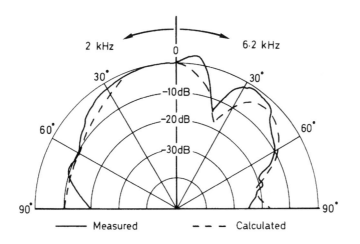

Figure 3.14. Directivity at two frequencies, 2 kHz and 6.2 kHz (after Philips [7])

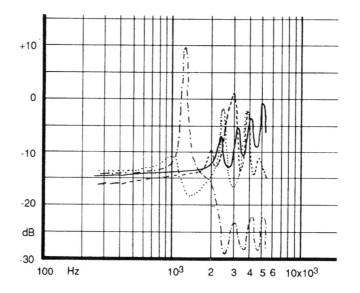

Figure 3.15. Comparison of three 100 mm cones. (With coil and surround) - - - - conical,
- - - - - - convex, - · - · - concave. ———— conical (Without coil or surround) (after Shindo *et al.* [6])

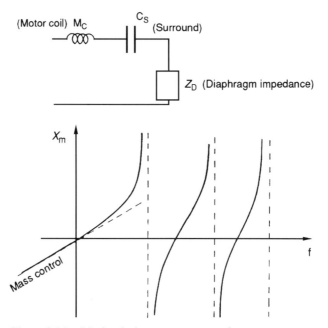

Figure 3.16. Mechanical reactance versus frequency

In theory, a diaphragm may be energized by a number of transducer methods, with the diaphragm characteristics (mass and area) being dimensioned to match.

Table 3.1.   Some examples of film-type transducers

Electrostatic
(1) Dayton Wright Electrostatic — An array of convex elements with a sulphur hexafluoride filling to increase dielectric strength and hence allow higher voltage operation. A horn super tweeter was often used with this system.
(2) Quad Electrostatic — Classic two-way system of adequate power capacity. Superseded by ESL63 of better directivity and matching. Multi-electrode concentric structure.
(3) KLH9 — Electrostatic multi-way system in a tall, vertical panel format with strip elements for the higher frequencies.
(4) Acoustat — A large angled panel full-range system with direct-coupled valve amplifier eliminating transformer/polarizer — other passive versions available. Figures 3.20, 3.21.

Magnetic
(1) Magneplanar — A large panel transducer with a mylar film diaphragm using copper wire bonded over its surface area.
(2) Strathearn — A magnetic drive film transducer with a transformer coupled single foil turn and strip magnets (mid and treble range only), Figure 3.22.
(3) Apogee — Differentially tensioned trapezoidal surface conductive film, plus true ribbon upper frequency section.

Film Transducers

The large surface area transducer has always had a following. If uniformly driven over its surface, a light low mass diaphragm will enjoy a dominant resistive air loading which will tend to suppress resonances. Theoretically such a transducer should also be efficient, due to the excellent radiation coupling. In practice, however, engineering limitations reduce the efficiency to a level at or below that of the average moving coil driver. The best known and longest lived example is probably the original Quad loudspeaker (Figures 3.17 and 3.18), and its successor, the Quad ESL63, which employs electrostatic drive to a light plastic film diaphragm [9]. This is a two-way, push-pull system with a high surface-resistivity diaphragm ensuring constant charge operation. The two elements of push–pull electrodes and constant charge impart linear operation up to fairly high frequencies and amplitudes, a performance not possible with the simple electrostatic driver employing a single fixed electrode.

The greatest problem with large area diaphragms is directivity. With increasing frequency the response angle narrows in the forward axis until it becomes too critical for normal use. As a general rule, for a frequency range up to 15 kHz, the radiating diameter of the source should not exceed 40 mm for reasonable directivity at the highest frequencies. Ideally the diaphragm should possess a cylindrical or, better still, a spherical contour, but this poses great constructional difficulties. Often the frequency range will be split between different sized drivers. A vertical strip element provides improved directivity in the horizontal plane, though often this is achieved at the expense

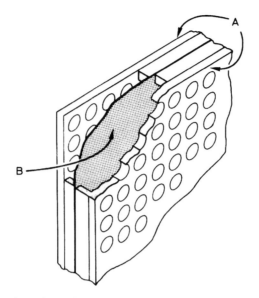

Figure 3.17. Section of an electrostatic transducer. A: Front and rear perforated electrodes (conductive). B: Very thin film diaphragm of high surface resistivity (10^{12} ohm/square) (after Jordan [20])

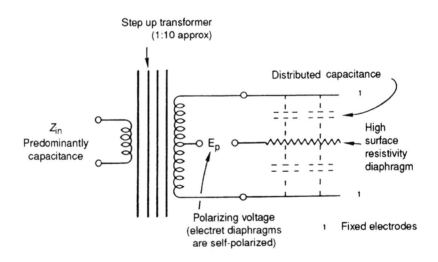

Figure 3.18. Simplified electrostatic driver circuit (conventional)

of the directivity in the vertical plane. For a very tall loudspeaker, e.g. over 2 m, the listener is placed in the near field and thus avoids this effect.

In addition, most electrostatic systems are run at a high polarizing voltage, 5–7 kV, and need substantial screening for safety and leakage reasons. Such precautions can impair the performance. The high audio driving voltage required may be achieved with wide band, high voltage, low ratio transformers fed from integral valve power amplifiers, but more usually a high ratio step-up transformer is employed. The system is fed from a

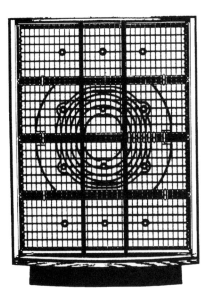

Figure 3.19. The Quad ESL63 (see Figure 8.19)
with grille removed

conventional low source impedance power amplifier. The input impedance characteristic of such a system is often regarded as difficult to drive,* approaching a short circuit at very low frequencies. The impedance is initially strongly inductive, and then becomes increasingly capacitative, sufficient on occasion to cause oscillation with some amplifiers.

Almost 20 years of research by one designer, Peter Walker [10], has resulted in a set of solutions which go a long way to solving a number of the above problems. In its latest form his ESL63 electrostatic system is conceived as essentially one large flat diaphragm $0.6\,m \times 0.8\,m$, approximately. The analysis is largely unaffected by the practical constraints which necessitated dividing the area into more easily handled sub-panels where the required rigidity of the fixed electrode structure can be more easily attained. The uniform surface resistivity radiating area is driven by an electrode structure arranged in the form of a set of concentric rings. A high voltage audio delay line of critical construction feeds the electrode elements, so energizing defined areas of the diaphragm with power of calculated frequency response, phase delay and amplitude. A phased array is thus produced where the directivity and response may be closely defined. In the case of the ESL63 the system is corrected both for edge discontinuity and the clamped boundary, and is also adjusted to produce an expanding wave front. This simulates a point source $0.3\,m$ behind the diaphragm. The plastic film diaphragm has a fine gauze material in loose contact to aid damping, and the whole is enclosed in an acoustically transparent plastic film enclosure to help reduce the damaging effects of dust and humidity on the system. Both these and the necessary metal mesh grilles inevitably impose some small compromise on the performance.

The success of this multiple concentric ring approach to planar speaker design may be judged from the amplitude frequency response taken at $2\,m$ ($1\,m$ is too close owing to proximity effects). First of all, the axial response is particularly smooth and well balanced while in addition the $15°$ and $30°$ off-axis responses show comparatively little

*See Figure 3.20.

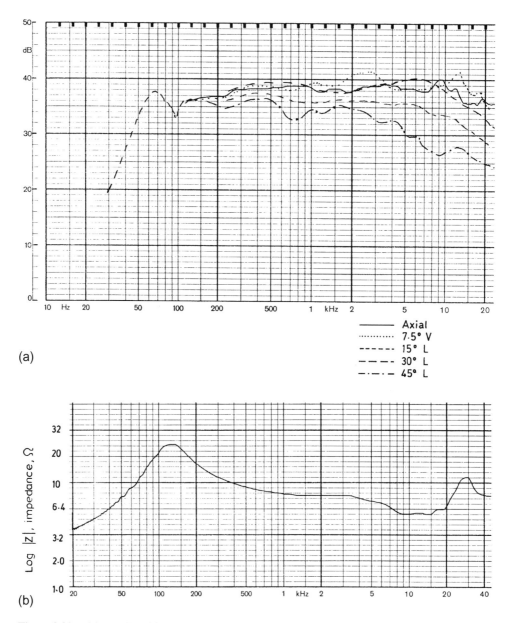

Figure 3.20. (a) Family of forward responses at 2 m anechoic. Note well damped bass resonance and the uniformity of off-axis responses. (b) The impedance curve falls at low frequencies due to the finite primary inductance of the matching transformer, and at high frequencies due to predominantly capacitative loading of the transducer

variation showing very good consistency over a 60° forward directed arc. By 45° off-axis, the fall off resembles a normal speaker in magnitude but shows commendable uniformity with no sudden changes in slope. Thus the usual boundary and directive effects have been well controlled in this design. As regards load impedance the plot of

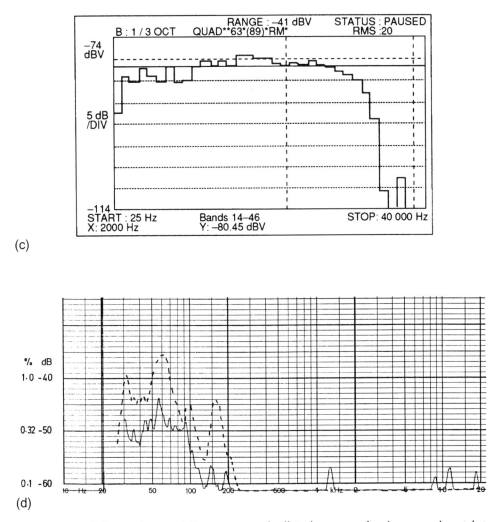

(c)

(d)

Figure 3.20 (c) Room Averaged Responseover the listening space showing a good match to room environment and good correlation with the anechoically derived trends. (d) Measurement of distortion for this push–pull electrostatic at 86 dB s.p.l. illustrating the low levels achieved over most of the range, particularly of third harmonic, typically −70 dB, 0.032%

impedance modulus versus frequency shows a fairly smooth characteristic not falling below 4 Ω from 50 Hz to 20 kHz and averaging 7 Ω over most of the range.

A di-polar speaker such as this excites side wall modes at low frequencies rather less than a normal box speaker and the fine room-averaged response at the listening position is a testament to the good interfacing ability of this model when placed in a medium sized, well proportioned listening room. The speaker was placed about 1 m from the back wall on 30 cm stands, the units angled inwards by about 12°. The low frequency distribution is particularly consistent down to the effective working limit of 35 Hz (see Figure 3.20(a), (b), (c) and (d)).

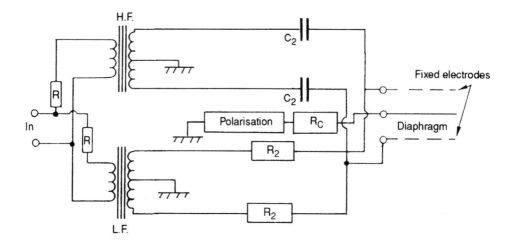

Figure 3.21. Double transformer drive with RC crossover improves the input impedance of electrostatic with a wider bandwidth (compare with Figure 3.18) (after Strickland [11])

In the early days of electrostatic speaker design, single plate electrodes with constant voltage polarization were employed. The force equation is non-linear with a second order term, $F \propto V^2$, producing second harmonic distortion. $F = Eq$ also governs the behaviour where $E = $ electric field and $q = $ static charge. Modern electrostatics set q constant by employing a high resistance to the diaphragm or using a high resistivity diaphragm, ensuring constant charge operation. In theory the drive equation is thus perfectly linear over the full diaphragm excursion, orders of magnitude superior to other systems. At 90 dB s.p.l. total harmonic distortion of 0.03% is possible at middle frequencies (see Figure 3.20(d)).

Sheathed Electrodes

Commonly electrostatic electrodes comprise rigid perforated metal or metallized plates. These must be free of self-resonance and of sufficient window area to minimize reflections and allow easy egress of the diaphragm output. Polarizing potentials above 5 kV are desirable to improve sensitivity but higher voltages generally result in breakdown under adverse climatic conditions (see Figure 3.21).

At the typical 5 kV used, contact with the fixed electrodes by a diaphragm under full excursion results in a discharge arc sufficient to burn a small hole, eventually impairing performance. Complex protection schemes may be devised to reduce this effect (Quad) or an alternative method of construction [11] may be used. Originally developed by Janzen (USA) the fixed electrodes may be formed by a grid of insulated or sheathed wires supported on an open cellular structure. The sheath dielectric properties are deliberately chosen (a grade of pvc) to provide a controlled high resistivity. These share a proportion of the total polarizing voltage with the air gap (see Figure 3.22).

Air gap gradients exceeding $1 \, kV \, m^{-3}$ are approaching the ionization limit, with increased noise, and in the sheathed electrode system the wire to diaphragm spacing is

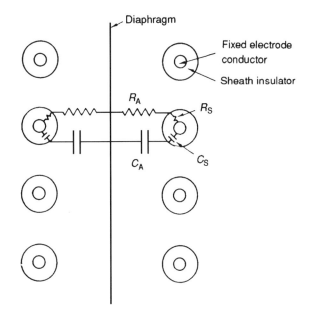

Figure 3.22. Sheated electrode construction. R_A=airgap resistance, R_s=sheath resistance, C_A=airgap capacitance, C_s=sheath capacitance (after Strickland [11])

held at 2.5 mm. An additional benefit is the electrical 'soft landing' provided when the diaphragm contacts the sheathed wire. The insulation remains intact and no arcing can occur.

Alternative Forms of Film Transducer

Other large area transducers have employed a variant of the moving coil whereby the 'turns' are laid out uniformly over the area of the film diaphragm, with the necessary static magnetic field supplied by an array of powerful magnets. Sufficient gaps must be provided in the steel structural backing plate to allow the sound energy produced to escape (see Figures 3.23 and 3.24).

It would be wrong to suppose that for a film diaphragm the beneficial air mass to diaphragm ratio will eliminate resonances entirely. In fact, a number of more or less well damped breakup modes are present which are akin to those occurring in a thin plate clamped at its edges. With some designs additional damping has proved necessary in the form of a local sympathetic vibrating membrane and/or surface diaphragm coatings. This is particularly true of the higher mass magnetic diaphragms which often employ nodal clamps (see Figure 3.25(a) and (b)).

Large Film Transducers

In an effort to avoid the particular subjective distortions present in cone-type speakers and their usual box enclosures, many high end designers have attempted open baffle

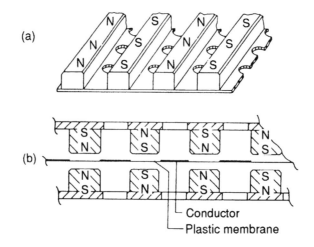

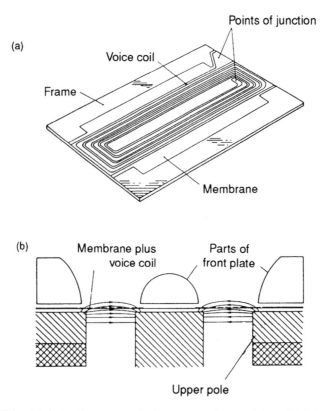

Figure 3.23. Construction of a leaf or pseudo ribbon tweeter with slotted or pierced plates

Figure 3.24. (a) A double aperture, leaf tweeter, pole assembly and (b) the stretched film element employed (after Nieuwendijk)

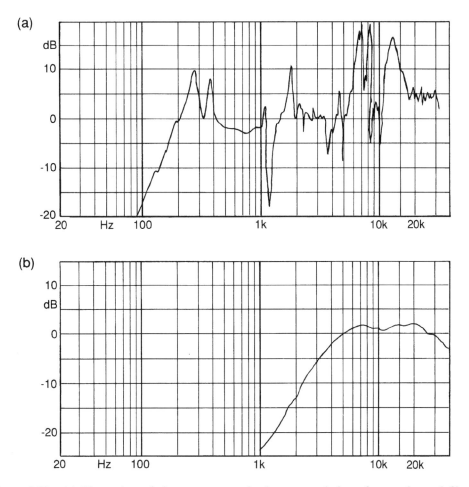

Figure 3.25. (a) Illustration of the resonant modes in a poor design of an undamped film 'ribbon' unit (nearfield measurement). (b) A ribbon-film HF unit with satisfactory damping achieved via a thin mat of very fine glass fibre (after Nieuwendijk)

designs. These generally avoid the use of a cabinet, opting for a more accurate sound by substituting large surface driven film elements for the transducer in place of small drivers. In some cases these film models have been described as ribbon designs, but this term should only be applied where the moving element is essentially freely suspended and can move as one in the defined axis of motion. Many of the so-called 'ribbons' are stretched-film transducers, clamped at all edges and subject to complex plate and similar resonance modes. Steps must be taken to control these if the whole point of the exercise is not to be lost. In one design series the element is a stretched mylar film with a copper wire conductor array bonded to it. A number of clamps are successfully applied at specific points over the diaphragm area to control major resonances (Magnepan). In another film design the diaphragm is horizontally pleated like a ribbon in order to inhibit harmonic bending modes. Its composite aluminium foil–Kapton film element self-dampens fairly well while its semi-trapezoidal tapered geometry also avoids the symmetry that would otherwise encourage resonant modes (Apogee bass section, see Figure 3.26). Despite major theoretical difficulties encountered with practically sized open-backed design, several design aspects have been exploited to overcome these, although the Q of the LF modes is often rather high.

In the example cited, the film panel operates up to 500 Hz, the range beyond covered by a tall, true ribbon element, 25 mm wide by 2 m high, mounted asymmetrically in an acoustically significant baffle section (see section on Line Sources). The ribbon consists of three vertical conductive paths of aluminium film bonded to a flexible Kapton foil substrate and a resistance of $3\,\Omega$ is achieved without recourse to a matching transformer. For the bass panel an array of bar magnets is bonded to a perforated steel backing plate to create a series of horizontal alternated magnetic fields. Employing single sided drive only, the pure aluminium diaphragm film is slotted horizontally in a zig-zag manner creating a long current path whose elements are correctly aligned over the magnetic field pattern. The bass radiating sheet is essentially surface driven as one element. With such large radiating areas the power handling capacity is very high and thermal compression effects are quite negligible. At normal s.p.l. the distortion is very low, comparable to good electrostatic design, though when driven hard the one-sided field results in a fairly harmless increase in second harmonic distortion or 'doubling' at low frequencies.

Figure 3.26. A selection of large magnetically driven two-way systems. Film LF–MF, true ribbon HF (see Figure 4.8)

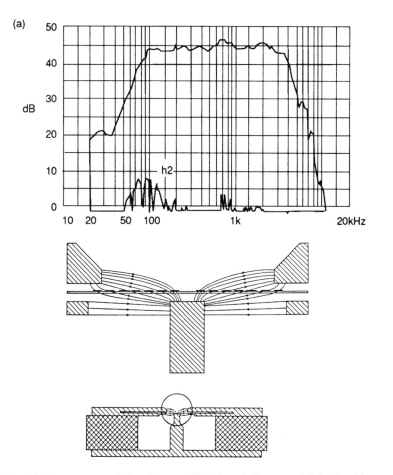

Figure 3.27. (a) Response possible with a well designed film-type 'ribbon' midrange. (b) Its construction involves a split-top pole section driving a single 10 μm polyamide element, 100×80 mm with a surface conductor pattern in 20 μm aluminium foil (after Nieuwendijk)

With careful system design the result can also be very close to linear phase over a wide bandwidth.

By moving from a circular or square shape to a tall, elongated form, especially one with an asymmetric, angled side, the rolloff at the baffle cancellation corner frequency is softened, and the resulting response smoother and more extended. In the L.F. region a clever technique of differential tuning is employed. The diaphragm tension is separately adjustable for the upper and lower sections of diaphragm. The upper part is tuned to typically 60 Hz filling in the otherwise declining range below 100 Hz, while the lower section tuning is set away from the harmonic multiple, at 37 Hz or so, continuing the response down to 27 Hz under listening room conditions.

The Q of the lowest mentioned resonance tends to be well above unity and further development of this type of design is anticipated using a degree of acoustic resistance for Q_t control. The sensitivity is typically 86 dB/W (2 m measurement) with a resistive load impedance.

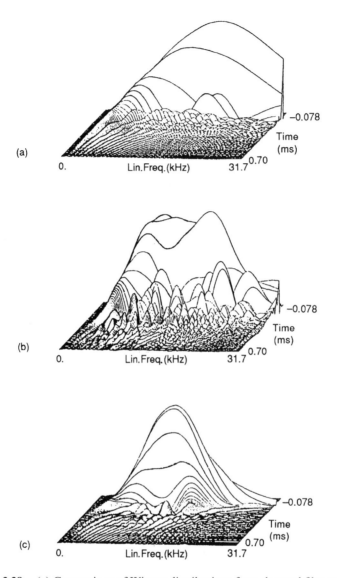

Figure 3.28. (a) Comparison of Wigner distributions for a damped film-type 'ribbon' midrange. (b) A 50 mm 'hi-fi' midrange dome. (c) An experimental 300 mm circular electrostatic midrange unit. Much information is conveyed here for further analysis while the fundamental behaviour of the drivers is confirmed (after Nieuwendijk)

To date, no full range film loudspeaker has proved to be particularly cost effective or electrically efficient. Due to a rigid constraint of field strength versus allowable diaphragm displacement, which restricts the resulting sensitivity, there are definite limits as to the maximum sound level that can be attained. For the magnetic types, the sheer cost of the high-energy magnets is prohibitive, and on the electrostatic side, the combination of voltage breakdown and the reactive input impedance of the capacitive element together act to restrict the maximum acoustic power output. Where high levels

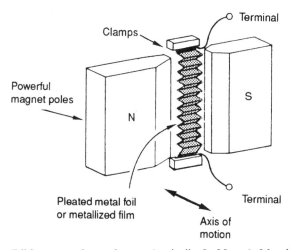

Figure 3.29. Ribbon transducer element (typically 5×25 mm). May be horn loaded to improve air coupling

are required, these large systems are also considerably more directional than well designed direct radiators, although with care the directivity may still be adequate over the important part of the range (see Figures 3.27(a) and (b) and 3.28(a), (b) and (c)).

In recent years, electret elements have been developed for use in headphone sets and microphones, where the polarizing potential is present in the form of a permanent charge akin to magnetization. Large drivers for loudspeaker applications may well be developed along these lines. Some designers have developed hybrid designs that combine moving coil bass systems with film transducers for the upper range.

The Ribbon

The so-called 'ribbon' is a variant of the film transducer and comprises a thin corrugated conducting foil placed between the poles of a powerful magnet [12], thus constituting a single turn, very low impedance coil (Figure 3.29). As with the normal solenoid type motor coil, the motion is at right angles to both the coil and magnetic axes, the ribbon equating to a diaphragm element clamped at each end. It is difficult to produce large ribbons, due to problems of generating a sufficiently high field strength over a wide magnet gap; the transducer may be horn coupled to achieve reasonable air matching and efficiency. A transformer is also required to match the amplifier to the very low impedance of the conducting element (see Figure 3.30)—unless a multi-conductor form is adopted.

A Stabilized Ribbon for MF–HF Duty

On the basis of some empirical private investigations carried out by the late Carl Pinfold, an interesting development of the medium size ribbon transducer has been designed and launched commercially by Celestion. Intended for mid and high frequency

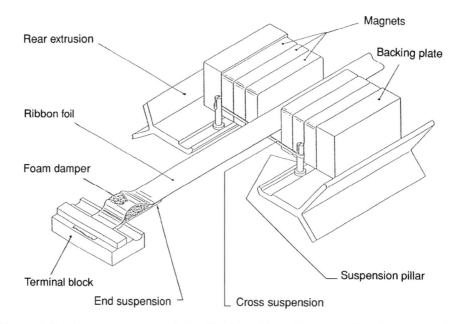

Figure 3.30. The construction of the Clestion ribbon (patent applied for). A number of interesting features are included (only one section of magnets is shown).

Rear Extrusion — This consists of a strong box section aluminium extrusion onto which all the main parts of the ribbon are fixed and which in turn is screwed into the cabinet.

Magnets — Five sets of strontium ferrite bar magnets provide the magnetic field across the ribbon diaphragm. The ribbon always see a linear magnetic field due to the magnets being much deeper than the usual maximum excursion of the diaphragm.

Backing plate — Two steel bars that hold the magnets as well as tuning the field for optimum strength and consistency.

Ribbon Foil — A 12 μm strip of aluminium foil which is corrugated for longitudinal stiffness. Since the foil is of low mass but with a large area, heat is easily dissipated so that the ribbon does not suffer from thermal compression problems greatly.

End Suspension — This is moulded from high temperature silicon rubber and provides the end termination and correct tension for the ribbon foil. Because the ribbon is not directly clamped at its end, mechanical fatigue in the foil is greatly reduced, therefore in normal use the ribbon will not suffer fracture at this point. This has been proved by the ribbon passing the same power tests that are used for moving coil treble units. The ribbon is tensioned so that any thermal expansion is taken up by the end suspension.

Foam Damper — This damps out any motion in the part of the ribbon foil between the point where it is attached to the end suspension and the end clamp. It is made from fire-retardant foam to ensure longevity.

Cross Suspension — To ensure that the ribbon diaphragm is central in the magnetic field and does not rub during large excursions, four silicon rubber cross suspensions stabilize the foil. These are made from the same material as the end suspensions.

Suspension Pillar — These hold the cross suspensions in place and provide the necessary adjustment by rotating the pillars.

Terminal Block — There is one of these at each end for holding the end suspensions and giving a point from which to take the electrical connection to the ribbon foil. The end suspension is held under a spring to enable a replacement diaphragm to be easily fitted and the end of the foil is clamped under a gold plated brass terminal to which is soldered the lead from the crossover board (after Celestion)

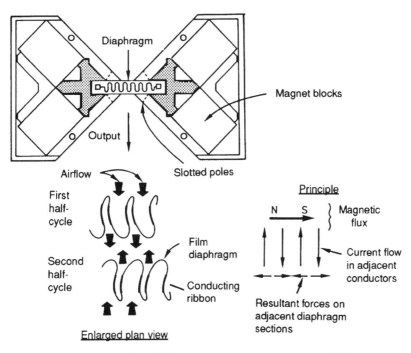

Figure 3.31. Heil 'air-motion transformer' (after ESS)

service the unit employs inexpensive ferrite bar magnets set at a spacing where gap geometry and natural fringing together result in a satisfactorily uniform distribution of magnetic flux across the ribbon. This is essential if the moving element is to act as one, and not fail prematurely due to bending fatigue. The horizontally pleated or corrugated, $12\,\mu\text{m}$ aluminium foil ribbon operates under relatively constant tension regardless of ageing and other effects, thanks to its thin silicon rubber suspension elements (see Figure 3.30). Normally such ribbons are self-supporting and they tend to slacken with age or when driven to high power levels. A small foam damper controls the self-resonance of the foil loop which provides for the electrical termination.

The alignment of the delicate ribbon within the long magnet gap is critical and any contact may damage the ribbon and cause severe buzzing and distortion. In consequence, many designers have resorted to wide gap clearances which reduce efficiency due to increased front to back leakage cancellation. In the Celestion refinement adjustable cross suspensions are employed, these of extremely low mass, made of compliant silicon rubber threads. Four such suspensions are used over the ribbon length (height) and these are screw adjustable to control relative tension and the precise alignment of the ribbon in the gap. This allows for tight gap tolerancing.

Working over a 1 kHz to 25 kHz frequency range the 0.52 m long ribbon has an essentially pure resistance of $0.14\,\Omega$ and is matched to a nominal $6\,\Omega$ impedance by a small 50% nickel–iron core transformer designed and wound with bi-filar windings for minimal leakage inductance. Used in a two-way system in conjunction with a 210 mm moving coil bass driver, an overall sensitivity of 87 dBW has been achieved and,

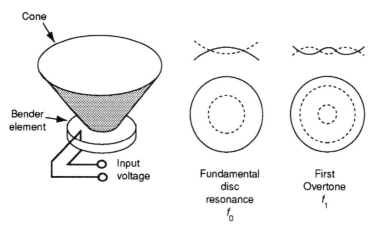

Figure 3.32. Piezo-electric horn driver element, working range between f_0 and f_1 (cone may be coupled to short slotted horn—not shown) (courtesy Motorola [6]). Direct radiator types are also made

unusually, this commercial realization operates the ribbon in monopolar mode, the rear radiation exhausted into a carefully designed non-resonant sealed sub-enclosure. While the natural frequency response of such a transducer is not level, due to the strong reinforcement of lateral diffraction effects in line sources such as this, its unusual placement of the corner or edge of the system/enclosure provides a great reduction in edge diffraction and also results in a better horizontal polar response directed towards the listener. Very good stereo focus has been reported. Frequently, large panel 'ribbon' systems can sound more diffuse than small direct systems, and this 'edge' placement of a line source is clearly a powerful tool in reducing line source diffraction.

Air-motion Transformer

The so-called 'air-motion transformer' is a device that was developed in the USA by Oskar Heil, and resembles a ribbon driver in construction (Figure 3.31). The vital element in this case is a pleated plastic film diaphragm with a metallized conductor pattern of normal $6\,\Omega$ or so resistance. The forward/aft motion of a 'ribbon' is not exploited but instead, the element is encouraged to expand and contract along its length like a bellows, pumping air in and out of the folds. This is claimed to result in a $5:1$ magnification of the air load presented to the diaphragm resulting in good damping and reasonable efficiency. Certainly the performance of the commercial examples available would tend to bear out this claim.

At present a 400 Hz to 20 kHz range is possible, and research is continuing on a larger mechanical version of the 'air pump' for bass frequencies, using a solenoid-type motor element. High internal loss materials such as polyethylene or teflon (p.t.f.e.) are used for the film in the present HF models as these inhibit self-resonances.

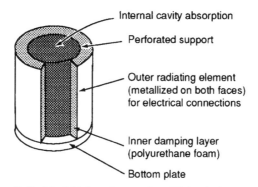

Figure 3.33. Cylindrical high-polymer piezo-HF unit (courtesy Pioneer [7])

Piezo-electric and High-polymer Transducers

Certain materials have the property of a dimensional change when an electric potential is applied across them, this being a reversible mechanism. Both crystalline substances such as Rochelle salt, and the so-called ceramics, such as barium titanate, demonstrate this piezo-electric effect. Some plastics, known as high-polymers, have also been developed which possess this property; incidentally, these are polarized internally like electrets. The piezo-electric effect in these polyvinylidene-fluoride 'high polymers' can be up to ten times that of quartz.

The high mechanical impedance of the mineral substances normally restricts their use to high frequencies; one inexpensive design employs a ceramic element, a horn loaded HF unit (Figure 3.32) developed by Motorola [13]. It uses a circular 'bender' 22.6 mm in diameter and 0.5 mm thick, and a small cone joined to the centre of the element, acoustically coupling the latter to a moulded semi-sectoral horn. The axial response covered is 4 kHz to 20 kHz, within ±3.0 dB limits.

Although high-polymer 'benders' may also be produced, another use for this material has been suggested. If a thin film (30 μm) of high-polymer plastic is formed into a cylinder (Figure 3.33) the dimensional changes under an applied electric field variation will cause the cylinder's radius to pulsate. Such a structure approximates to a theoretical cylindrical source [14].

With additional damping and support elements, cylindrical section mid and treble range drivers have been constructed with a voltage sensitivity similar to that of typical moving coil drivers. The input impedance is high and predominantly capacitive.

Hybrid Digital Loudspeaker Driver

Philips patented a hybrid loudspeaker for digital and analogue drive signals (US 4,555,797). This patent was filed in 1983 and notes that available technology limited the maximum resolution to 8 bits for direct digital drive for a single multi-section loudspeaker voice coil. If efficiency is the primary goal, then this remains a practical proposition provided that the remainder of the dynamic range is covered by an alternative technique. In Figure 3.34, eight of the windings are fed bit weights up to the

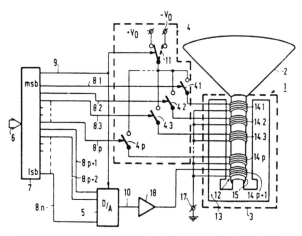

Figure 3.34. Amongst numerous proposals for digital loudspeakers this one is more practical than most. Low levels are dealt with using a small linear amplifier and a digital-to-analogue convertor. High levels are efficiently handled by direct conversion with summing achieved at the motor coil assembly (Philips patent, filed Netherlands 1983)

eighth (not all are shown). Assuming 16-bit audio coding the final lowest 8 bits are fed to a digital-to-analogue convertor run in the best, upper part of its dynamic range, followed by a relatively small amplifier. This need only be sufficient to cover the output range from 60 dB downwards; s.p.l. of 0.1 W is sufficient to drive the ninth coil section. At first, this scheme looks practicable, until the requirements of high quality, wide response systems are considered. Thus it can be seen that the extension of this hybrid digital approach is difficult to execute where multi-way designs with good directivity characteristics are involved.

Magnetic Circuits and Motor Coils

The most widely used motor system is the cylindrical coil immersed in a strong radial magnetic field, usually generated by a permanent magnet, acting on a narrow radial gap.

The force on the coil is given by $F = BlI$, where B = the flux density, l = the length of wire immersed in the magnetic flux and I = the current (Figure 3.35). The coil/cone velocity is represented by $U = Z_M$ where Z_M is the total mechanical impedance reflected on the coil (see Section 2.3).

The transducer action is reversible and hence the cone motion gives rise to a corresponding current in the coil. This is the so-called 'back e.m.f.' and is reflected in the motional impedance of the coil.

Although Z_M is a mechanical impedance it may be transformed into an electrical equivalent, Z_{EM}, via the motor relation to give

$$Z_{EM} = \frac{B^2 l^2}{Z_M}$$

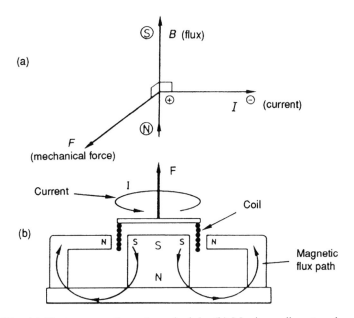

Figure 3.35. (a) Electromagnetic motor principle. (b) Moving coil motor element

The total input impedance of a moving coil driver also includes the coil resistance and associated inductance (Figure 3.36); hence

$$Z_{in} = j\omega L_c + R_c + \frac{B^2 l^2}{Z_M}$$

Figure 3.37 shows variations in input impedance of a typical moving coil driver mounted in free air. The impedance curve may be readily measured via a constant current source and gives much useful information about a driver.

For the example curve in Figure 3.37 typical values for a bass driver would include a moving mass equivalent to $200\,\mu F$, a system compliance equivalent to $40\,mH$, a $6.2\,\Omega$ resistance, a system Q of 0.5 and a motor coil inductance of $1\,mH$. R_{ac} never quite equals R_{dc} mainly due to eddy currents in the pole structure and in the coil former if it is conductive.

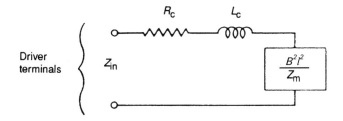

Figure 3.36. Moving coil driver equivalent circuit (electrical side), where Z_{in} is the total impedance and Z_M is the total mechanical impedance, i.e. moving system and air load [21]

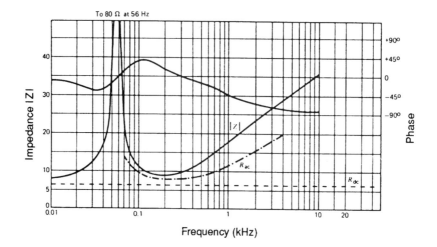

Figure 3.37. Typical impedance curve for a moving-coil driver (infinite baffle mounted) (see also Chapter 6, Figure 6.12)

3.8   MOVING COIL MOTOR LINEARITY

Ignoring the contribution of the suspension or surround, the magnetic motor will itself possess non-linearities due to the magnetic gap flux deviating from a theoretically uniform radial field. This is attributable to fringing effects at the edges of the poles, which result in the magnetic field extending beyond the gap. For maximum flux utilization most short-throw coils are overwound by a factor of some 15% in length, in order to make full use of the available flux including the fringing field.

The output is fundamentally dependent on B times l, the product of the flux density and the wire length immersed. If a coil undergoes considerable excursion, as is required to reproduce low frequencies at a reasonable power level, then it will move from one region of flux density to another while the coil portion in the usable flux may also change, this in combination producing non-linear drive and consequent distortion. Such a driver, under considerable excursion at low frequencies, will also produce intermodulation with the mid-range, the latter output, showing an amplitude variation due to the modulated Bl factor. Another secondary effect concerns the coil inductance which may be a significant part of the driver's input impedance at mid and high frequencies. This inductance may be coil position dependent, in which case it will vary in sympathy with the excursion, thus adding a further modulation factor.

There are two obvious methods available to improve linearity. One consists of lengthening the coil to twice or possibly three times the magnet gap length, while the other utilizes a short coil fully immersed in a longer gap. Both solutions allow a given length to occupy a constant flux within the designed excursion limits, but they are costly in terms of efficiency. In the case of the long coil, a large proportion of the input power is wasted in coil resistance, as only a percentage of the turns are active. For the short coil only a small proportion of the available flux is used one time, thus necessitating an overlarge magnet system.

On economic grounds, the long coil presents the most attractive solution and is widely used for long throw, medium efficiency, low frequency units. The increased coil area results in an improved thermal power handling but gives an increase in both coil mass and inductance, these tending to curtail the output in the mid and upper range, which is no problem unless bass-end coverage is required.

Where standards are high, another solution must be adopted since flux leakage asymmetry will provide an ultimate limit to the linearity of conventional magnet systems. The best motor coil thermal capacity is obtained by making sure it remains immersed in the gap, in close proximity to the heat dissipating pole faces. Hence the requirement is to restrict the driver excursion and use a fairly short coil, well immersed in a symmetrical field gap.

Instability

Examination of the variation of average flux-linked turns (or Bl factor) for a coil under large excursion shows that Bl reduces at the amplitude peaks where the coil begins to leave the gap. In the region above the fundamental resonance where the coil driver is under mass control, the motional impedance is still considerable and the back e.m.f. at the driver terminals is both correspondingly high and proportional to $(Bl)^2$. As the coil moves away from the central rest position, the back e.m.f. is reduced due to the lowered Bl factor, and hence more power is drawn from the amplifier (assuming a normal constant voltage source). This drives the coil proportionately farther, the process accelerating until the coil is constantly displaced away from the zero position, this resulting in extreme distortion [15]. The moving coil system is thus fundamentally unstable in this respect. High compliance systems with a large cone mass are particularly susceptible and some poor designs, thermally rated at 100 W, have been known to 'lock out' with as little as 6 W input at 100 Hz. The mid-point shift or bias is progressive with input level due to a d.c. component of the varying magnetic field generated by the rectifying effect of the fringing non-linearity. Increasing the suspension stiffness to control the excursion is the main cure, and long coil design for extended throw drivers is another aid.

Clearly long coil or long gap design helps to reduce the Bl variation and minimize the effect. Good suspension design can ensure that the system mechanically 'soft limits' before the jump-out excursion is attained. This intentional suspension non-linearity is to be preferred to the almost 100% half-wave distortion which occurs during jump-out. Investigation has shown that with reflex loading the effects are more severe than with sealed box. This is partially due to the lack of box air stiffness helping to restore the diaphragm/coil to the linear range, but more importantly to the high sensitivity of a reflex system to variations in damping or Bl product.

Another solution proposed for 'jump-out' is the addition of two high conductivity rings placed at each end of the motor coil [16]. These act as eddy current brakes, firmly halting additional excursion as the rings enter the magnetic field. As such, the braking can be sufficiently sudden to produce an audible knocking sound. R. Walton of Elac (UK) has suggested a revised geometry for the rings adjacent to the coil winding to produce a more progressive action. Deep teeth or serrations here would smooth the transition. Alternatively the rings could be tapered to a thinner and less conductive section nearer to the winding.

Distortion Compensation

It has been suggested that suspension non-linearity may be balanced against magnetic instability since the former is negative (waveform compression with increasing excursion) and the latter positive (waveform expansion with increasing excursion). Obviously with a given system this may be taken to a certain power level beyond which the cancellation fails, producing gross distortion [17].

By using a slightly oversized coil with a suitable choice of suspension and surround, the power handling for a given distortion level may be increased by 3 dB to 5 dB above that possible with a conventional unmatched suspension. Incidentally, the non-linearity of the air stiffness may also have to be taken into account. The degree of distortion will again be limited in conventional magnet systems by fringing asymmetry.

Combined Suspension and Motor Design

The benefits of integrating the design of the motor system and suspension are now apparent. In some cases, a similar coil/gap length may be chosen for optimum efficiency. The suspension stiffness and total non-linearity may be used to both extend the power output by nearly 5 dB and prevent jump-out. The choice of a sensible coil length also results in sensible values of inductance and mass, thus improving the upper range output and reducing the position dependency of the coil inductance.

A driver designed along these lines is likely to exhibit a rather higher fundamental resonance than the typical long throw low frequency unit; for example, 35 Hz as against 20 Hz and consequently reflex loading is more usual with these optimized medium resonance drivers.

Coil Inductance and its Suppression

Several methods have been suggested for suppressing coil inductance which are all essentially the same in principle. These involve placing a shorted turn in the magnetic coil system, such as a conducting motor coil former or a second coil, or the use of copper shielded or plated magnetic poles. The circulating current induced in the copper 'turn' will reduce the motor coil inductance and will also shunt higher frequency eddy currents from the magnetic poles, a factor responsible for residual harmonic distortion in some drivers. If a shielding method is not employed, then the latter effect may be overcome by laminating the poles, as is done in wide-band audio-transformer cores.*

The 'shorted turn' will also have an effect on the back e.m.f. in the motor coil at low frequencies, particularly at the driver's main resonance, as it will reduce the magnitude of resonance slightly, due to eddy current loss. However, the effective coupling factor at low frequencies is small, and in consequence it can be ignored.

One disadvantage of such a copper turn or sleeve arrangement is the reduction in magnet efficiency, since to be effective the copper needs to occupy 20%–30% of the gap thickness, with a consequent reduction in flux density. The ring may then be placed below the coil.

---

*Woodman of ATC has demonstrated magnet systems with pole faces of high permeability and high electrical resistance; the powder based matrix gives good control of eddy currents and distortion resulting.

The effective improvement in rise time or transient high frequency response due to coil inductance cancellation has been incorrectly described as a major advantage for bass units, where the inductance of a large, long throw coil may be considerable. However, this is a false assumption, as the system transient performance and rise time is largely determined by the highest frequency unit in the system, i.e. the HF driver, and has no connection with the bass unit. The LF transient response is a function of the low frequency resonance and its damping, which are unaffected by considerations of coil inductance.

Some magnetic distortion arises from the modulation of total pole flux by the motor coil flux variations. A conductive sleeve placed in the magnet system somewhat away from the motor coil can have a significant effect in reducing this distortion, notably odd order harmonics (see subsection on Magnet design in Section 3.10).

Dual Coil L.F. Unit

Watkins (USA) developed a method for extending the LF range by rematching the speaker impedance in the resonance region using a double motor-coil. One coil, of nominal impedance, is driven over the usual range and provides an overdamped alignment; the second, lower resistance coil, is fed via a bandpass filter at low frequencies providing the extra power to augment the low frequency output. Effectively the driver is impedance rematched around the LF resonance.

3.9   INFLUENCE OF MAGNETIC FIELD STRENGTH ON LOUDSPEAKER PRESSURE RESPONSE

With loudspeaker diaphragm assemblies falling within a prescribed range of mass for good colouration and response, the other major design factor responsible for efficiency is the magnetic field strength or flux density, B.

Figure 3.38 shows the influence of B on a typical driver (LF design will be considered in detail in Chapter 4). Curve 1 illustrates an optimum magnet strength. 3 shows the effect halving the field and curve 2 that for doubling the field. The uniform reference response band shows expected changes of 6 dB, i.e. the output is proportional to the square of the flux change.

At low frequencies near resonance the results are perhaps the reverse of what might be expected. The larger magnet has actually halved the LF power at resonance through overdamping, and since the mid-band has risen 6 dB, the effective bass loss totals 12 dB. By contrast the smaller magnet has reduced the damping to below optimum, giving a 6 dB rise in bass, exacerbated by the accompanying 6 dB fall in mid-band level and resulting in an effective bass boost of 12 dB.

This example demonstrates that magnet strength may not be indiscriminately increased or reduced. There will be an optimum Bl value for any design of enclosure, driver and loading (see Chapter 4). Precise control of flux density is a vital part of loudspeaker design and manufacture.

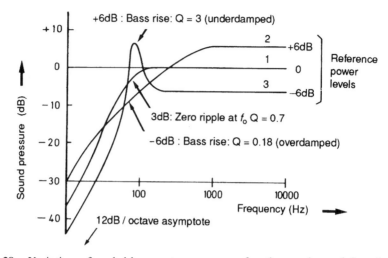

Figure 3.38. Variation of sealed-box system response for three values of flux density: (1) maximally flat Butterworth flux level, (2) flux density doubled (initial rolloff at 6 dB/octave) and (3) flux density halved (+ 12 dB 'bass boom')

3.10   MAGNET SYSTEMS

The high sensitivity of a system's output to flux strength variations means that most magnet designs are operated with the poles saturated in order to attain a more constant field in production. This necessarily implies a wastage of available flux. Further loss occurs in the leakage paths and at the physical interfaces between the magnetic components. It is not uncommon for poorly designed commercial magnet systems to utilize as little as 10% of the total flux in producing acoustic power.

The gap geometry (Figure 3.39) must be dimensioned so that the motor coil assembly, with the required wire length, may be accommodated with sufficient clearance for normal scrape-free movement. It must also include provision for manufacturing

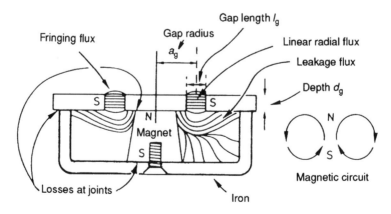

Figure 3.39.   Typical alnico 'pot' magnet structure

tolerances and aging drift in the cone suspensions. Clearly a short-throw coil will require less gap margin than a long-throw version and hence will allow greater utilization of gap flux.

The ratio of gap to pole dimensions will determine the degree of magnetic leakage and hence how much wasted flux is left outside the working gap. This factor may vary between 2 and 10 for various pole and magnet geometries. The relevant parameters for a magnetic material are B and H, respectively flux density and coercive force. The optimum working point is usually where the $B \times H$ product is at a maximum, in other words, at the knee of the curve.

The equation

$$H = \frac{B}{\mu}$$

where μ = magnetic permeability, is analogous to an electric circuit (i.e. $V = I/S$, where $S = 1/R$ and is the conductivity). Electrically,

$$\text{power or energy} = VI$$

and for the magnetic equivalent

$$\text{energy} = BH$$

Magnet Design

The following relations govern the design and specification of the energizing magnet and the magnetic circuit required.

The fundamental energy factor in a moving coil loudspeaker is the Bl product, where l is the length of coil wire immersed in a magnet flux density B. The geometry of the gap, i.e. coil or centre pole radius a_g, gap thickness or depth d_g, and the gap width l_g together with the required flux density B_g, will determine the other parameters.

For the gap

$$H_g k_B = H_m \qquad \text{and} \qquad H_g = B_g A_g$$

where H_g = total gap flux, k_B = a loss factor, H_m = total magnet flux, B_g = gap flux density, and A_g = gap area. Also

$$H_m = B_m A_m$$

where B_m = magnet flux density and A_m = magnet area. Hence the magnet surface area is

$$A_m = \frac{k_B B_g A_g}{B_m}$$

Substituting the gap dimensions for A_g

$$A_m = \frac{k_B B_g d_g . 2 a_g}{B_m}$$

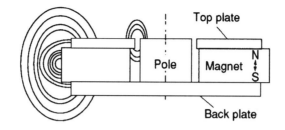

Figure 3.40.   (a) Typical ceramic magnet structure illustrating height/CSA ratio

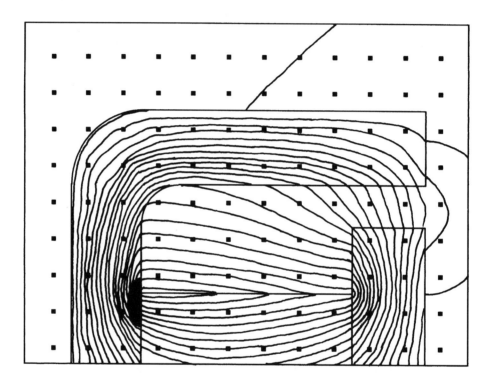

Figure 3.40.   (b) The use of Finite Element Analysis for modelling field distribution for a cup type magnet. The darkest areas show maximum flux density in the steel pole, line spacing shows density and distribution in the gap. These development tools are helpful in obtaining maximum magnet linearity and efficiency (Elcut)

The magnetic potential across the gap equals that across the joints, pole paths and magnet; this is analogous to a static electric circuit where the generator potential equals the sum of the potentials across each series circuit element. The potential across each magnetic element depends on its reluctance, which is a function of the material and geometry. A loss factor k_H of between 1.1 and 1.3 is associated with the potential circuit. Thus

$$H_m L_m = H_g L_g k_H$$

and hence the magnet length is

$$L_m = \frac{H_g L_g k_H}{H_m}$$

If the flux density of the magnet is fixed, then the magnet dimensions may be found using the above equation.

An Alnico or Alcomax magnet will have a typical B value of $1.0\,T$ and H values of 0.5, giving similar proportions to those illustrated (Figure 3.32). In contrast, a ceramic magnet will have B close to $0.25\,T$ and H at 0.12 which, as can be seen from the above equation, will result in a magnet with low profile proportions (Figure 3.40).

The magnetic circuit efficiency is defined as $100/R_H$ and is the percentage of total flux used in the available gap flux. Operation of magnets at the knee or BH maximum implies they are virtually saturated, as is the case in practice, thus improving consistency. Loudspeaker drivers are usually assembled in the demagnetized condition to reduce the risk of gap contamination. Magnet 'charging' is carried out at a final stage and the magnet design must take this requirement into account.

Recently it has proved possible to produce modest cost, controlled excursion, bass drivers with short motor coils and a modified gap geometry to improve fringing symmetry. An inevitable consequence is a reduction in gap flux due to earlier saturation of the thinner 'T' section of the centre pole employed.

JBL the US audio company, [18] have investigated this in some depth for high performance units in a general analysis of motor linearity, which they refer to as 'SFG' or symmetrical field geometry. A useful reduction in distortion may be obtained with the additional help of a shorting ring. In Figure 3.41 the linearity of alnico and ferrite structures are compared.

T-shaped magnet poles can provide significant reductions in distortion at modest cost and the examples show two similarly sized and priced loudspeakers driven to a modest 86 dB s.p.l. at 1 m, plotted for second and third harmonic distortion. Speaker A has a purely average distortion of 0.3%–0.4% in the mid range, while B with a T-shaped centre pole offers a dramatic reduction, particularly in regard to the magnetically

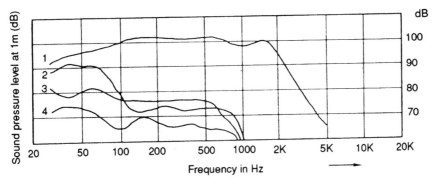

Figure 3.41. Distortion due to different magnet types (after JBL company brochure); (1) loudspeaker, (2) Alnico, (3) ferrite, (4) ferrite with T-shape pole

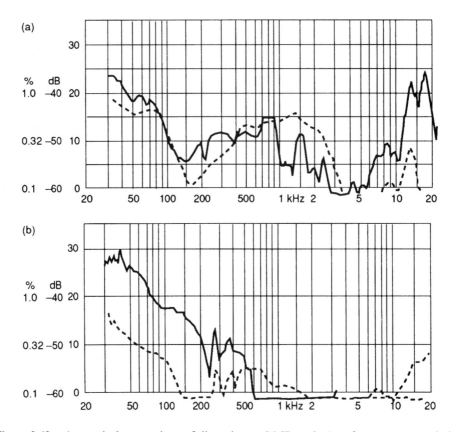

Figure 3.42. A practical comparison of distortion at 86 dB s.p.l., 1 m, for two magnet designs. (a) Inexpensive two-way moving coil system. (b) Similar design to (a) but with a 'T' cut centre pole fitted to the LF unit. Key: —— second harmonic, - - - - third harmonic

related third harmonic (see Figure 3.42(a) and (b)). Incidentally, proof that a large area, push–pull planar system offers inherently low distortion is given by the Quad 63 electrostatic, also driven to 86 dB s.p.l. (Figure 3.20(d)). Above 200 Hz the distortion falls below the 0.1% base line and typically remains at the 0.03% level for both second and third harmonics, an excellent result.

Both ribbon and large area planar designs, including magnetically driven types, can also offer very low distortion, far superior to the usual moving coil performance.

Neodymium

Neodymium alloy magnets are showing increasing popularity because of (a) a fall in price and (b) to meet the need for very small magnet systems to provide small diameter HF units for centre channel speakers. Here the unit is flanked by a pair of bass units in close formation to try to maintain reasonable directivity in the driver plane. Initially of controlled price and availability due to patent restraints, and seen in micro headphone

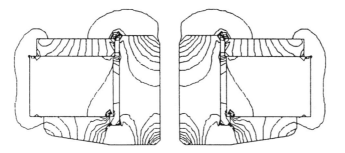

Figure 3.43. Typical loudspeaker magnet assembly using an external ferrite ring permanent magnet

and pickup cartridge applications, these high power magnets (NeFeBo; neodymium–iron–boron alloy) are likely to fall in price thereby expanding their scope.

Their BH_{max} product is almost 10 times that of ferrite or 'ceramic', while the remanence exceeds the already high value of ferrite by 2.5 times. In practice the area required to carry the flux could be halved, and considering that the coercivity of this new material is seven times that of AlNiCo (750 kA/m as opposed to 50 kA/m) very little height is required, virtually a thin disc. Tweeter magnets are typically 3–4 mm thick and 20–22 mm diameter in this material. One advantage is the low stray field of the cup type of pole used with neodymium. This is compared with a ferrite-based assembly in Figures 3.43 and 3.44.

For large volume applications Aura Inc (USA) produce neodymium cup magnet systems which place an array of strip magnets that line the inside of one of the gap faces ensuring maximum utilization for available flux. The result is so effective that even a low-cost driver can benefit from a long gap–short coil motor, e.g. 15 mm of linear gap height for a 6 mm coil and with a screened assembly. So far it has not been made available to lower volume hi-fi makers who arguably will derive the most performance gain from such a magnet design.

FEA Applied to Magnets

Comprehensive FEA packages can deal with magnet systems, which can be considered a subset of thermodynamic problem-solving. Shareware is available, for example the Russian ELCUT (windows) which, although limited to 200 nodes in its basic form, is nonetheless surprisingly effective. If valid material parameters are entered, interactive

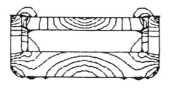

Figure 3.44. Typical loudspeaker magnet assembly using an internal NeFeBo disc permanent magnet

design can provide worthwhile improvements in linearity and flux density while making better use of the available material volume. Variations in pole shape may be modelled, often with surprising results, while regions of premature saturation may be identified and corrected.

Magnetic Shielding

The popularity of home theatre systems involving the use of loudspeaker drivers or speaker systems in close proximity to a television CRT or monitor, necessitates a low value for stray magnetic field.

Pot magnet systems are inherently self-shielded, for example the old iron alloy magnets using an Alnico or similar centre pole and a thick steel enclosure for the return flux. The new generation of neodymium magnets, while of low profile, are also of the pot type and have a low stray field. However, the most popular type of larger magnet using an open ring of ferroceramic has a substantial stray field which may disturb the colours on a CRT from up to a metre away. The first line of attack is to place a second ceramic ring, preferably of similar power, in reverse polarity and bonded to the back plate of the existing assembly. This has the effect of bending the stray field inwards and has the further minor advantage of increasing the effective flux in the gap by a few per cent. This practice may be sufficient for smaller assemblies such as high frequency units. For bass drivers it is usually necessary to install, in addition, a deep drawn steel can of around 0.5–1 mm thickness completely enclosing the magnet system and preferably with its rim in contact with the front plate of the magnet. Magnetic shielding may also be seen as a benefit in terms of the design of a speaker since the interaction between the fixed and varying fluxes of the drivers and with the cables and crossover network is thereby eliminated.

REFERENCES

[1] Kates, J. M., 'Radiation from a dome', *J. Audio Engng Soc.*, **24**, No. 9, 735–7 (1976)
[2] Yuasa, Y. and Greenberg, S., 'The beryllium dome diaphragm', *Proc. Audio Engng Soc. 52nd Convention*, October–November (1975)
[3] Suzuki, H. and Tichy, J., 'Radiation and diffraction effects by convex and concave domes', *J. Audio Engng Soc.*, **29**, No. 12 (1981)
[4] Harwood, H. D., 'New BBC monitoring loudspeaker', *Wireless World*, March, April and May (1968)
[5] Sakamoto, N., *et al.*, 'Loudspeaker with honeycomb disc diaphragm', *J. Audio Engng Soc.*, **29**, No. 10 (1981)
[6] Shindo, T., *et al.*, 'Effect of voice coil and surround on vibration and pressure response of LS cones', *J. Audio Engng Soc.*, **28**, Nos. 7/8 (1980)
[7] Frankort, F. J. M., *Vibration and Sound Radiation of Loudspeaker Cones*, Philips Research Reports Supplements, No. 2, Eindhoven, Netherlands (1975)
[8] Barlow, D., *et al.*, 'The resonance of loudspeaker diaphragms', *J. Audio Engng Soc.*, **29**, No. 10 (1981)
[9] Walker, P. J., 'Wide-range electrostatic loudspeakers', *Wireless World*, May, June and August (1955)

[10]   Walker, P. J., 'New developments in electrostatic loudspeakers', *Audio Engng Soc. 63rd Convention Preprint* 1472 (D-10) (1979)

[11]   Strickland, J., 'Acoustat design paper', Acoustat, Florida, USA (ca. 1981)

[12]   Nakajima, H., *et al.*, 'Direct drive ribbon tweeter', *J. Audio Engng Soc.*, **29**, No. 10 (1981)

[13]   Bost, J. R., 'A new type of tweeter horn employing a piezo-electric driver', *J. Audio Engng Soc.*, **23**, No. 10, 796–801 (1975)

[14]   Tamura, M., 'Electroacoustic transducer with piezo-electric high polymer films', *J. Audio Engng Soc.*, **23**, No. 1, 21–6 (1975)

[15]   Barlow, D. A., 'Instability in moving coil drivers', *Proc. Audio Engng Soc. 50th Convention*, London, March (1975)

[16]   SEAS Fabrikker AS, 'Dynamic Damping', Data Sheet (1978)

[17]   Harwood, H. D., 'Loudspeaker distortion associated with low frequency signals', *J. Audio Engng Soc.*, **20**, No. 9, 718–28 (1972)

[18]   Gander, M., 'Moving coil loudspeaker topology as an indicator of linear excursion capability', *J. Audio Engng Soc.*, **29**, Nos. 1/2 (1981)

[19]   Bank, G. and Hathaway, G., 'A three-dimensional interferometric vibrational mode display', *J. Audio Engng Soc.*, **29**, No. 5 (1981)

[20]   Jordan, E. J., *Loudspeakers*, Focal Press, London (1963)

[21]   Wright, J. R., 'An empirical model for loudspeaker motor impedance', *J. Audio Engng Soc.*, **38**, No. 10 (1990)

BIBLIOGRAPHY

Baxendall, P. J., 'Electrostatic loudspeakers', Ch. 3, *Loudspeakers and Headphones Handbook* (Ed. Borwick, I.), Butterworth (1988), 2nd edn (1994)

Cunningham, W. J., 'Non-linear distortion in dynamic loudspeakers due to magnetic effects', *J.A.S.A.*, **21**, No. 3 (1949)

Frankort, F. J. M., 'Vibration patterns and radiation behaviour of loudspeaker cones', *J. Audio Engng Soc.*, **26**, No. 11 (1978)

Fryer, P. A., 'The holographic investigation of speaker vibrations', *Proc. Audio Engng Soc. 50th Convention*, London, March (1975)

Gayford, M. L., *Acoustical Techniques and Transducers*, Macdonald and Evans, London (1961)

Geaves, G. P., 'Design and validation of a system for selecting optimised midrange loudspeaker diaphragm profiles', *J. Audio Engng Soc.*, **44**, No. 3, 107–17 (1996)

Hladky, J., 'Holography of loudspeaker diaphragms', *J. Audio Engng Soc.*, **22**, No. 4, 247–50 (1974)

Kates, J. M., 'Analysis of decoupled cone loudspeakers', *J. Audio Engng Soc.*, **25**, Nos. 1/2 (1977)

Kelly, S., 'Loudspeaker enclosure survey', *Wireless World*, 552–8, November (1972)

Kelly, S., *Loudspeaker and Headphone Handbook*, Ch. 2, (Ed. Borwick, J.), Butterworth (1988) (2nd edn 1994)

King, J., 'Loudspeaker voice coils', *J. Audio Engng Soc.*, **18**, No. 1, 34–43 (1970)

Kinsler, L. E. and Frey, A. R., *Fundamentals of Acoustics*, Wiley (1982)

Lian, R., 'Non-linear time delay distortion', *Proc. Audio Engng Soc. 47th Convention*, Copenhagen (1974)

Merhaut, J., 'A horn loaded electrostatic loudspeaker', *J. Audio Engng Soc.*, **19**, No. 10, 804–6 (1971)

Millward, G. P., 'The isodynamic principle', *Proc. Audio Engng Soc. 50th Convention*, London, March (1975)

Nieuwendijk, J. A. M., 'Compact ribbon tweeter/midrange loudspeaker', *J. Audio Engng Soc.*, **36**, No. 10 (1988)

Sakamoto, N., *et al.*, 'Frequency response considerations for an electrostatic horn tweeter using electret elements', *J. Audio Engng Soc.*, **24**, No. 5, 268–373 (1976)

Scott, J. and Kelly, J., 'New method of characterising driver linearity, *J. Audio Engng Soc.*, **44**, No. 4 (1996)

Suzuki, K. and Nomoto, I., 'Computer analysis and observation of the vibration modes of a loudspeaker cone', *J. Audio Engng Soc.*, **30**, No. 3 (1982)

Whelan, R. C., 'A novel planiform loudspeaker system', *Proc. Audio Engng Soc. 50th Convention*, London, March (1975)

4

Low Frequency System Analysis: Room Environments and 2π Theory

It is relatively easy to measure the low frequency parameters of a drive unit. These relate to the most consistent and well-defined area of operation where the diaphragm may be considered a true piston with hemispherical radiation properties.

Armed with the characteristics of the driver — moving mass, suspension compliance and the like — an equivalent circuit may be readily constructed for the driver and its enclosure, whether sealed or ported (reflexed). The analysis brings out useful ways of looking at the system as a whole and the theory can be remarkably successful in predicting the low frequency performance of real loudspeakers provided that the practical details of manufacture, assembly and measurement are reasonably well controlled.

4.1 GENERAL CONSIDERATIONS

The initial theory of LF loading has been covered in Chapter 2, where the piston was mounted in three ways; in free space (finite baffle); on an infinite baffle; and on the end face of an infinite cylinder. In this chapter only the lowest octaves will be considered and the directional characteristics at higher frequencies, noted in Chapter 2, Section 2.1, thus will not apply. The impractical form of 'finite baffle' loading has minimal coverage, since the effective reproduction of low frequencies requires the use of an inordinately large structure (the large film diaphragms are a special case, see Chapter 3). However, it is worth noting that with an open baffle the rate of rolloff is first order, at 6 dB/octave [9], the most gentle slope of the LF loading methods. If required a 6 dB/octave slope may be obtained for a box speaker over a more limited frequency range via an overdamped alignment, or by electronic equalization.

The theory presented in this chapter is based on analyses which generally assume that the speaker enclosure is mounted in a large wall, with the system driving a large room volume whose dimensions are considerable compared with the lowest

103

wavelengths under discussion. In accordance with the LF radiation pattern examined in Chapter 2, coverage is hemispherical, i.e. into a half-space (2π steradians). However, when the finite size of a typical listening room is taken into account, the situation is more complex. The effective radiation space will vary with frequency, thus influencing the axial pressure response.

The term 'infinite baffle (IB) mounting' may be a source of confusion, as the surface of such a baffle is metaphorically brought round to the rear of the driver to form a closed or sealed box. The only practical approximation to a true infinite baffle consists of fitting a driver into a dividing wall between two rooms, a technique rarely employed for obvious reasons. In fact, the IB enclosure as we know it is more correctly termed a 'closed box'. Infinite baffle radiation theory does not hold true for a box system unless its front panel is flush mounted in a large wall. In practice this condition may be achieved only with small bookcase enclosures as the larger models are likely to be used in floor or stand mounted positions, located at some distance from an adjacent wall. Free standing location is employed for many high quality loudspeaker systems today and hence the relationship between 'IB' theory and the sound radiation from an enclosure in a room must also be taken into consideration. (Reflex systems may be considered as a sub-division of closed boxes.) Half space represents a first approximation to room drive; full space is clearly incorrect.

Responses in Rooms

Much confusion exists concerning the ideal response of practical loudspeakers in practical rooms, both in terms of the desired response or responses, and the method of assessing that performance in a room. Han [36] covers this subject in some depth by critically reviewing the available published material. Han refers back to the sound field present at the microphone at the moment of recording and goes to the heart of the matter by pointing out that the reproducing loudspeaker should aim to reproduce that original incident sound field in the listening room. If this argument is followed through to its conclusion, then one arrives at the view that virtually all present loudspeakers, which it must be conceded have been designed largely on empirical grounds, are fundamentally incorrect on grounds of their erroneous directivity. The discussion is complicated by considerations of proximity effects which alter perceived timbre and frequency response; for a human listener these are significant at 1 m or less. Other relevant factors include more effective analysis of the acoustics of domestic sized rooms.

Primitive 'live-versus-recorded' tests made under semi-anechoic or anechoic conditions aim to assess the ability of a speaker to mimic a small, simple acoustic source. Success here helps to define standards for transducer colouration and also for the accuracy of the forward region frequency response, but does not define in any way the ability of such a speaker to reproduce a complete sound field, i.e. create a facsimile of the acoustic energy arriving at a recording microphone at a real event such as a concert. Han points out that the correct requirement for a sound field reproducer is one where at a point normally occupied by the listener's head the speakers can produce both a flat direct frequency response and a flat reverberant field, i.e. the level *difference* between the direct and reverberant fields must be constant with frequency.

By virtue of their design nearly all box-type direct radiator loudspeakers have the opposite characteristic to that which is required. For most speakers their power response reduces with frequency, worsened by the natural increase in room absorption with increasing frequency.

Without degrading stereo focus this discrepancy could be largely countered, by avoiding omnidirectional box-type low frequency systems, employing instead gradient or dipole designs. If necessary provide additional energy at higher frequencies by adding auxiliary mid and treble drivers to the other surfaces of a box speaker, e.g. the top or rear panels. Related considerations involve control of the listening room acoustics so that a given loudspeaker has a better chance of generating a more uniform reverberant characteristic.

Han also notes the theoretical colouration present in a stereo image due to the comb filter effect, this resulting from the acoustic combination of the two sound sources — the left and right channels. Under controlled conditions, listening with one ear only, these resulting colourations are easily heard yet with two ears some psychoacoustic function exists which subjectively suppresses that colouration. Nevertheless the character and timbre of a single speaker reproducing monaurally in a listening room does not compare well with a phantom mono source reproduced by a stereo pair of the same speakers. When designing speakers it is absolutely essential that the stereo route is taken as early as possible in the listening development process.*

Active Room Compensation

While research has proceeded on the active control of studio acoustics, for example by using transducer systems, driven so as to cancel low frequency standing wave energy, recently such devices have become available for domestic use. Designed as a self-contained entity these active room compensators are placed at or near the maximum energy points of low frequency room resonance, generally near a boundary. The device acoustically monitors the incident energy, inverts the phase and outputs sufficient acoustic power to reduce or even totally cancel the incident energy. Such methods can be fairly wide band, as found in industrial noise cancellation systems, and are in effect a kind of acoustic 'negative feedback', smoothing the peaks and troughs in the overall listening room acoustic. Assessors have reported greater uniformity of low frequency response with their use together with a subjectively improved low frequency transient performance of the loudspeaker system and room to which the technique was applied. A 10 Hz to 200 Hz working bandwidth is usual; at higher frequencies room modes are diffuse, and cannot be 'trapped' in this manner.

Room Interaction

Take the case of a medium sized system designed to produce a uniform axial pressure response under anechoic conditions (true free-field and not conventional small anechoic chamber measurement). If it is placed against a wall, at frequencies where the enclosure is substantially omni-directional (below 200 Hz or so), the radiation in the forward

*See Appendix Design Type 2

plane will be doubled due to the addition of the reflected rear directed energy. Hence a + 6 dB step up in the sound pressure will occur below 200 Hz. If we were to add a floor to this hypothetical wall and place the enclosure at the boundary, then the effective radiation space is halved again, resulting in a further 6 dB lift at low frequencies. If a second wall were to be added, forming a corner, then once again the axial pressure level will increase by 6 dB. Inspection of the radiation paths to the listening position shows that each adjacent wall is acting as an acoustic mirror, redirecting off-axis sound back into the forward plane. While this situation may be desirable on grounds of efficiency, the reflected images may cause interference with the direct path radiation and disturb the stereo effect. Furthermore, the greater the number of wall surfaces acoustically coupled to the reproducer, the stronger will be the excitation of the eigentones or room standing wave modes, a corner being the worst position in this respect [1].

For many so-called 'free space' systems it has become customary to use an open structure, rigid floor stand to elevate the enclosure. By careful placement it is possible to adjust the three critical path differences in the listening room, namely to the floor, to the back wall and to the side wall, so that they are non-coincident. By this means the successive coupling of the room boundaries is less serious in terms of their combined effect on the perceived low frequency response. Below 200 Hz the reduced aural transient discrimination imples that, for local boundaries, the ear cannot distinguish the reflected low frequency radiation from that emanating from the source. At low frequencies, the local room boundaries must be considered part of the loudspeaker system design, particularly when the latter is intended for semi-free space positioning (see Figure 4.1 which shows the power rather than the pressure increase).

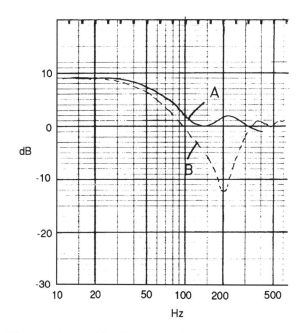

Figure 4.1. Theoretical gain at low frequencies due to two specific speaker placements. (A) Near ideal with the radiating centre 0.3 m from wall, 0.7 m from floor and 1.0 m from side wall. 0 dB is the free field reference. (B) Worst case with all spacings equal at 0.5 m (after Adams [38])

Optimum System Design for Low Subjective Colouration

So far the most satisfactory subjective results have been obtained in the smaller listening rooms through minimizing the coupling to the nearby walls or floor. This is achieved by employing an open stand about 0.4 m high on which the enclosure is placed, the latter spaced clear of adjacent walls by 0.6 m or more. Unavoidably there will be some irregularity in the response at the listener position due to remaining room reflections, but subjectively these do not appear to be too serious. By removing the enclosure from immediate contact with the floor, the frequency band at which severe local reflection begins is sufficiently low to minimize the colouration in the more important mid-band (300 Hz and upwards).

LF Limitation Due to Room Size

Below 200 Hz the enclosure will be less than a wavelength away from the floor and thus may be assumed in contact with it. At some lower frequency it will be joined by the nearest wall, and so on until the room dimensions themselves have been taken into account. This state of affairs partly explains why uniform low frequency reproduction in modest listening rooms is difficult to achieve. This does not mean to say that an extended low frequency response should not be aimed for, but does help to explain why larger rooms have a less coloured bass sound.

Integrated Room/Speaker Design

One solution suggested by Allinson involves the deliberate exploitation of the low frequency radiation properties of the loudspeaker and listening room. He has shown that there exists an optimum driving point for the most uniform low frequency response, and that a system may be tailored both physically and electrically to make best use of this position. His paper illustrates a worthwhile improvement at low frequencies but gives indications that the mid-range quality may be affected [2]. This is precisely why the free space stand method described above has been widely adopted for high quality monitoring speaker installations.

It has been shown [1,2] that it is possible to simultaneously cope with the need for close coupling at low frequencies and offer dispersed semi-free field radiation at higher frequencies. The latter requirement is essential for the formation of precise unambiguous stereo images and also to minimize the type of colouration which makes small source sounds, such as the human voice, sound unnatural.

The technique consists in first choosing an appropriate crossover frequency between the low and mid frequency units — typically between 200 and 350 Hz. Secondly, the enclosure/system is so designed that the MF unit is well clear of the floor and rear wall boundary, thus ensuring that the interference dips which are reflected on that driver's output actually appear below its crossover frequency. These occur at between 120 and 160 Hz for a driver mounted between 0.4 and 0.6 m from the nearest boundary. Finally the LF unit is located in close proximity to the local boundaries, for example on the side of the lowest available section of the floor mounted enclosure. The small spacing ensures that the boundary interference effects are located between 500 and 700 Hz outside of the LF driver's passband. By this means the problem of first-order boundary interference is neatly side-stepped and no longer appears as an irregularity in the system's acoustic output in a conventional listening room (see Figure 4.3).

Some Subtleties in Room Acoustics

Architects have long known, regardless of the care with which new auditoria are designed, and taking into account the overall volume and enclosed shape as well as the effect of the fittings, surface finishes and treatments, that the resulting sound may still not conform to expectations.

Acoustic modelling is complex and may stretch even the most powerful computer analysis available whether using finite element analysis (FEA) or boundary techniques. Some of this variability and uncertainty is associated with second-order effects, the subtle underlayer of perceived quality of the room or hall acoustic.

First-order acoustics covers aspects such as reverberation, its complex relationships with frequency and time and the general approximations often applied to get the concepts into a workable form. The absorption or its converse, the reflectivity of surfaces and materials, are explored — wood, plaster, etc — and applied to designed absorbing structures. Likewise, the geometry and size of the acoustic space defines the three-dimensional standing wave modes which may be energized, and ideal spaces can be designed which aim to provide the most even distribution of sound energy, in theory leading to the most natural sounding acoustics.

Even when all these factors are taken into account, a listening room or hall may still have a recognizable aural colour, which may be more or less pleasant and which may impair listening accuracy and enjoyment, whether for live or reproduced music.

Borrowing from experience with loudspeaker cabinets, specifically the influence of panel resonances on sound quality, while noting that these are relatively strong in loudspeakers due to mechanically coupled vibrations from the speaker drive units, it is worth considering the concept that the room is just another loudspeaker box in which we sit, significantly energized from within by our sound sources.

In addition to the fundamental acoustic properties of the room we must add the effects of the structural resonances in free or clamped panel surfaces, and the effects of coupled vibration from the sound sources to structure, such as the floor.

Energy absorbed by room elements will be delayed and re-radiated according to the mechanical, vibratory and acoustic properties of the element, and will impart a smearing energy decay and tonal error to the room acoustic. It is typical for these effects to be significant; some way into the acoustic reverberance they die away and it is difficult to find techniques that can separate these effects from the broader acoustic. While the measurement technique is insensitive, the ear has no problem in identifying these problems.

In the author's new listening room a decorative array of tall acoustically dispersive shelving/bookcases was installed on the side walls in the speaker region and these extended for several metres. Notes on the quality before and after installation were taken. The mixed array of CDs, books and ornaments on the shelving was effective in diffusing the side wall reflections and major standing waves, without sucking out too much energy or drying up the sound in the mid range. But there was also a serious loss in low frequency energy, subjectively some 2–3 dB from an estimated 30 Hz to 150 Hz. After some tapping tests on the bookcases the problem was narrowed down to the thin ply backing to the prefabricated shelving units. Spaced typically a centimetre or two from the wall, in the bass it was acting as a large area membrane absorber.

Fortunately, a cure was not too hard to achieve. Small holes were drilled in the panels, and using a suitably small nozzle, hard-setting expanded polyurethane foam was injected into these thin cavities, bonding the panels to the wall.

As the aerosol foam cured, the panels were stiffened and the bass energy originally defined by the chosen loudspeaker returned.

Similar effects can result from an excessive number of pictures hung on the wall, from large furniture units or cabinets, particularly those with thin doors or backs; from heavy upholstered seating, especially if present in excess. An open layout, a variety of surfaces and furnishings kept to a minimum and a completely clear region in the vicinity of the loudspeakers are all helpful in improving room sound.

At low frequencies an enclosure floor mounted, and fairly closely positioned to the rear wall will have an acoustic load corresponding to π steradians rather than the 2π assumed in the theory for low frequency analysis. Some loss is to be expected due to wall and floor flexure and typically an overall gain of 2 dB is possible below 100 Hz. This adds further evidence to support the suggestion that the speaker system Q factor should be on the low side, 0.5 rather than 0.7, to offset increasing room gain.*

Exploitation of Boundary Effects

Allinson has assessed the effect of room boundaries and has devised commercial systems whose frequency response is adjusted for specific locations. One such system, a small two-way design, is conceived as a cube of 300 mm sides with a front mounted HF unit and a top mounted 200 mm bass mid-driver. Intended for wall positioning the LF unit is thus brought close to the wall boundary which acts as a good 2π acoustic mirror right into the crossover range.

Proof of the effectiveness of the technique is given by a room response measurement using the multiple averaging technique or room averaged response. Over a 25 Hz to 20 kHz range the output in the region of the listeners is shown as curve B for the correct wall position and curve A for a conventional off-the-wall, stand-mounted location (Figure 4.2). The former is an impressive result for a room response taken for a wall mounted speaker, while the poor off-the-wall result correlated well with the poorer results of a second listening test. Many conventional speakers are often spectrally balanced for a wall mounting but cannot achieve optimum smoothness unless more effective steps are taken to control the interface between the speaker system and its immediate boundary, for example by using specific driver placements on the enclosures to minimize comb effects in the working frequency range.

While the bulk of this chapter draws on valuable established theory concerning the analysis of systems ideally mounted in a 2π infinite baffle, this is a relatively poor approximation to real life use in domestic rooms.

---

*This assumes solid building construction. For timber-framed houses, low frequency losses are greater and Q values of 0.7 to 0.8 may be appropriate.

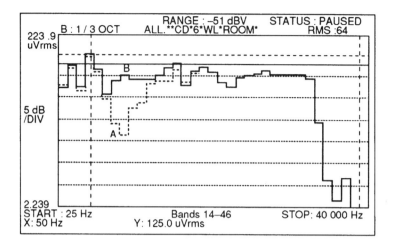

Figure 4.2. A boundary matched loudspeaker, the room averaged responses at the listener position: free space (curve A), close wall mounted as designed (curve B). In free space the output notches severely at 120 Hz due to the symmetry of the top mounted bass unit relative to the local boundary

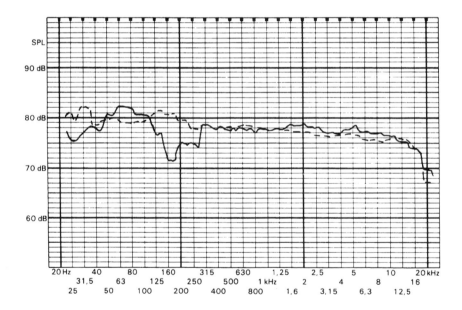

Figure 4.3. Comparison of two curves taken by multiple microphone position averaging in a domestic listening room. Both curves are for speaker systems with similar drivers. The dashed response applies to a cabinet with a close coupled LF driver and crossover separated mid range. The solid response is that of a conventional system with front mounted drivers in the same room location, i.e. 1.6 m from the side walls, rear of cabinet close to rear wall, and mid units at near ear level. The improvement in low frequency uniformity shown is typical for a number of microphones and speaker positions (after Holl [28])

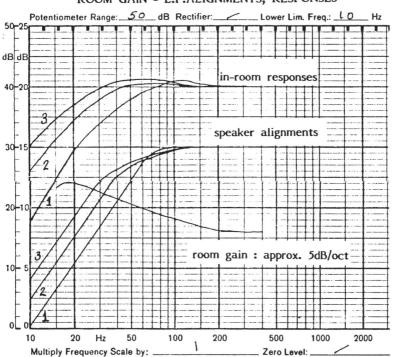

Figure 4.4. (a) Room gain curve at low frequencies; (b) three new LF alignments to take account of domestic room gain at low frequencies

It is clear that a better set of alignments is required to tailor speaker response more closely to room conditions. Figure 4.4 shows the typical room gain at the listening position for optimally placed, free standing loudspeakers. A group of three possible low frequency response shapes is shown. The initial rolloff is progressively overdamped with reducing f_c (see $-3\,\text{dB}$, 45 Hz curve) to provide a better room match. Room gain has then been added to produce the final output, as heard. An anechoic $-6\,\text{dB}$ at 30 Hz results in a room response of $-3\,\text{dB}$ at around 20 Hz, depending on ultimate room losses.

Loudspeaker Placement

Unavoidably the loudspeaker is strongly affected by the listening room both in terms of local boundaries, their reflections, and the overall room acoustic. It makes sense to make the best of the room and to maximize the design and performance of the speaker. Ill-considered placement can make the resulting sound substantially poorer than it need be. On the other hand, sensible placement can help make the best of what is in essence a combination of speaker and room. With speaker systems of generally good fidelity the room may be responsible for up to 50% of the ultimate sound quality. That means that a poorer loudspeaker, if well matched to a given room and well placed, can in practice be made to sound significantly better than the case of a superior loudspeaker which has been poorly located.

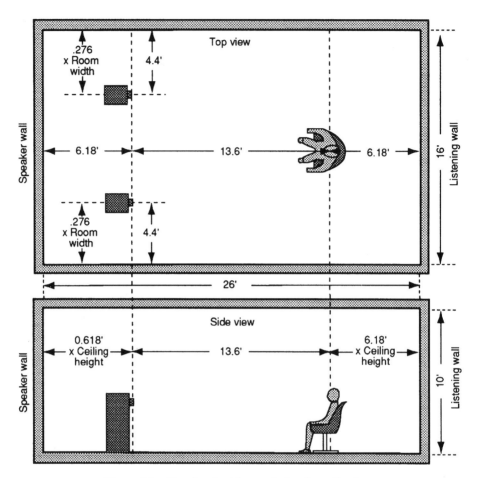

Figure 4.5.   Reprinted with permission from Cardas

With a high performance speaker it is crucial that the listening room be studied carefully and practical trials must be carried out using a wide range of programmes in order to ascertain the best location both for the listener(s) and for the loudspeakers.

In certain circumstances a chosen loudspeaker may prove unsatisfactory in a given room owing to an unfortunate coincidence of the bass characteristic inherent in the speaker and a dominant room mode, this resulting in excessively boosted loudness in the low frequency range. This can occur despite the circumstance of a good room and speaker married to a carefully tuned and optimal set-up aimed at the most uniform perceived frequency response and the most favourable stereo image. Since it is usually difficult to change the room, an obvious solution is to change the loudspeaker for one with different low frequency characteristics.

It is not constructive to direct criticism at rooms or simply to take a theoretical view of their inevitable resonant modes with their consequent impact on the final frequency response of an installed audio system. We live and work in rooms and we are acclimatised to their essential acoustic properties. We recognize and deal with the acoustic character of rooms under normal conditions such as conversation. We know from theory that the character of someone's voice is modified by the room they are in and by their position in it, yet we have no problem recognizing those voices and, unless the acoustic is particularly bad, understanding them. Likewise, as audio listeners we must accept that listening rooms do change the sound of loudspeakers, but generally there is no reason why we cannot deal with this subjectively. While the ragged graphs showing speaker response in a room may disturb us, we know from experience that satisfying reproduction of speech and music is possible a the real world domestic room (see Figure 4.5).

Wilson Placement Method

Making the best of the room acoustics can make a large contribution to sound quality. The American company Wilson Audio are key exponents of a speaker placement technique which begins with a good listener making a subjective assessment of the room acoustic, especially the reflecting boundaries close to the proposed loudspeaker position. You begin by clearing the region between the loudspeakers and the nearby walls of any potential acoustic obstacles, including the proposed loudspeakers themselves. In one method pioneered by David Wilson, he reads aloud from a neutral text, employing a level monotone, while at the same time pacing over an imagined grid defined by the speaker and listening regions. The reader listens for changes in the character of his or her voice, in particular the early reflections. In most rooms it is found that neutral zones exist present as bands located about 1–2 m from the side walls and running parallel to them. Depending on the absorption of the local wall and that opposite, these neutral regions may be between 0.3 and 0.8 m wide and are identified by the neutral sounding reverberation returning to the listener when he stands in these regions. The operator is looking for an absence of colouration, where the sound is neither boomy nor chesty, lacks roughness and the words sound most clearly articulated. The underlying principle is based on the interchangeability of source and observing listener. The assessment may be beneficially guided by a second listener who would be located in a fairly neutral position, located centrally in the listening area. The quality of aural discrimination is improved with practice. To find the optimum listening position in a given room a first approximation is given by defining the initial speaker positions, 1.5 m from the rear wall, where each speaker has a median axis placed centrally in its neutral zone, so defining the space into the side walls (Figure 4.6).

The next step is to triangulate from the speakers to a central point where the distance to each speaker is about 15% greater than the spacing between the speakers. Additional text may be read out at each nominated speaker position whilst assessing the listening position. This may be fine tuned nearer or further from the speakers, listening for the clearest and most natural speech sound.

For the final stage of the Wilson procedure the loudspeakers are installed but not locked or spiked. Trials are then conducted in stereo with a high quality wide range programme. The sound quality must be assessed quickly taking immediate notes before the ear/brain has a chance to acclimatize and make unwanted compensations. Those first opinions are the basis for adjusting placement of the speakers, forward and back, for the most even perceived frequency response. Trial and error will quickly show the

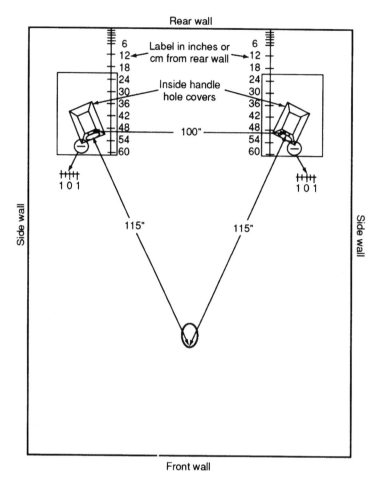

Figure 4.6. After Wilson

surprisingly small adjustments which may be influential on perceived spectral balance (Figure 4.6).

For critical applications there is a final test Wilson calls 'vowelling'. Here a piece of simple programme with a complex harmonic timbre may be used in single channel working for the listener, optimally seated, to assess subtle shades of mid-band clarity, this controlled by the critical adjustment of the speaker with respect to its nearest boundary. The use of calibrated masking tape temporarily fixed to the floor can aid this calibrated adjustment which is done for left and right channels individually. When correctly carried out, the sound acquires a further dimension of precision focus and articulation in the stereo image.

Free-space Listener Placement; Wall Mounted Speakers

Some loudspeakers are designed specifically for placement against a main boundary, typically the rear wall, and adjustments are made in the acoustic design to account for the way in which the room is driven. Account still needs to be taken of the change in sound resulting from the variation in spacing to the side walls and perhaps the floor where the speaker height is not defined by the designer. Generally the listener sits in a relatively free field position out in the room.

At first hearing wall mounted speakers can be disconcerting since the room acoustic is more strongly excited, and that can give an impression of reduced stereo depth and an altered tonal balance. Often the bass is superior owing to the effective inclusion and control of one and sometimes two local boundaries, namely the rear wall and the floor.

With continued listening it is generally found that the ear becomes familiarized with the acoustic condition and depth is heard once more in the stereo image. There are obvious advantages to wall or boundary placement. The speakers are more easily incorporated into a domestic living room, although for the best results the wall where the speakers are placed should have no significant reflecting cabinets or structures to disturb the sound fields radiating from the speakers. Structural walls are best.

Free-space Loudspeaker Placement

In an interesting reversal of the above, Joachim Gerhard of Audio Physic has proposed that the listener be placed on the boundary, usually working across the shorter dimension of the room and with the loudspeakers placed slightly wider than the equilateral triangle would define and placed substantially away from the room boundaries, typically 3.5 m from the side walls and 1.5–3.0 m from the nearest rear walls (Figure 4.7). The effect, particularly in the mid range, is of an almost anechoic, low colouration quality to the sound which is dominated by the speakers and not by the room–speaker combination. A partial disadvantage is the very rapid changes in overall tonal balance, bass and low mid-range weight which are perceived with relatively small changes in the spacing of the listener's head from the rear walls, e.g. 5–40 cm. Where a room is particularly difficult this technique tends to free the loudspeakers from the room problem and may offer a practical solution. It is worth noting that when the longest dimension of a room is used for stereo and results in uneven low frequency performance, compromising with a smaller stereo sound stage and working across the smaller room dimension can often be helpful. Flexibility in matters of home furnishing is also essential if the keen listener is to succeed in getting the best possible sound quality.

Comparisons of Anechoic and In-Room Responses

While much discussion and criticism is made of significant irregularities in the in-room frequency response of practical speakers, it remains true that well-designed speakers of all types still work better in real rooms than do poorly designed systems.

Well balanced design, confirmed by unsighted panel testing, can be shown to correlate well with a sensible approach to anechoic measurement combined with spatially averaged measurements taken in the listening area of a worthwhile room.

Two divergent examples are chosen, one a modern compact two-way design of 22 litre internal volume intended for free space positioning on stands. The other is a more

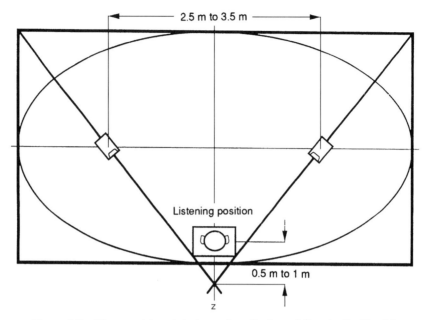

Figure 4.7. Placement to minimize early reflections (after Audio Physic)

complex system, namely a large magnetic di-polar using a long vertical ribbon element above 400 Hz in combination with a surface driven, stretched-film diaphragm below 400 Hz. Such a two-way panel is bi-directional and it can be quite difficult to achieve smooth, well balanced amplitude-frequency responses from such a design. The axial measurement response shown for the panel model is marred by proximity effects due to the relatively small anechoic chamber used but nevertheless gives an idea of the axial character (see Figure 4.8(a)). In-room the averaged measurement is rather striking and this was no fluke for it was also matched by a larger design in this series. With the exception of a rather poorly damped fundamental diaphragm resonance mode at 30 Hz, this partially magnified by the room, the energy drive to the listening area is particularly uniform, and sounded like it (see Figure 4.8(b)).

Conversely, the moving coil compact loudspeaker is worthy of attention. Careful system design shows how well integrated the acoustic outputs of even a conventionally mounted pair of drive units can be. No time delay compensation or specific low diffraction precautions are present. On axis, using a high resolution slow sine sweep, the anechoic result is smooth and well balanced, both with and without the grille (see Figure 4.8(c)). No smoothing has been used on this graph. In fact this speaker achieves essentially ± 1 dB limits from 100 Hz to 20 kHz and is a testament to high quality moving coil driver technology and system design. Below 100 Hz the low frequency response is tailored, with a damped rolloff slope reaching down to 50 Hz, then increasing to 12 dB/octave below this point. Tuned to 33 Hz the final rolloff rate is 24 dB/octave with this ported design. The deliberately 'tailored' low frequency response allows for a useful 87.5 dB/W sensitivity and aims for a good match in conjunction with the bass augmentation provided by typical rooms. The second measurement describes the speaker's output in the forward direction. The axial result at 2 m (solid line) is

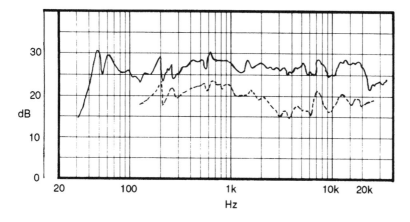

Figure 4.8. (a) Axial response of medium sized, two-way planar magnetic dipole with tall ribbon HF. The upper trace is at 2 m and does not change greatly up to 5 m. The lower trace is taken at 1 m and shows the effect of proximity unbalancing the relative outputs of the bass and treble sections (see Figure 3.26)

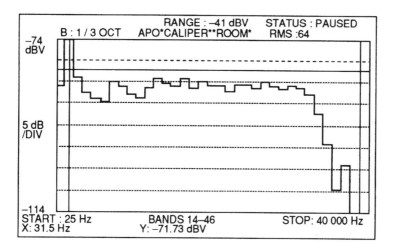

Figure 4.8. (b) Room averaged response for a large magnetically driven planar, 'line' dipole (two-way). The Q at system resonance is excessive (30 Hz). This response shows how skilled design can achieve remarkably uniform room responses judged both by measurement and by sound quality

compared with that at 15° vertically above the system, and with those responses for 30° and 45° in the lateral direction (see Figure 4.8(d)). For a spaced two-unit system the on- and off-axis uniformity is judged to be very good. (These may be compared with the predictably excellent off-axis responses obtained with a true concentric two-way driver in a system of similar size (see Figure 5.18).)

Finally, our example two-way box speaker is used to drive the listening room. Allowing for a mild 6 dB dip at 100 Hz due to the floor reflection a ±3 dB in-room response has been achieved from 30 Hz to 10 kHz (see Figure 4.8(e)).

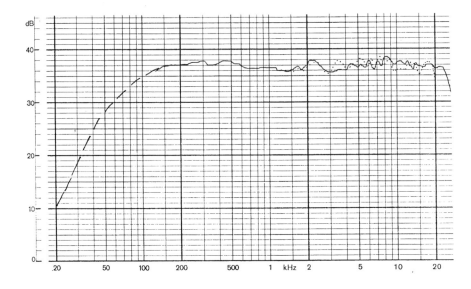

Figure 4.8. (c) Reference response, 1 m axial, high resolution for the system in Figures 4.8(d) and (e). Noteworthy for the accurate response which can be achieved with standard rectangular enclosures. The dotted line shows the effect of fitting the grille assembly. In this example the effect is present but of mild degree. With many commercial systems the grille exerts a major effect on the response

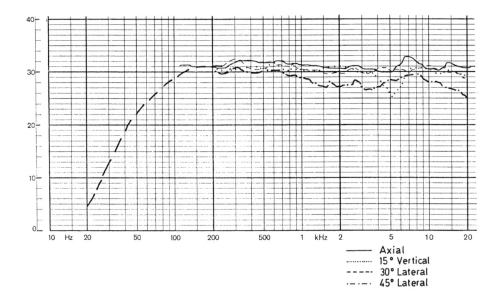

Figure 4.8. (d) The forward response set; axial, 15° vertical, 30° lateral, 45° lateral taken at 2 m anechoic with $\frac{1}{3}$ oct. weighting for the system in Figure 4.8(c). Note the good off-axis uniformity shown for this conventional, separated driver system

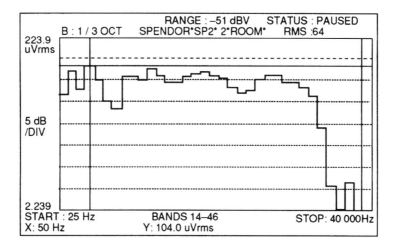

Figure 4.8. (e) Room averaged response for the two-way compact enclosure. Stand mounted and well positioned with respect to local room boundaries. The mild dip is due to floor cancellation. Sensible extension to 30 Hz is achieved in practice from this well aligned 22 litre bass reflex system. This energy weighted response shows good correlation with the sound quality and the set of forward frequency responses in Figure 4.8(d)

One element of its separated two-unit nature may be seen in the mild dip in energy between the upper mid range and where the high frequency unit takes over. The transition is from a 200 mm bass to a 20 mm HF at 2.5 kHz and nevertheless shows how well the gap may be traversed.

4.2 LF SYSTEM ANALYSIS

Modern thinking is best illustrated by the many fine papers on the subject, the majority of which have appeared in the *Journal of the Audio Engineering Society* (AES) or else been read at their conventions. The classic work on reflex systems (vented boxes) is that by Thiele [3], which was first presented as early as 1961. This represented both an effective summary of work to date as well as providing new viewpoints. A decade or so later Small [4–9] produced a series of papers on loudspeaker analysis which were republished in the *Journal of the Audio Engineering Society*, the first appearing early in 1972. Those readers seeking a detailed treatment of the subject are strongly advised to acquire a set of these articles, as a single chapter in a book cannot do justice to all the available material. In the main, this material assumes 2π loading.

Fundamental Performance Limitation

The performance limit of any simple box system [10] (excluding horns and the like) may be described by the following equation for efficiency, where η_0 is the efficiency in the level reference region:

$$\eta_0 = k_n f_3{}^3 V_B \qquad\qquad (4.1)$$

where k_n is an efficiency constant, f_3 is the $-3\,\mathrm{dB}$ low frequency rolloff point, and V_B is the box internal volume. The system is assumed to be driving a standard 2π free field.

Of fundamental importance, this equation shows that for a given type of system and driver, any alteration of volume or cutoff frequency will also affect the efficiency. For example, given a bass response to 70 Hz at $-3\,\mathrm{dB}$ in a sealed box system, the efficiency may be doubled only by doubling the enclosure volume. Alternatively, if for the same starting efficiency the response was to be extended by an octave, for example to 35 Hz, then the volume would have to be multiplied by a factor of 8.

f_3 and V_3 are not of course wholly independent parameters, and equation (2.7) defines this relationship accurately. This explains why compact, wideband systems are so inefficient and also suggests that the pursuit of an extended low frequency response for its own sake is likely to prove extremely costly and hence represents unbalanced engineering. Since neither broadcast nor disc programme contains significant energy below 35 Hz, the desire for a substantial LF extension has proved rather pointless. More recently two factors have emerged to change this view. Digitally mastered and replayed programme is becoming increasingly common, these systems possessing a recorded bandwidth to near d.c. Significant recorded output to below 20 Hz is no longer improbable. In addition recent psycho-acoustic research suggests that the phase shift, and more importantly the group delay associated with high-pass filters at low frequencies, is more important than, for example, that due to low-pass (high-cut) filters near the audible band edge. A steep rolloff at 40 Hz can no longer be regarded as an acceptable basis for system design at the higher subjective quality levels.

Magnitude of k_n

k_n is an efficiency factor that depends on the following: the class of system, i.e. sealed or ported; the system losses; the required response shape or alignment; and the driver–enclosure compliance ratio.

Maximum values may be estimated, with practical limits, for a well designed system calculated to be $k_n = 2 \times 10^{-6}$ for a reflex loaded system* and 1×10^{-6} for a sealed box (MKS units) [10].

Referring to the example on reference efficiency in Section 2.4, the driver examined is suitable for sealed-box loading in a 30 litre enclosure, and will possess a $-3\,\mathrm{dB}$ frequency at 50 Hz. Using the limitations set by equation (4.1) above, we get

$$\eta_0 = 50^3 \times 30 \times 10^{-3} \times 10^{-6} = 0.37\%$$

which compares well with the reference efficiency calculated using the driver constants. It is, however, important to note that the driver size does not appear in the equation. Theoretically, by ignoring practical factors such as diaphragm excursion limits, any driver size could be used in any practicable enclosure. With respect to low frequency reproduction, there are no intrinsic special properties possessed by large cones as opposed to small. Simply, larger drivers are ultimately capable of radiating greater acoustic power for a given distortion level because they move less.

---

*i.e. a reflex may be 3dB louder than a sealed box design of comparable $-3\,\mathrm{dB}$ cutoff.

High-pass Filter Analogy

A driver-box assembly may be represented as a high-pass filter. The sealed box equates to a second-order network with a 12 dB/octave rolloff below the cutoff frequency, while the reflex system is a fourth-order network, with a 24 dB/octave rolloff. A third-order response, 18 dB/octave, may be obtained by the addition of a series high-pass elements in the electrical circuit of a closed box system. An approximation to a third-order rolloff over a limited range (about 1.5 octaves) may be obtained by resistively or otherwise over-damping a fourth-order reflex system. (Multi-bore or fabric blocked ports can provide the box resonance damping.) A double cavity enclosure system fed electrically via a first-order bandpass filter [27] can offer an overall third order 18 dB/octave bandpass response over three octaves, this relevant to sub-woofers.

Infrasonic Overload

While the high-pass analogy extends an octave or so below f_3, the extreme LF behaviour must be taken into account, since audio signals may impart infrasonic energy due to disc warps, etc. In the case of a sealed box, the driver motion continues to decrease with reducing frequency below cutoff. The sealed box plus third-order element, the latter being a series capacitor, also follows this pattern. However, the ported reflex systems do not. The idealized resonant augmentation of low frequency output with reduced driver excursion relates only to the frequency range close to f_3, and at lower frequencies the port output shifts out of phase with the driver resulting in the 24 dB/octave rolloff and there is little control of driver excursion.

In this lower range as the acoustic load presented to the driver is small and since the air cushion of the sealed box is absent, the driver will have little motional resistance to infrasonic excitation. In many domestic vinyl disc playing systems, the LF driver in a reflex enclosure may be seen to be in continual oscillation at subsonic frequencies thus impairing both linearity and the system power handling capacity.

It can be argued that infrasonic energy should be removed by the addition of suitable filters in the amplifier, but as does exist in practice, the problem cannot be ignored in system design.

With digital replay there often remains some sub-low frequencies due to auditorium noise, traffic rumble and air conditioning for example.

Subjective Considerations and Transient Response

When choosing the desired low frequency characteristic for a high performance system, subtle subjective considerations may well prove important. For example, it is known that a low frequency response with a relatively high cutoff (60 Hz) and a slow rate of rolloff (initially 6 dB/octave), may be preferable in terms of colouration and perceived bandwidth to one where a wider response is uniformly maintained to 45 Hz, but which rolls off quickly at 24 dB/octave. From the viewpoint of an engineering specification the latter would appear to represent the ideal case, in fact, subjective judgement indicates the former to be preferable. Why this is so is not fully understood, although it may be associated to some extent with the poorer transient response and inevitable delayed ringing present to some degree with typical fourth-order designs (reflex systems). The rapid phase shift as the system enters the resonance range may also be a contributing factor. Although covering a range of reduced aural sensitivity, an overall response extension to say 35 Hz seems to be more important than a flat response with a higher cutoff. Such aural preference is also in agreement with preferred LF alignments designed for better interfacing with the room.

High Acoustic Outputs

It was mentioned earlier that the attainment of high acoustic powers requires a large driver diaphragm. The greatest excursion is demanded at low frequencies where the volume of air displaced is the quantifying factor. Thus

$$V_{A_{max}} = S_D d_{pk}$$

where S_D = piston area, and d_{pk} = the allowable peak excursion. The maximum power output for a given bandwidth over the reference frequency range is given by

$$W_{max} = k_p f_3{}^4 V_{A_{max}}^2 \quad (k_p = \text{power constant})$$

with the drive adjusted to keep the unit's displacement within d_{pk} over the working passband. k_p depends on the spectral energy distribution in the programme and also on the system and its response shape. For typical speech and music programmes, k_p approximates to 0.85 for a closed box, and 3.0 for a reflex enclosure [10].

In the case of the 30 litre enclosure discussed earlier, let us compare the alternatives of reflex and sealed driver loadings as designed for the same cutoff frequency. If the maximum acoustic power was equivalent to 100 dB s.p.l. for the sealed box, then the reflex equivalent for the same diaphragm excursion would be 105.5 dB, assuming that the driver could tolerate the increase in input power, and that non-linearities in the vent were not excessive at this sound pressure level. k_p values for 2π systems are rather generous if the true acoustic space loading is accounted for as in normal rooms.

Air Non-linearity

Another potential source of distortion is the excessive compression of air in the box. If this exceeds 5% by volume, harmonic production from this source may be significant. In practice this is unlikely to occur as the output in the case of a medium sized enclosure would be approaching a very loud 120 dB s.p.l. at 1 m. If greater levels are required, then it would be expedient to increase the box size. Since a larger driver with a high power rating will undoubtedly prove necessary, a larger box will in any case be employed.

Incidentally, such distortion is a problem with horn systems, particularly high output mid-range drivers where the effective air compression in the horn throat is considerable.

Compact Enclosures

While it is self-evident that for most applications a loudspeaker system should be as small and unobtrusive as possible, the fundamental efficiency equation (4.1) dictates that if a given system is reduced in size, then a sacrifice must be made in terms of either LF extension or efficiency. If the latter consideration is inconsequential thanks to the ready availability of high power amplifiers, then the bandwidth and output power may be maintained in a small box. However, a stage will ultimately be reached where either:

- the thermal power handling limit of the motor coil is reached, or

- the non-linearity of port or air volume becomes significant, or

- the size of the magnet system required to energize the increasingly long throw coil becomes too large and costly.

The most influential factor is undoubtedly the cutoff frequency f_3, and as the efficiency is proportional to the third power of f_3, a sacrifice in bandwidth is a logical step. Merely by moving the cutoff frequency one-half an octave higher, e.g. from 38 Hz to 50 Hz, the enclosure volume may be reduced by nearly 2.5 times without any significant loss of efficiency or maximum acoustic power output.

Low Frequency Sensitivity of the Ear

Another relevant point concerns the human ear's low frequency characteristic and its maximum acoustic output.

Reference to the Fletcher–Munson hearing sensitivity curves will show the ear's poor sensitivity at low frequencies. Clearly there is no point in designing a speaker system whose response extends to 35 Hz if it cannot generate sufficient acoustic levels for these frequencies to become audible (see Figure 4.9).

Taking the case of a 90 dB sound level at 1 kHz, and assuming this to be the maximum output of a certain system, suppose that the programme carries an orchestral bass drum note at 40 Hz, which appears at a level 30 dB below peak. Noting the maximum system s.p.l. at 90 dB the drum note will be reproduced at 60 dB, which is just on the threshold of perception, and hence the attainment of a 35 Hz cutoff in such a speaker with an overall 90 dB s.p.l. maximum is almost worthless. To do justice to modest level bass sounds the system s.p.l. maximum would have to be raised to at least 100 dB. This fact virtually rules out frequency responses extending below 50 Hz from compact enclosures with low maximum output levels, regardless of the miracles of engineering which are claimed to exist inside. Hence for effective low frequency performance a high quality speaker must be capable of producing an adequate acoustic output over the designed range. It is precisely because the ear's low frequency sensitivity is poor that bass musical instruments are the largest in the orchestra; their size is indicative of their ability to produce sufficiently high energy to be heard at low frequencies.

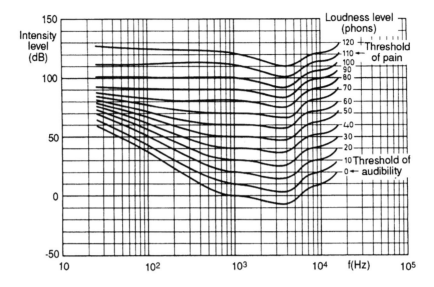

Figure 4.9. Hearing sensitivity versus frequency and level (after Fletcher and Munsen)

Assessment of LF Responses

Assessment of the bass performance from standard response plots is complicated by two factors which require that greater attention be paid to small detail than might be expected.

The first concerns the ear, generally described as having a logarithmic sensitivity allowing the use of the familiar dB scale. This is true in the mid and treble ranges, but at low frequencies the hearing response is close to the linear characteristic (Figure 4.9). Inspection of these curves shows that in the common working range of 60–90 dB s.p.l. the close spacing of the curves shows a near linear characteristic at frequencies significantly below 200 Hz. A linear characteristic implies that the ear has a greater sensitivity to changes in level and hence irregularities in bass response are more audible than the usual logarithmic response presentation suggests.

The second factor concerns visual interpretation and an understanding of numbers. If we consider that the range 20 Hz to 120 Hz spans 100 Hz, thinking of the mid range say 1 kHz, then this does not seem very much. But we must remember that music is scored in octave intervals each containing tones and semi-tones, sharps and flats. 440 Hz, middle 'C', runs an octave to 880 Hz, i.e. 8 full notes. Consider then the low frequency range, an octave from 120 to 60, another octave from 60 to 30, and still a few notes more until you reach 20 Hz.

For these reasons every 10 Hz of clean, uniform bass extension is worth having, is musically important, and adds to the value of the system. Designers need to focus more clearly on low frequency system design [39].

Programme Spectral Content

The previous example citing the bass drum at 30 dB below the mid-band maximum level is not uncommon, and in general the average bass content of normal programmes falls rapidly below 100 Hz. This implies that below this level the maximum undistorted acoustic power capability of a system may be compromised without audible deterioration. Such a concession will in fact allow a typical system possessing a cutoff at 50 Hz to be operated at a maximum overall programme level 6 dB higher than the actual 50 Hz overload point.

Alternatively, where a system is designed with bass boost equalization [30] in mind, there will be approximately 6 dB of available headroom at 50 Hz for equalization, when referred to the nominal reference band power level. (When a speaker system is to be used specifically for low frequency reproduction such as a solo electric bass guitar, this headroom will need to be reduced, and the system designed accordingly [11].) Home theatre woofers also operate on less headroom owing to the high effects levels present at low frequencies.

Damping

Both Thiele and Small have indicated that there is no theoretical need for acoustic damping in a system at low frequencies [3,4]. (Absorption is of course necessary for the higher frequencies; see Chapter 7.) Generally speaking, it is only those systems which are poorly designed, often with insufficient electromagnetic damping, that require significant additional absorption in the form of a dense volume filling or resistive structures in the port. Such enclosures invariably have low values of k_n, the power constant; much lower than the practical values suggested earlier in this chapter. In addition, k_p, the power handling factor, is usually below optimum.

Damped systems generally reflect a designer's decision to make use of an existing driver, possibly chosen for its mid-band rather than LF properties. Making the best of a theoretically poorly matched driver/box combination by smoothing the resulting output irregularities through resistive damping is inevitably wasteful of bass power and system efficiency — though is much used in practice.

Until recently, much LF analysis ignored the fact that the use of lumped, single-value parameters was at best an approximation. Highly accurate measurements of systems, particularly vented types, show small but significant discrepancies with the theory. In terms of response aberration the effects may be considered negligible, but it remains a fact that these artifacts may add colouration to the final system. In such cases some measure of damping may well assist in absorbing the higher modes, so helping the resulting lossy system to conform more closely to the lumped simplification. Absorption inside the enclosure is essential in practice to avoid colouration arising from standing wave modes.

Low Frequency Parameters

While elegantly simple programmes are available that deal with the basic maths for low frequency analysis, these working on high-pass filter theory, it is worth remembering that this approach relies on the assumption that lumped parameters are valid for the acoustical quantities. Such low frequency analysis is a reasonable approximation. It generally assumes that higher order acoustic radiation effects are negligible and rarely is note taken of wave properties inside the enclosure. The basic theory is satisfactory for small enclosures of low efficiency, e.g. below 93 dB/W and 50 litres. With higher sensitivity design (96 dB plus) and larger enclosures the acoustic load and related radiation factors play their part, while in box sizes of 150 litres and over the air volume can no longer be regarded as a simple spring. More complex models have to be devised involving higher order equations governed by the radiation properties of the cones and enclosure as well as the size and proportion of the cavity bounded by the enclosure. Some more recent computer programmes do take some account of these factors.

Low Frequency Alignment

View point — What is an optimal low frequency Q?

Should the Q value of a system be based on classical 2ii theory defined or measured against the pistonic reference band level as is customarily done, or is it more accurate to reference it to the overall performance of the speaker system, the actual mid-band averaged level? Consider the real and theoretical low frequency Q for an example $2\frac{1}{2}$-way system, one with a bass unit and a bass–mid unit operating in parallel. If the alignment for the upper bass–mid driver is adjusted for optimum mid efficiency, then it is usual for the reference range, and consequently the bass region, to be depressed relative to the actual, in use speaker sensitivity. For that upper section the level shift could amount to 3 dB relative to the reference range, so a low frequency cut-off of say 60 Hz is now -6 dB instead of 3 dB, and is effectively of lower Q, typically down to 0.55 from the original 0.7. Next take account of the bass section of this system whose job is to fill in the depressed bass region, in a sense restoring the system Q and, according to design practice, extending the bass response to still lower cut-off frequency.

What matters subjectively is the bass alignment relative to the final response, the transfer function of the whole speaker system, and not some arbitrary notional calculation for a bass driver based on 2ii theory.

In any case the boundary effects in a real room complicate the issue so much that targets for system Q need revising according to the cut-off frequency and the intended room placement.

Undue reliance on computer-aided design can result in inadequate speakers which interface poorly with real rooms. An overview is essential to guide computation and make sure that it does relate to working acoustic conditions.

Adaptable Low Frequency Design

Active loudspeakers often have adjustments for their low frequency characteristics to suit placement and room acoustics, both full range and sub-woofer systems.

In principle there is no reason why passive loudspeakers should not be adaptable also and there are a number of possibilities.

One good reason for the facility is the varying bass acoustics found in houses across the globe, which means that a given room matched alignment in one location may be off target when auditioned in another, and this despite careful positioning and an optimal listener placement.

An intriguing example of adjustability is the Sonus Faber Extrema, a two-way stand mounted compact system for semi-free space use. Employing a rear panel mounted ABR for the reflex alignment the system is tuned to be critically damped if the ABR were to be mechanically blocked. In this instance the ABR is a large bass unit in its own right complete with motor system and it is electrically connected to a rotary switch and an array of moderate power resistors. Wired across the motor coil this provides an elegant method for tuning the Q of the ABR and thus the alignment of the whole system, over a range from a near sealed box Q of 0.6 to a maximally flat Q of 0.7, full reflex.

On test it proved versatile in room matching and, in addition, allowed the performance profile to be adjusted according to use and programme type. For example, moderate loudness classical might sound best with a fairly high Q while heavy rock played at high levels will sound faster and better controlled with a lower Q.

For some three-way systems using rather low crossover points, e.g. 150 Hz, the large electrical network to the bass unit constitutes a point where the low frequency response shape, and hence the effective alignment, may be altered via the network Q or perhaps using a power resistor placed across the motor coil (the latter would have little or no effect on a direct coupled bass unit or where a higher crossover frequency is involved).

Another technique involves making an allowance at the system design phase for user adjustable variations of the port. Ports may be interchangeable, adjustable for length. Alternatively, or additionally, one or a range of porous foam plugs may be provided — an easy way of adjusting the port Q factor or efficiency. With sensible design a useful range of alignments can be made available to the purchaser for fine tuning the speaker to the relevant local environment. A foam liner for the port is yet another useful point of variation.

At the price of potentially greater signal impairment, some designers have advocated small, active low frequency equalizers, generally with switch selected response shapes. Such a unit would be placed on the tape loop of an amplifier or preamp. In practice their addition generally results in a small but disappointing loss in quality, often resulting in their ejection from an enthusiast's audio system.

Where a woofer is built with a double motor coil, the second coil may be used for low frequency control; this feature also allows the low frequency response to be tuned electrically. A design could be conceived using adjustable filter elements which would offer some control of the low range and hence the effective bass alignment.

4.3 CLOSED-BOX SYSTEM

The closed-box system is an attractive form of loading due to its simplicity of construction and ease of manufacture. It may be considered theoretically as an infinite baffle mounted driver with an additional stiffness component added to the existing suspension compliance due to the springiness of the air volume trapped in the enclosure. Clearly a smaller box will have a greater stiffness contribution than a larger one, and in sealed-box systems this air restoring force is normally made dominant compared with that of the driver suspension [13].

Overmuch has been made of the high linearity and of this 'air spring' loudspeaker. In fact the air stiffness in the phases of compression and rarefaction differs and if the volume change is significant by comparison with the total value (i.e. more than 5%), then the resulting distortion may be obtrusive. Harwood states that in a typical $60\,\text{m}^3$ listening room, to produce a sound level of 105 dB, a sealed-box speaker with an f_3 cutoff frequency of 40 Hz must be at least 65 litres in volume, if a 3% second harmonic distortion criterion is specified. With a 10 litre enclosure and a 60 Hz f_3, the listening room sound pressure cannot exceed 96 dB [8]. In practice some additional allowance may be made for the reducing LF content of normal programme. While the larger infinite baffle speakers will certainly supply enough undistorted output for most domestic and medium level monitoring purposes, the distortion described may well prove to be a limitation for some high level IB designs.

Analysis

The complete acoustical circuit (electromechanical) for a sealed-box system is illustrated in Figure 4.10. Due to the poor efficiency of such speakers the air load impedance is very small and may be neglected. Likewise resistance losses in the cabinet (leaks, etc.) may be assumed negligible. The circuit now simplifies to Figure 4.11. Here

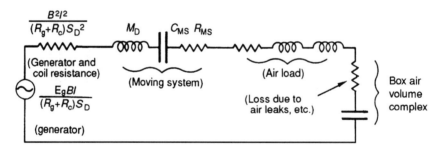

Figure 4.10.   Electromechanical equivalent circuit for a driver in a closed box

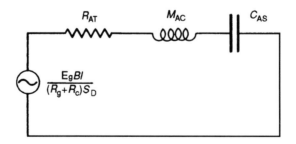

Figure 4.11.   Simplified form of Figure 4.10 (like components are summed forming a simple tuned circuit)

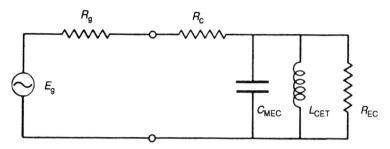

Figure 4.12. Voltage or impedance equivalent form of Figure 4.10 ($M_{AC} \equiv C_{MEC}$, $C_{AT} \equiv L_{CET}$ and R_{EC} represents the mechanical losses)

R_{AT} represents the total series resistance, M_{AC} the total mass including diaphragm and adjacent air mass, and C_{AS} the total compliance of both driver and air spring components. This acoustical circuit may be transformed into its electrical equivalent to facilitate analysis, as in Figure 4.12.

The system response is that of a damped single-resonant circuit. Two reactive components are present, hence the transfer function $G(s)$ is second order, and demonstrates a 12 dB/octave ultimate rolloff below resonance:

$$G(s) = \frac{s^2 T_c^2}{s^2 T_c^2 + \dfrac{s T_c}{Q_{TC}} + 1}$$

where

$$T_c^2 = \frac{1}{\omega_c^2} = C_{MEC} . L_{CET}$$

defines the resonant frequency f_c. Since Q_T = total driver Q at f_c, for a normal low output impedance driving amplifier

$$Q_T = \frac{Q_E Q_M}{Q_E + Q_M}$$

where

$Q_M = \omega_c C_{MEC} R_{EC}$, the mechanical Q

$Q_E = \omega_c C_{MEC} R_c$, the electrical Q

$$C_{MEC} = \frac{M_{AC} S_D^2}{B^2 l^2}$$

$$R_{EC} = \frac{B^2 l^2}{(R_{AB} + R_{MS}) S_D^2}$$

A minimum phase characteristic is shown by the system and the parameters of phase and amplitude/frequency are sufficient to describe completely its behaviour.

Response Shape

Figure 4.13 illustrates the possible range of useful response shapes for a normalized cutoff frequency f_c, when Q_{TC} lies between 2 and 0.5. The latter figure corresponds to the critically damped alignment where the response is $-6\,dB$ at resonance, and the initial rate of rolloff is slow at $6\,dB/octave$, down to approximately half the cutoff frequency. The transient response will show no ringing and little overshoot.

The Butterworth 'B2'* is a popular alignment where $Q_{TC} = 0.7$; this places the $-3\,dB$ point at f_c. The response is maximally flat and still possesses a satisfactory transient characteristic. Putting Q_{TC} equal to unity provides a greater bandwidth since the $-3\,dB$ point appears at approximately 0.8 of f_c, but this is attained at the expense of a 1.5 dB response lift above cutoff. The impulse response possesses a noticeable overshoot accompanied by a cycle or so of ringing. Increasing values of Q_{TC} do not serve to further extend the response, and the overshoot and response peak at resonance continue to increase (see Figure 4.13(b)).

If a 2 dB peak in the response is acceptable, corresponding to the 'C2 Chebychev' alignment where Q_{TC} is 1.1, then it results in the optimum efficiency alignment for a sealed-box system and offers a nominal 1.8 dB increase in sound pressure over the B2 alignment for the same $-3\,dB$ cutoff frequency. (In essence, the smoothness of the 'B2' alignment has been traded for a higher reference efficiency.)

A C2 alignment is permissible for a small system of limited bandwidth, e.g. with an f_c of 65 Hz and above. With increasing low frequency bandwidth the best subjective results for stand mounted systems will be attained also by progressively reducing the Q value. For an f_c of 50 Hz a Q_{TC} of 0.6 is suggested, while for 40 Hz a Bessel alignment $Q_{TC} = 0.52$ is appropriate. These alignments, adapted for domestic room matching, show that efficiency need not be unduly sacrificed in the search for better low frequency extension. For a given box volume and system resonance an alignment with $Q = 0.5$ will provide a 3 dB increase in reference band efficiency over the B2 response.

Sensible alignments will provide better all round performance but at present many boomy underdamped designs remain popular.

Enclosure Volume and Efficiency

A table has been drawn up of the maximum reference efficiencies for various box sizes (C2 alignment). Figure 4.14 shows, for example, that the maximum efficiency even a large 120 litre sealed-box enclosure can achieve is limited to 1%, for a 35 Hz cutoff frequency. If the latter were increased to 45 Hz (a not unreasonable figure), then the theoretical efficiency may be seen to double to 2%. A conventional 40 litre system with a typical 40 Hz cutoff cannot exceed 0.5% efficiency (f_c about 48 Hz with the C2 alignment).

The section on LF equalization in this chapter explains how the effective efficiency may be increased by choosing an over-damped alignment with a high reference efficiency, and then restoring the $-3\,dB$, f_c point through bass lift in the accompanying amplifier, provided that sufficient headroom is available.

*This and the other alignments mentioned are from Thiele [3]. These are respectively Butterworth alignment 2 and Chebychev alignment 2.

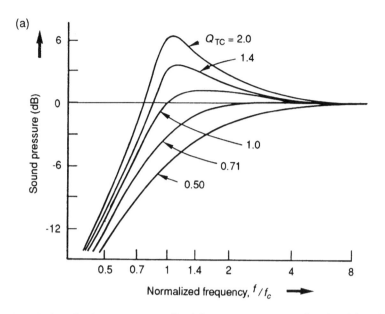

Figure 4.13. (a) Amplitude versus normalized frequency response of a closed-box loudspeaker system for several values of total system Q_{TC} (after Small). See Figure 3.34. (Note the reference levels for each case are normalized)

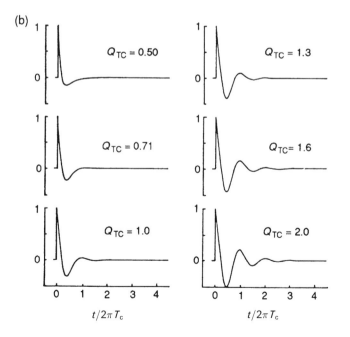

Figure 4.13. (b) Normalized step response of the closed box loudspeaker system (after Small [7])

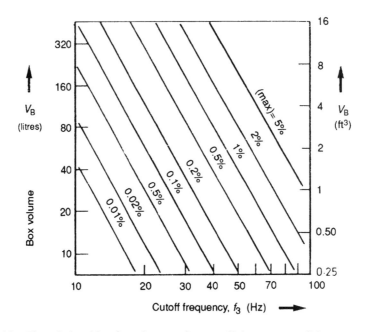

Figure 4.14. The relationship of maximum reference efficiency to cutoff frequency and enclosure volume for the closed-box loudspeaker system; f_3 is the cutoff frequency (after Small [7])

Box Filling or Damping

A volume filling may offer an apparent air volume increase of up to 15%, and additionally may add a mass component due to the physical movement of the filling at the lower frequencies. The combined effects lower the system resonance and must be accounted for in the design. With a light diaphragm and a dense but flexible filling, the effective cone mass increase may be as much as 20%. Very dense fillings will increase the frictional losses in the enclosed air volume and augment the damping. If the system is designed correctly, then such damping is not required, but may help to control a system where the Q_{TC} is too high, perhaps due to inadequate magnet strength. Movement of the filling is undesirable because it is usually non-linear and impairs subjective rhythm.

Design Example

To assess its performance in a range of enclosure sizes, Small [10] examined an LF unit with the following properties: $f_0 = 19$ Hz, $Q_m = 3.7$, $Q_E = 0.35$, $V_{AS} = 540$ litres and $Q_T = 0.32$. Using

$$\eta_0 = \frac{4\pi^2 f_0^3 V_{AS}}{C^3 Q_E}$$

the reference efficiency was found to be 1.02%. The effective piston radius was estimated at 12 cm, so $S_D = 45 \, \text{cm}^2$, the piston area.

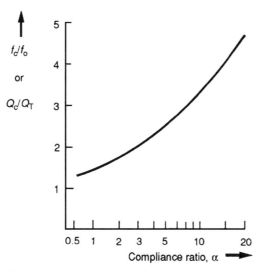

Figure 4.15. Ratio of closed-box resonance system f_c and Q_{TC} to driver resonance frequency f_0 and Q_T, as a function of the system compliance ratio α (after Small [7]). (Note that the system Q increases with α, i.e. with reducing box size)

The system damping is invariably lower than that of the driver alone (i.e. Q_{TC} is greater than Q_T) and the ratio of the two depends either on the ratio of the enclosure compliance to that of the driver, or alternatively on the ratio of the enclosure volume to the equivalent compliance volume of the driver. The ratio is denoted α, where $\alpha = V_{AS}/V_B$, and since the box should represent the controlling restoring force in a closed-box system, α is rarely below 3. Since the response shape is proportional to the total system Q, and this depends on α, both shape and cutoff frequency for a given driver are controlled by this ratio. Figure 4.15 shows this relationship, and alignments for the driver example are given below.

The lowest Q_{TC} (0.72) is attained with the driver using the largest enclosure volume worthy of consideration (135 litres). Despite this, Q_{TC} remains more than double the driver Q alone (0.32) which indicates that not only should the driver compliance be high for sealed-box designs, but also the driver Q should be sufficiently low. This alignment approximates to B2 and hence in this example f_c (the box resonance) and f_3 (the -3 dB cutoff) are the same, i.e. 42 Hz, approximately twice the driver free air resonance.

A smaller 60 litre cabinet offers a compliance ratio equal to 9. The system Q rises to unity with a small peak at resonance (60 Hz) and this peak extends the -3 dB point to 47 Hz. If the cabinet volume were to be further reduced, then the Q would exceed unity but could probably be adequately controlled by additional damping offered by a dense volume filling. Many drivers available on the o.e.m. (original equipment manufacturer) market are not too well suited even for sealed-box use, due to their excessive Q. Many have Q_T values between 0.6 and 1, which implies that virtually any box design used with them will have an excessive system Q and a resulting rise at resonance, unless heavy damping is applied in the form of a dense volume filling and/or thick felt layer applied over the rear chassis of the driver. This resistively loads the local air movement adjacent to the diaphragm but efficiency is bound to be impaired as a result.

In his series of papers Small [4–9] comprehensively covers an example design of driver and enclosure to meet a given specification.

In this present driver example let $Q_E = 0.19$ and $V_b = 100$ litres, a realistic size. Now

$$\eta_0 \simeq 2\%$$

$$\alpha = \frac{540}{100} = 5.4 \quad \text{giving} \quad \frac{Q_{TC}}{Q_T} = 2.5$$

$$Q_T = 0.2 \quad \text{with} \quad Q_{TC} = 0.5$$

$$f_c = 2.5 \times f_0, \quad \text{or} \quad 47\,\text{Hz}$$

which for a Q_{TC} of 0.5 will be the $-6\,\text{dB}$ point.

Alternatively, if moving mass were to be traded for sensitivity and η_0 was set at 1% then the moving mass M_D can rise by $\sqrt{2}$. Adjusted to the same Q values, the $-6\,\text{dB}$ point f_c is now 37.5 Hz (the latter is a preferred alignment, given in Figure 4.13). With normal room gain the $-6\,\text{dB}$ response of this system *in situ* is typically 20 Hz.

Constant Pressure Chamber (Isobarik)

In the early 1960s, a French designer produced a speaker system which had further low frequency drivers placed close behind those on the front panel. A small sealed chamber is present in the intervening space. Driven electrically in parallel, advantages of response extension, linearity and power handling were all claimed. In the early 1980s, the UK Linn company patented a similar device, called an Isobarik. Their description was based on the idea that the driven rear loudspeaker takes care of the acoustic drive to the main sealed chamber thus endowing the front radiating driver with constant pressure operation (see Figure 4.16); the atmospheric pressure loading is said to be

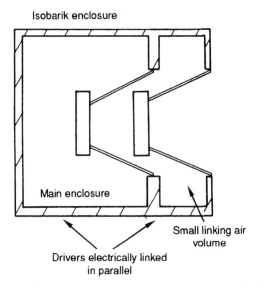

Figure 4.16. Series operation of LF drivers in a sealed-box enclosure (Isobarik). The units are electrically connected in parallel and work in cascade to drive the main chamber volume

maintained on both the front and the rear faces. It is then suggested that the natural, unbaffled low frequency range is provided.

In my opinion, a more convincing explanation goes as follows. The small air chamber between the drivers is essentially incompressible at frequencies below 150 Hz. Hence the diaphragms may be regarded as close coupled, as if by a lightweight rod. Now the analysis is simple. Conventionally connected, with the motor coils wired in parallel, the composite dual driver has the following characteristics if compared with the single device: twice the moving mass, half the compliance, half the impedance for which it draws twice the input current and hence double the power. In a sealed box the rear chamber air spring is much less compliant than the driver suspensions, and is the dominant restoring force. Hence the main Isobarik result is a reduction in system resonance of nearly $\sqrt{2}$ or 1.4. Typically, if the single driver-in-box resonance was 40 Hz, -3 dB, then this technique would offer a reduction to typically 30 Hz, -3 dB, with some improvement in power handling and linearity. An alternative route would be mass loading of the single driver, at risk to its dynamic performance and also its response extension to higher frequencies. On a positive note, the double driver 'composite' may well reduce the audibility of standing wave reflections in the main enclosure.

Resistive Chamber Coupling

While the theory of low frequency analysis suggests that for most properly designed systems acoustic resistance is not required, it is often used in practical designs where limited availability constrains the choice of bass driver and its low frequency parameters.

In these cases acoustic resistance may be desirable, even essential, to achieve a good result. Many commercial designs have used designed acoustic resistances, for example low Q vents in an enclosure wall, felt pads placed over the back of an LF unit to reduce Q_m, or placed in the division between the inner compartments of a multi-chamber enclosure. The performance of the acoustic resistance is often variable with frequency and unwanted secondary 'flapping' resonances are common which complicates the analysis.

In the case of the SBL and IBL loudspeakers,* a two-chamber bass loading scheme has been adopted. With the bass driver mounted in a smaller upper enclosure, a lower, larger volume is coupled to the first via a carefully designed non-resonant acoustic resistance, a finely divided multi-slot moulding of high precision (see Figure 4.17).

Height mode standing waves within the structure are suppressed, the response and impedance variations in the system resonance region are smoothed, while the resulting tapered response is better at matching the room, particularly in view of the against-the-wall position intended for these designs. The power handling approaches that defined by the restraining compliance of the smaller box volume, while the overall low frequency extension approaches that defined for the larger box volume. The head is decoupled from the box.

*Naim Audio Ltd.

(a)

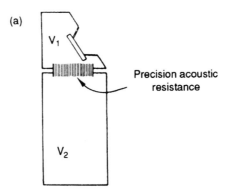

Figure 4.17. (a) A sealed-box, coupled-cavity system employing a precision, purely resistive element for coupling the cavities (after Naim). Here the technique allows for a small, separate low colouration 'head' enjoying the loading of a larger enclosure

(b)

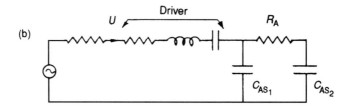

Figure 4.17. (b) The equivalent mobility circuit for the resistively coupled cavity system

4.4   REFLEX OR VENTED ENCLOSURES

The idea of venting or porting an enclosure to augment and extend its LF response by exploiting the Helmholtz resonance is quite an old one. The first reflex patent can be attributed to Thuras and dates from 1930. Since then numerous simplified and improved methods of analysis have been developed, culminating in the review of techniques presented by Small [8,9] in the early 1970s and the several computer-aided design programmes that have derived from his work [14,15].

For a given engineering expenditure, venting offers the following advantages compared with a sealed box.

1.   Greater maximum acoustic output (up to 5 dB increase for a typical programme drive).

2.   Higher efficiency (+3 dB for similar ripple-free responses).

3.   Lower cutoff frequency (30% lower than with the standard B2 alignment).

The hybrid form using an auxiliary bass radiator (ABR) may be a little more expensive owing to the cost of the additional moving element (with a normal reflex the extra radiating element is simply the plug of air trapped in the port or tube). The ABR

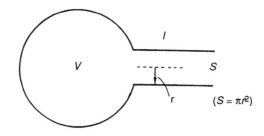

Figure 4.18. Helmholtz resonator: effective length $l_e = l + (\pi r^2/2)$

method [9,27] gives a cleaner sound as the windage noises and possible pipe resonances associated with small long ports needed for compact enclosures are thereby avoided. The ABR undoubtedly extends the scope of reflex design right into the compact enclosure range traditionally dominated by sealed-box enclosures.

Transient Response

The fourth-order, 24 dB/octave final rolloff for a reflex system is associated with a fundamentally poorer transient response than a sealed box. However, lower-than-optimum efficiency alignments such as QB3 or SC4* may be chosen whose characteristics are quite similar to the sealed box while retaining the reduced diaphragm excursion and consequent power handling advantage of the reflex.

Opinions concerning the subjective transient response of reflex systems may be influenced by the boomy character of the many poorly designed examples in production today, some of which have Q_{TC} values greater than 1.5. Reference to typical room reverberation characteristics indicates that passband ripple up to 1.5 dB at low frequencies is likely to be subjectively inaudible, this criterion permitting the use of the response-extending Chebychev alignments.

The time response given for the standard B4 alignment, maximally flat Butterworth, does not tell the whole story, nor has the failure of the system values concerned to act as finite lumped parameters been taken into account. For example, ducted ports have strong transmission modes at higher frequencies which effectively adds delayed resonances to the energy/time curve or transient response. If the analysis is to be carried out to a high degree of accuracy, then it will be found that with typical enclosures the lumped box air compliance parameter is subject to error above 50 Hz and the port mass component is uncertain above 150 Hz. Internal cabinet proportions and port geometry have a strong influence here and account for a significant proportion of the subjective differences and the response errors observed between otherwise similar systems.

It is worth noting that the original Helmholtz resonator (compare with Figure 4.18), designed to operate clearly at a single frequency, i.e. to follow a lumped parameter target, employed a spherical cavity with the ducted port inside and its extremity placed

---

*These and other alignments mentioned are from Thiele [3]. These are respectively 'Quasi Butterworth 3' and 'Sub-Chebychev 4'.

at the single mode or standing wave minimum. Pure tone reflex resonators were well understood by 1890!

Do not assume that other acoustic systems are free of resonance mechanisms, as investigation of the upper-range transducers used in many loudspeaker designs or alternatively the operation of a number of high performance microphones, will show. In both cases, the use of controlled acoustic resonance techniques to extend and flatten the response is widespread.

Analysis

Omitting the preliminary stage by considering Figure 4.11 we may add the mass component of the port (ABR or air plug) to the equivalent circuit to obtain Figure 4.19. The complete ABR analysis is more complex due to the finite compliance of the passive radiator suspension.

Whereas in Figure 4.11 the enclosure compliance was combined with the driver, in this case a separate circuit is necessary, comprising the addition of the enclosure port resonance, the total forming a fourth-order, high-pass network. As with the closed-box case, the compliance ratio α is

$$\alpha = \frac{C_{AS}}{C_{AB}} = \frac{L_{CES}}{L_{CEB}}$$

The full derivation is beyond the scope of this book, and system engineers are recommended to follow the references quoted [3,4]. Small's alignment chart for the basic C4, B4 and QB3 alignments is given in Figure 4.20, with the accompanying responses, Figure 4.21.

To a first approximation the enclosure losses may be neglected (high losses are caused by badly leaking cabinets or a theoretically unnecessary volume filling of absorbent). On the chart the values of k and B specify the C4 and QB3 alignments. f_B is the Helmholtz resonance of the enclosure; f_0 is the driver free air resonance; h is the tuning ratio equal to f_B/f_0; and Q_T is the total driver Q.

For classic B4 alignments a rule of thumb is to ensure $\alpha = 1$, so the port mass equals the driver's total moving mass. This natural equivalence offers equal acoustic output from the port and driver, and thus a well-tuned uniform response.

Design Example, Vented Box

Different alignment factors apply for different values of box loss; typically $Q_{box} = 7$. Noting that the chart reproduced (Figure 4.20) is for a lossless system and choosing a well damped QB3 response (Figure 4.21) an example can be tried.

A small 110 mm frame unit has $f_{s(0)} = 38\,Hz$, $Q_M = 6$, $Q_E = 0.33$, $V_{AS} = 23.6\,L$, $Q_T = 0.3$ for a nominally low source and cable resistance. With α set at 3 (left-hand scale for Q_T), the B factor appears at 2.2 (upper scale); h is 1.3 and $f_3/f_s = 1.5$ (right-hand scale).

The enclosure volume $V_B = V_{AS}/\alpha = 24/3 = 8\,l$. The alignment requires

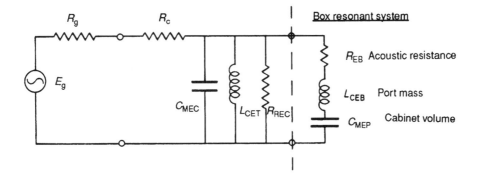

Figure 4.19. (a) Equivalent circuit of driver and ported (reflexed) cabinet

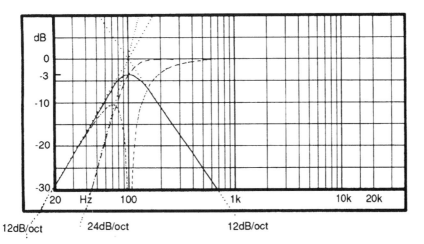

Figure 4.19. (b) Theoretical responses for a QB4, fourth-order maximally flat response for a bass reflex system showing response asymptotes and scaled to a vent resonance at 100 Hz. The port output reaches a maximum of -3 dB relative to the reference level. The driver output goes through a null as the port reaches to maximum, presenting the maximum value of acoustic impedance load to the rear face of the driver cone. The combined level of driver and port extends to 100 Hz, -3 dB, the rolloff slope at 24 dB/oct., the sum of the 12 dB slopes of the port output and the driver. (Key: ——— port, - - - - driver, – – – summation, ·········· asymptotes)

$$f_B = hf_0 = 1.3 \times 38 = 50 \text{ Hz},$$

and the vent resonance is 3 dB down at

$$f_3 = 1.5 \times f_0 = 56 \text{ Hz}.$$

Using a sensible 35 mm diameter ducted vent the Thiele–Small nomogram (Figure 4.22) gives a duct length of 110 mm. If in this design the example box was sealed, the -3 dB point would move up to 70 Hz, with a loss of power handling in the 45 to 75 Hz range.

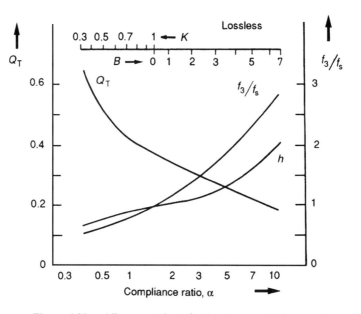

Figure 4.20. Alignment chart for lossless vented-box system

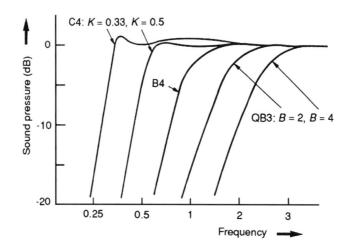

Figure 4.21. Normalized response curves for B4 and selected C4 and QB3 alignments of vented-box loudspeaker system (after Small [8]). (B4 maximally flat Butterworth)

The efficiency η_0 is

$$= \frac{4\pi^2 f_0{}^3 V_{AS}}{c^3 Q_E}$$

$$= \frac{4\pi^2 \times 38^3 \times 0.0028}{345^3 \times 0.33}$$

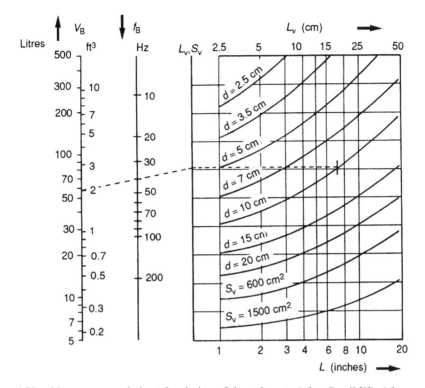

Figure 4.22. Nomogram and chart for design of ducted vents (after Small [8]). (f_B = enclosure resonance, L_v = vent tube length and S_v = vent area, no tube required.) The dashed line shows an example route for a 58 litre, 40 Hz vent, 10 cm diameter by 18 cm long

thus

$$\eta_0 = 0.45\%$$

(Note, in SI units, $4\pi^2/c^3 = 9.64 \times 10^{-7}$.)

Vented-Box Distortion

Harwood has investigated typical situations and shown that for a 60 m³ listening room, to produce medium sound levels of 100–105 dB with an f_3 in the region of 40 Hz, the distortion from quite small vents of 50 mm and upwards in diameter will not prove subjectively obtrusive [12]. While this is true in a general sense for orchestral programme and the more gentle pop material, the evolution of more powerful and percussive sounds in modern pop suggests that for critical listening this rule may not be adequate. In the 96–100 dB range distortion from a 50 mm vent diameter may be heard as a chuffing or 'windage' effect. Under these transient, turbulent conditions the port factors of loss, effective length, effective diameter and consequently vent mass are strongly modified. This produces a non-linear transient response. Subjectively the low frequency range may also exhibit compression effects.

Gander [37] investigated both vent and thermal compression effects in some depth. As regards temperature, a 380 mm bass mid-driver fitted with a 63 mm motor coil was tested, a unit intended for musical instrument applications. The response was compared for a 1 W and a 100 W input (100 dB and 120 dB s.p.l.). No compression was evident at the sealed-box system resonance of 50 Hz; elsewhere a significant 2.5 dB of compression was seen at the higher power, almost halving the effective rated power input.

Vent compression was illustrated in a 125 litre enclosure tuned to 42 Hz by one of two choices of vent. The first was a 110 mm aperture in the 19 mm thick front panel, the second a pair of ducted 110 mm ports with 125 mm tubes, 173 cm^2, twice the area. At 96 dB (1 W) the vent efficiencies were the same. However at 116 dB (100 W) the smaller vent lost a surprising 4 dB in its working range while the loss from the larger held to just 1 dB.

The use of an ABR of sufficient dynamic range can reduce distortion at high powers and can also block the egress of standing wave energy from within an enclosure at higher frequencies.

With large boxes the internal absorption can be very effective; an accompanying reasonably sized vent offers negligible colouration and distortion, and may well be capable of higher maximum acoustic output than the ABR. It is worth noting that the precise area of an ABR is almost immaterial; the high volume velocity of a well designed vent may exceed that of a practicable ABR by several times, and the expression 'increased radiating area' as applied to an ABR is misleading. Ultimately the reflex system will be limited by the non-linearity of the air volume in the cabinet, in the same way as with a sealed box.

For a given vented enclosure quite striking improvements in subjective and objective performances may be achieved by improving the smoothness of airflow in the vent. The velocity of air in the vicinity of a port may be very high, well above the value required to produce turbulence, and this is particularly troublesome with the smaller ports below 60 mm in diameter.

In some cases the use of a tube or duct helps by allowing an increased port diameter, but there are drawbacks, with Harwood indicating that a length to diameter ratio of more than 2:1 is inadvisable. Fluid flow theory indicates one immediate difficulty with 'ducted ports', namely that the ease of air flow is greater for the input or enclosure volume compression part of the cycle than the output or expansion part. This is because in the former the air flow is essentially from a hemisphere over the front panel down the tube, whereas in the latter the internal air stream approaching the exposed end of the port tunnel is divided, reducing the local pressure gradient and thus the volume flow. In consequence over a few cycles of low frequency drive in the region of enclosure resonance a net increase above atmospheric pressure is built up within the cabinet. Distortion, already produced by this rectification mechanism, is now further augmented as the back pressure pushes the driver diaphragm out of its linear working region.

Tunnel or tube ports are also acoustic structures in their own right and exhibit self-resonance or standing wave modes, a departure from lumped parameter behaviour. If coincident with modes within the enclosure the level of unwanted port output may be considerable at higher frequencies, thus contributing to system colouration.

With small ports, two practical ideas have been tried with significant measurable improvement. One consists of streamlining the important edges, for example for a simple aperture the inside and outside edges may be radiused. One model showed a

40% improvement in distortion at 50 Hz at a nominal power level by the addition of a 12 mm annulus of soft polyurethane foam at the port opening. Under stroboscopic examination its mode of action was found to be one of turbulence damping, the port liner edge moving with and thus controlling the 'windage', particularly reducing the higher 'chuffing' harmonics (this addition due to Spendor).

Bose have developed quite sophisticated port design in a small reflexed enclosure using fluid flow theory to establish a complex moulded streamlined form. Others have tried related methods such as a matrix of fine parallel pipes in the tunnel to aid laminar air flow. These are often realized as packed straws or a moulded honeycomb structure.

One solution to the ducted port modes is to incorporate an absorptive lining, the first suggested by Spendor as an extension of their foam liner. Here the entire length of the duct is lined, for example with 12 mm thick polyurethane foam, and forms a low-pass acoustic filter (see also [33]). Duct air-flow symmetry may be greatly improved by the addition of a small baffle or rim on the inner extremity of the duct. In one solution the standard cylindrical duct was replaced by an asymmetric triangular form, reducing self-resonance. In another, the duct and matching radial slot aperture was built into the floor of the enclosure, inhibiting resonance and imposing air-flow symmetry. Conversely, with the latter location, the internal duct entry cannot be easily located near the nodal region of the enclosure; this is a negative aspect. While taking account of driver location and the surface needs of wide dispersion treble units and the like, the port should be located towards the centre of the cabinet and not tucked away in a corner. A convenient location may well be the cabinet rear, behind the treble driver. Some designers have found also that a slant cut on the inside edge of the duct reduces self-resonance.

Twin Bass Units

Complications arise from twin bass drivers sharing a single reflexed enclosure. When using double bass units in a single sealed box the pressures and powers sum quite well and the results are according to expectation. Each driver sees close to half the box volume.

When the enclosure is reflexed the summation may not be perfect and the drivers may share the single port unequally, perhaps due to individual variations or a lack of symmetry in the placement of the port and/or the drivers. With smaller boxes the predicted Q for the port is not attained and its output at resonance may be so weak that it is dominated by the upper resonance of the driver pair in the enclosure volume. In one example the expected 33 Hz system resonance was entirely disguised and the port measured as a bandpass with -3 dB points covering a wide 20 Hz to 85 Hz.

Where possible, in such cases it is good practice to subdivide the enclosure so that each bass driver can be tuned individually. This also gives scope for differentially tuning the two sections to provide a more extended, tapered low frequency alignment, more suited to room boundary matching.

Multiple Ports and Resonance Distribution

Some designers have suggested the use of multiple ports of varying lengths to spread the box–port resonance. In practice, the lumped parameter approach remains valid and the box volume is still tuned to a single frequency, effectively 'integrating' the port group into a single vent. An approximation may be gained by averaging the port lengths according to their contribution to the total port area. Distributed port lengths may well be helpful in reducing the effect of the organ pipe modes by distributing them over a frequency range.

Resonance distribution can be achieved by using additional coupled box volumes associated with their own tuning ducts. Bandpass enclosure design shows just how sophisticated this subject can get.

Port Length

Typically, calculations for ports are based on the Helmholtz resonator principle, namely the effective mass of air in the reflex port of the duct acting with the compliance of the trapped air spring within the chamber or enclosure.

If the port length is zero, i.e. a plain hole in a thin wall enclosure, then there is still an air mass effectively present related to the diameter. The effective port length is $0.85 \times$ diameter. If the aperture has a finite duct length, then formulae are available which incorporate similar end correction factors to arrive at value for air mass (also see Figure 6.22).

Port Shape

In terms of area, square and circular ports are equivalent, but rectangular ports are not equivalent in area. Augspurger [40] suggests an end conversion factor of 1.13 times the smaller dimension times the fourth root of the aspect ratio, valid for aspect ratios between 1 and 10.

For a circular port the usual end correction is its diameter; this is added to the measured length. A square port is similar with a correction of 1.13 times a side. If multiple vents are used, then assess the end correction for an individual vent.

Some designers ascribe special characteristics to the shape of the exit aperture. While this may affect blowing or chuffing noises at high air velocities, taking into account local cabinet detailing, shape does not affect tuning or performance. Tests on slots, ellipses, rectangular forms and the usual convenient and inexpensive pipes show no low frequency difference, although the distribution of harmonic resonances in a longer duct may be altered by profile. A particular shape may well sound cleaner in a given enclosure design for this reason.

Port Location

Because there is generally some unavoidable spurious output from a port its location is influential. In some cases, particularly smaller boxes, it may be used constructively as part of the overall balancing of the design. A front panel location may be adopted. Conversely it is more common to locate the exit facing away from the listener to reduce the audibility of the unwanted sounds, duct blowing and resonances and acoustic leakage from within the enclosure.

With a floor standing design it may be possible to locate the aperture in the base of the enclosure, the necessary clearance from the floor being achieved with a matching base, or by the appropriate choice of spikes or feet.

Box Filling

Vent length calculations assume a predictable value for box volume, V_B, and textbook theory for low frequency analysis assumes minimal box damping 'said to maximize efficiency'. In practice, box interiors have a damping lining of 2 cm to 4 cm and/or some have a volume filling. These additions effectively increase the box volume by increasing the thermal mass of the interior. In an example tested by Augspurger, fibre glass wadding was used to a minimal fill of 50% of the surface area, thinly covered. This effectively increased V_B by 5%. A full lining 25 cm thick increased V_B by 10% and began to add resistive loss estimated at around 1% of the vent output. Moving on to a full volume fill with a clear space left around the driver and port gave a working V_B 20%–25% greater, while stuffing losses significantly decreased the vent output by 3–4 dB. However, this might even be useful in more difficult system designs.

It is more common to use Dacron or a similar fibrous wadding of lower density than glass for a self-supporting volume fill, and the V_B increase is unlikely to exceed 10%–15% unless very tightly stuffed.

The accurate calculation of box volume includes subtraction of the driver and duct volumes and any internal bracing of significant size. Thus, in practice a moderate fill tends to restore the original plain box volume used for the original calculations.

Port Modes and Measurement

As the port length is increased the resulting duct becomes an open pipe capable of unwanted organ-type resonances way above its intended bass operating range. The first mode is at the effective quarter wavelength, e.g. for a 14 cm duct it is around 450 Hz, allowing for end correction. Often it is the second harmonic at 900 Hz which is dominant and may lie only 10 dB or 15 dB below the intended port output. Unwanted resonances and colouration from ports do have a significant influence on overall sound quality. When assessing the likely effect of these resonances it is helpful to compare with the port blocked, making allowance for the attendant loss in bass output from the system. In measurement, note that a nearfield reading for the port will require scaling according to its radiating area to assess its impact on the primary response of the system. For example, the factor is in the square of the ratio of the two diameters, the bass driver and the port, taking care to use real and not specification figures. Thus, if a port is 70% of the diameter of the bass unit, then its pressure response measurement must be scaled down by half, or 6 dB, to align with the comparable nearfield trace for the bass unit. Phase plays a part and curve splicing remains an approximation.

If several mics are available — a practical proposition at low frequencies where ordinary units are often well behaved — their outputs may be summed in real time with a simple mixer, their individual outputs scaled according to the radiating areas. The total will represent the practical complex summation of the signals, still in the nearfield equivalent anechoic conditions (more correctly a consequence of the effective signal to room reflection ratio).

Port Modes Used Constructively

In one perverse commercial example, a 12 litre two-way system was tuned by a large 4.5 cm duct, 12 cm long, and the overall tonal balance was configured for an accepted and admitted proportion of port output adding energy in the lower mid range. An attempt to 'cure' the 'problem' by increased internal absorption and the substitution of low resonance port design turned out to unbalance the design, and in this case impaired the overall fidelity. This example design has enjoyed critical acceptance and goes to show that careful subjective assessment is a vital part of design and may throw up unexpected contradictions. When in doubt believe your ears first and look for explanations second.

Some Examples of Tailored LF Alignments

The following examples are for conventionally sized enclosures, closed and sealed box, fitted with well specified 210 mm (8 in) bass units, drawn from the Peerless line.

Take unit 850131, which has an f_s of 28 Hz, Q_{TS} of 0.37 and V_{AS} at 77.4 litres. Three reflex options are shown: first, a 30 litre box, strongly tuned to 29 Hz with a 5 cm diameter port 20 cm long, -3 dB at 42 Hz, -6 dB at 35 Hz (Figure 4.23). The output peaks fractionally at 80 Hz, although box losses would probably remove this theoretical lift. Approaching maximally flat Butterworth this alignment is probably too high a Q considering the local boundary gain. Leaving the port unchanged, raising the box volume to 40 litres results in a new tuning of 25 Hz; this small shift significantly lowering the Q for a better boundary match. The -3 dB point is barely altered while at 25 Hz there is a 2 dB improvement. Further increasing the volume to 50 litres, box 3, increases the damping a little, gives a still lower rate of rolloff and altogether a 4 dB raise in power at 25 Hz. Potentially the room gain may reach 6–8 dB, thus restoring more or less level output to this frequency.

For the next illustration I took driver 850137 which has an f_s of 26 Hz, Q_{TS} of 0.32 and V_{AS} at 92.2 — on the face of it, not very different. A good result can be reached in 30 litres reflex tuned to f_s, i.e. 26 Hz using a 25 cm long by 5 cm diameter duct. -3 dB is at 45 Hz, the -6 dB point comes in at 36 Hz, and the rolloff approaches the classic 24 dB/octave (Figure 4.24). Next compare with the small sealed box option 3, also near maximally flat but of more limited range. Now -6 dB is at 48 Hz. It does have a slower rolloff and hence a 'faster' transient response, while the near 12 dB/octave slope gives no perceptible advantage in terms of extreme bass extension, even down at 15 Hz.

The surprise alignment is for the larger 50 litre box, tuned to a low 21 Hz by a 15 cm long by 4 cm diameter port. This is nicely over-damped at the upper resonance, box 2. The nominal -3 dB point is 48 Hz, but the slow rolloff means that the -6 dB frequency is located at a desirably low 31 Hz, and down to this point the slope is very gentle, approximating to 6 dB/octave, quite unlike the usual criterion for a reflex design. In fact, the slope remains less than 12 dB/octave right down to 18 Hz.

Alignment flexibility such as this is a key factor in balancing the bass response of modern loudspeakers to give a natural sound in realistic living room acoustics. These examples also show that classic alignment theory should not be taken for granted and that a number of bass loading techniques can be adjusted to give good results provided that driver Q values are low enough and sufficient box volume is available when needed. In recent years there has been a tendency to crowd bass drivers into inadequate volumes.

4.5 BANDPASS ENCLOSURE DESIGNS

These are becoming available from several manufacturers, while the theoretical analysis has now expanded to include a whole heirarchy of more complex designs [34].

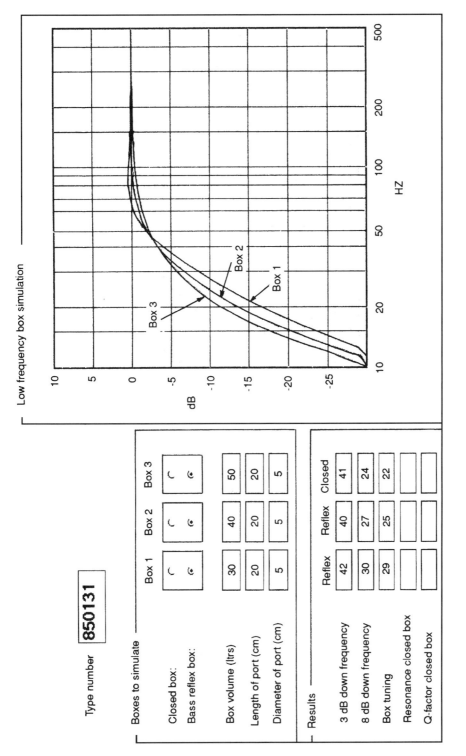

Figure 4.23. $f_s = 28\,\mathrm{Hz}$; $Q_{TS} = 0.37$; $V_{AS} = 77.4$ litres

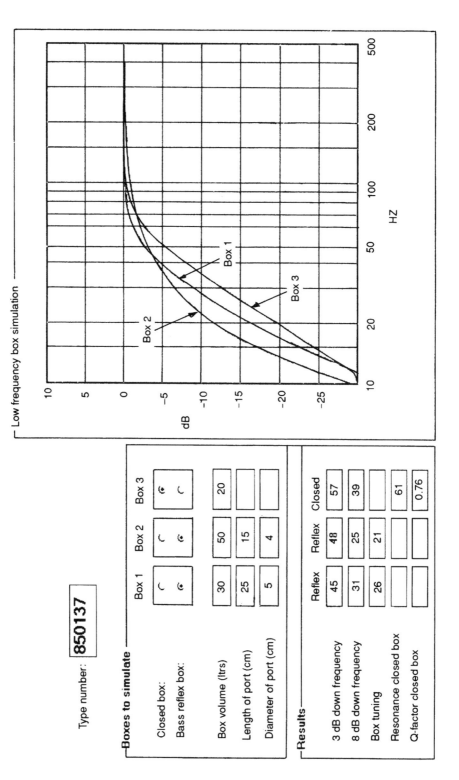

Figure 4.24. $f_s = 26\,\text{Hz}$; $Q_{TS} = 0.32$; $V_{AS} = 92.2$ litres

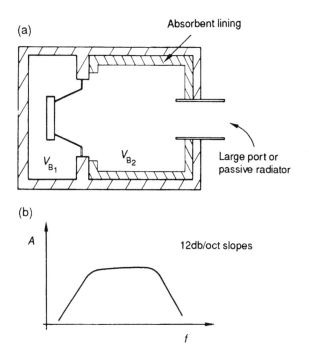

Figure 4.25. (a) The basic configuration for a bandpass loudspeaker enclosure commencing at fourth order. A sealed chamber back loads the internally mounted driver. The frontal output drives V_{B2}, the second cavity which is tuned via the large bandpass port or vent. (b) A 12 dB/oct, second-order slope is present for the high-pass and low-pass rolloffs. [US pat.; Lang 1954, France; Leon 1969]

Bandpass enclosures represent designs where the acoustic output is electro-acoustically filtered by a combination of acoustic and electrical low-pass functions to produce a defined acoustic bandwidth. This technique is particularly useful where a low crossover frequency is required in a multi-way system. Below 450 Hz passive crossover design is troublesome due to the unwanted interaction between the electrical low-pass filter and the reactive components of the motional impedance of the bass driver.

In a bandpass enclosure all the acoustic output exits via a large port which vents an inner chamber which is itself fed acoustic power by an internal drive unit or units. In turn, these are back loaded by a second chamber, often smaller than the first. Exponents of bandpass enclosure design cite improved filter slope characteristics, improved system efficiency, better low frequency extension, reduced cone excursion and, consequently, better linearity and power handling (Figure 4.25).*

The presence of acoustic low-pass filtering reduces distortion by attenuating harmonic products from the low frequency range. However, one drawback is the potential for upper range resonances in the main enclosure to be heard via the large port. In a given design, a compromise must be found between high absorption linings and a consequent loss of

---

*In fact the efficiency improvement relates to the control of bandwidth and not to a fundamental shift of the theory.

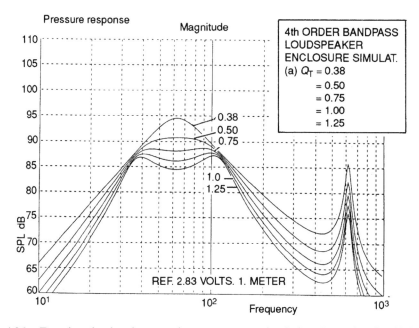

Figure 4.26. Fourth-order bandpass enclosure response simulation (second-order high pass, second-order low pass). The second peak results from a simplification of a quasi-distributed representation and is more complex in practice. A combination of electrical low-pass filtering and cavity linings will suppress these higher order modes. The curves show the effect of varying the drive unit Q_T from 0.38 to 1.25, a value of 0.7 close to maximally flat (after Geddes [34]) (see Figure 4.25)

efficiency. In addition, a port design free of windage and harmonic resonance modes must balance the need for a satisfactorily wide bandwidth (Figure 4.26).

A fourth-order alignment is common and corresponds to the first type described. Porting the inner chamber results in a 'sixth'-order type while the further addition of a second-order low-pass filter augments the low pass section to fourth order.* For the latter type, a surprisingly smooth, wide bandpass may be achieved (Figure 4.27). Double bass drivers may be used (Figure 4.28) while the designer may provide user adjustability via an optional active filter (e.g. KUBE). The use of bandpass enclosures for low-cost sub-woofers is also developing.

A detailed analysis is beyond the scope of this book, but for reference, the equivalent circuits and general form of equation are reproduced here for the second order, the third order, and for a passive electrical filter augmented, fourth order (Figures 4.29–4.31).

Computer-aided design (CAD), LF

Computer programmes can give the following:

(1) An indication of the range of theoretical solutions.

*i.e. 8th order overall for the bandpass filter.

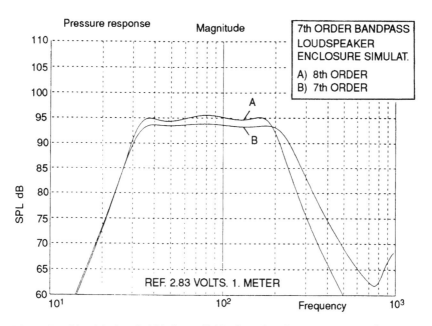

Figure 4.27. Considerable bandwidth is available from bandpass systems as these curves for higher order simulations show (Geddes). The 'eighth' order is achieved by a double ported design working with an electrical second-order low pass filter; the high-pass response is 'fourth' order, though the final rolloff will ultimately correspond to the 12 dB/oct. second-order slope of the lowest tuned vent*

(2) An indication of the possible practical solutions.

(3) An indication of the relative costs of (2).

(4) Quantitative results to enable practical values to be assigned to the variables in the system design equations

(5) Quantitative details of driver parameters, e.g. moving mass, compliance, free air resonance, etc.

A programme was based on the established LF loading equations to generate the following data. From the results obtained, curves could be drawn up relating the various parameters and hence allowing the rapid choice of useful solution sets (Fig. 4.27) (see Margolis and Small[32]). Basic criteria were sensitivity, −3 dB frequency, box volume and LF alignment. (See also CAD appendix).

Odd Order Alignments IB and Reflex

So far, even order second- or fourth-order systems have been considered, but with passive systems a further group may also be obtained, notably the third and the fifth, by

---

*There is no free lunch here, a characteristic 'sound' is often associated with bandpass systems of increasing 'order'.

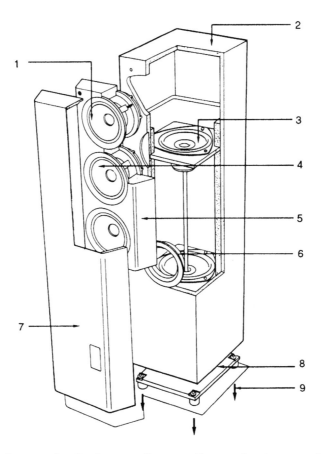

Figure 4.28. An example of a four-way floor standing monitor incorporating a fourth-order bandpass coupled cavity system (R105 III). The mid-treble enclosure is decoupled from the bass cabinet to minimize colouration. There are three cavities, one driven by the upper unit, one driven by the lower unit while the middle enclosure is tuned by a large 125 mm diameter vent with anti-turbulence flaring. R105 III (courtesy KEF Electronics). Key: 1—Low mid-driver; 2—Enclosure; 3—Bass unit; 4—Concentric MF-HF driver; 5—Solid, decoupled baffle; 6—Flared port/duct; 7—Grille; 8—Base panel; 9—Floor coupling spikes [29]

the inclusion of a large capacitor in series with the loudspeaker. Added to a sealed-box system we have third order with an 18 dB/octave ultimate rolloff, while for the reflex we obtain fifth order at 30 dB/octave. For both, the additional element provides a final subsonic rolloff of input power to the system, particularly helpful for the reflex type which otherwise has no final limit on excursion.

For a sealed-box system the additional design flexibility given by a series capacitor confers fourth-order type improvements in system efficiency; for a given f_3, and a lower Q_E, a higher sensitivity driver may be used. The technique may also be used to provide some tailoring of the extreme low frequency response below f_c, for example, to produce a better match to the listening room. Typical values are 200–400 μF and these need to be of decent voltage rating with high power systems; usually special non-polar electrolytics are employed, bypassed by smaller values for high performance designs.

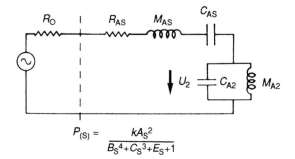

$$P_{(S)} = \frac{kA_S{}^2}{B_S{}^4 + C_S{}^3 + E_S + 1}$$

Figure 4.29. The simplified equivalent circuit for the basic bandpass system, second-order low pass and high pass. Acoustic components of the first cavity have been lumped with the driver. M_{A2} is the vent mass, and C_{A2} its associated chamber compliance

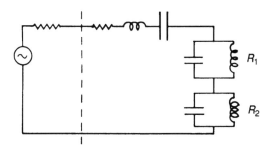

Figure 4.30. Shows the 'sixth'-order equivalent, a dual port design

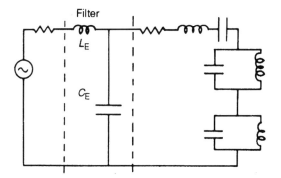

Figure 4.31. 'Fourth'-order high pass augmented to fourth-order low pass by the addition of a second-order passive electrical filter. (8th order bandpass)

It should be noted that with capacitor feed the full power handling augmentation of the fourth-order reflex is not provided by third order. For this fifth must be used and the production problems of tolerancing such a system often rule it out. Third order, used with a low Q long-throw driver, has proved worthwhile in some recent UK designs. With the right alignment an 'undersized' series capacitor provides the means to step the low frequency range in compensation for wall mounting a system.

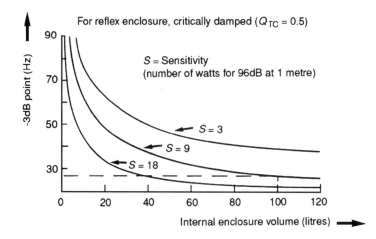

Figure 4.32.
Sensitivity: 86.0 dB at 1.0 m for 1.0 W input
System internal volume = 100.0 litres
Bass cutoff frequency 26.8 Hz
Bass unit parameters:

EL	V	F_s	Unit	Power	Bl^2/R_E	M_{MS}	C_{MS}
−23.0	100.0	26.8	130.0	50	0.0	0.0	0.0
−23.0	100.0	26.8	170.0	50	0.0	0.0	0.0
−23.0	100.0	26.8	200.0	50	7.17	16.6	21.1
−23.0	100.0	26.8	250.0	50	17.49	40.6	8.7
−23.0	100.0	26.8	300.0	50	0.0	0.0	0.0
−23.0	100.0	26.8	380.0	50	0.0	0.0	0.0

RE(1.1.1.1) = 000000
RE(10, 3, 5, 2) = 10.5361

The above extract follows insertion of data and shows two variable solutions: the one chosen is the 250 mm driver with a 50 mm pole, a moving mass of 50 g and a compliance of 8.7 × 10.4 m/N. The Bl^2/R_E refers to the driving force developed in the voice coil and magnet assembly and imparted to the cone. The programme continues to give details for the driver such as correct sizes for the magnet, the wire, top plate, throw, and the number of coil layers and turns. (This is an example of CAD-generated data)

For analyses a computer program deriving from the papers of Thiele and Small was published by the AES in the early 1970s [14,15] and many manufacturers actively pursue the various other aspects of computer-aided analysis; for example, the fast Fourier transform of loudspeaker impulse responses (see also Reference 23).

LF Alignments, Equalization

It has been indicated in the section on LF loading that certain alignments may require equalization. For example, take the case of the optimal system response in Chapter 3, Figure 3.38. If a +6 dB increase in efficiency were required, obtained by doubling the magnet strength, the system would then be overdamped and require equalization.

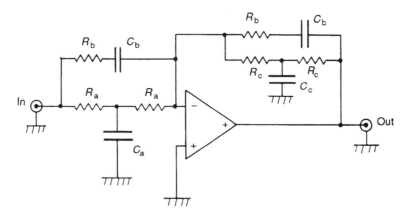

Figure 4.33. Versatile closed-box equalizer (after Greiner and Schoessow [30]). Any desired closed-box response may be obtained with almost any driver regardless of Q. (Equalizer should be suitably buffered in and out)

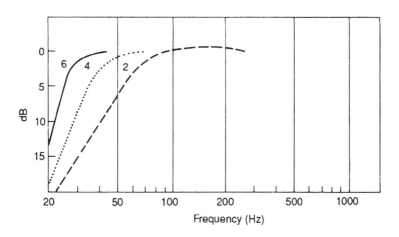

Figure 4.34. Low frequency performance of a 0.3 m frame, high compliance LF unit with $f_0 = 22\,\text{Hz}$, $Q_{ES} = 0.28$, $V_{AS} = 414$ litres; '2' is the second order response in a 40 litres closed box, $-3\,\text{dB}$ at 63 Hz; '4' is the fourth order QB_3 response in a 114 litre vented box tuned to 33 Hz, $-3\,\text{dB}$ system response at 33 Hz; '6' is the sixth order response, same cabinet as '4' but tuned to 26 Hz and with second order bass lift, $-3\,\text{dB}$ point now 26 Hz (after Keele [26])

Reference to the alignment pertinent to the new system will show the degree of bass lift necessary. In this case, the initial slope is 6 dB/octave, which may be applied at the input of the matching amplifier. An additional rolloff is desirable below the system cutoff frequency, so that overload at subsonic frequencies does not occur as a consequence of the otherwise continuing equalization (Figure 4.33).

The degree of equalization which may be applied is dependent on the spectral content of the programme employed, together with consideration of the excursion and thermal rating of the LF driver. The converse equalization is also possible, i.e. correction of an underdamped response. This is achieved by either a suitable 'dip' network ahead of the

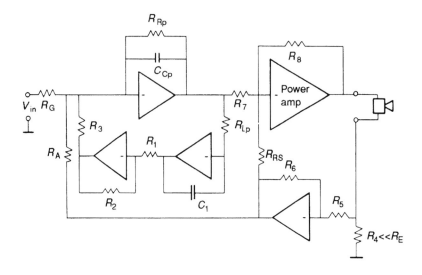

Figure 4.35. Realization of ACE-bass amplifier useful for experiment purpose (after Stahl [24])

$$R_s = \frac{1}{R_{R_s}} \frac{R_4 R_6 R_8}{R_5} \qquad C_p = C_{C_p} \frac{R_A R_5 R_7}{R_4 R_6 R_8} \qquad R_p = R_{R_p} \frac{R_4 R_6 R_8}{R_A R_5 R_7}$$

$$L_p = R_{L_p} \frac{R_1 R_3 C_1}{R_2} \frac{R_4 R_6 R_8}{R_A R_5 R_7} \qquad G = \frac{1}{R_G} \frac{R_A R_5}{R_4 R_6}$$

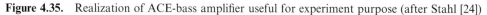

amplifier or alternatively by adjusting the output impedance of the amplifier to a negative value at low freqencies, thus increasing the overall electromagnetic damping [24,25].

The sixth-order alignment may be exploited to produce new performance standards for a given enclosure size. Keele [25] and Stahl [24] examine the classic Thiele analysis on vented enclosure design with reference to the sixth-order alignments, these relying on an auxiliary second-order filter usually ahead of the power amplifier. Thiele's B6 alignment 15 is chosen on which to base a new set of alignments, which allow a much wider choice of driver compliance. The second-order auxiliary filter is applied as a boost to the power input to the vented enclosure system and the whole provides high acoustic output with minimal excursion (Figure 4.34).

The effective Q_{TC} of this alignment may be adjusted to account for the practical values of room gain.

Both high compliance and low compliance drivers are amenable with suitable choice of alignment, if necessary with an allowance for some passband ripple.

Stahl [24] has shown how sixth-order alignment may be applied to the design of a sub-woofer, the latter referring to an enclosure added to an existing pair of speakers which then provides an octave or two of extreme bass often missing from smaller full range systems. Stahl extends the concept of utilizing an amplifier with a negative output resistance to control loudspeaker Q by introducing a complex negative impedance at the amplifier terminals connected to the loudspeaker, Figure 4.35. This dominates the characteristics allowing electrical control of the apparent mechanical characteristics of the connected driver, enabling the designer to utilize existing drivers and synthesize an exact desired alignment. An example is quoted [24] where a 20 Hz, $-3\,dB$ response was required from a 50 litre cabinet at 100 dB, 1 m (2π) at 5% or less distortion. Two special 170 mm drivers in push–pull were used which possessed a high force factor and large excursion.

The required mechanical parameters were synthesized via the amplifier impedance characteristic, and involved a moving mass increase from 0.038 to 0.26 kg, damping from 26 to 58 kg/s and a compliance reduction from 0.45 to 0.25 mm/N. The reference band efficiency remains unchanged at the high level set by the drivers themselves.

Motional Feedback

The idea of motional feedback [17] is not a new one and dates from the 1930s. Recently it has resurfaced in several commercial designs, incorporating modular transistor power amplifiers, where the convenience of the latter as compared with the earlier valve units must be a major contributing factor. The driver moving system is coupled to a sensor that generates a signal which may be fed back to the power amplifier and can be used to correct any motional error or non-linearity. The error signal will only be valid as long as the cone is operating as a piston and hence, in general, the correction can only be operative over the lower octaves. At very low frequencies the correction must be rolled off so that the unit is not driven out of its allowable mechanical excursion limits and, in consequence, feedback control is only possible over a specified bandpass. Within this range, however, the driver non-linearity may be considerably reduced, both in terms of pressure response uniformity and distortion. The technique allows poor quality, high resonance LF units to produce remarkable results in very small, sealed enclosures. Notwithstanding, the colouration problems associated with such enclosures still remain.

It is important to place the technique in its proper perspective. The compensation of major deviations in response through feedback involves equally large power demands to meet the deficiencies. A typical example employs a 280 mm driver in a 20 litre box, with an underdamped system resonance at approximately 110 Hz. The unequalized response at 35 Hz is more than 18 dB below the reference level, the latter specified at 200 Hz. While the application of motional feedback results in an f_3 point at 35 Hz, a 120 W amplifier is needed to achieve this, even taking into account the reducing headroom requirements at the lower frequencies. While the distortion of this example is under 2% from 40 Hz to 200 Hz, at 93 dB, 1 m, there is little evidence that distortion values of less than 5% at these lower frequencies are essential, even for a high performance speaker. Furthermore, if this driver example had possessed reasonable linearity the equalization could have been equally performed by a simple bass boost network in the driven amplifier, which would probably have resulted in a similarly satisfactory subjective result. Low efficiency is the penalty to be paid for extended bass response in small boxes. An efficient unit in a compact enclosure will require considerable equalization to achieve a wide uniform response. Consequently, in both the cases of equalization and feedback, a similarly high-powered amplifier will be required.

The only significant technical advantage of motional feedback lies in the improved amplitude linearity, but in any case, LF distortion would not appear to be much of a problem with conventional, well designed loudspeakers. While it is certainly true that poorly engineered systems may be 'rescued' by a feedback arrangement, when overload does occur it manifests itself with greater severity than in a non-feedback system. This

(a)   Servo loudspeaker system-layout

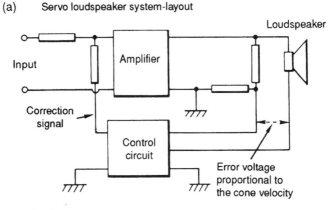

Figure 4.36. (a) A feedback system utilizing the variation in motional impedance. Alternatives include cone motion sensing accelerometers (courtesy Servo Sound)

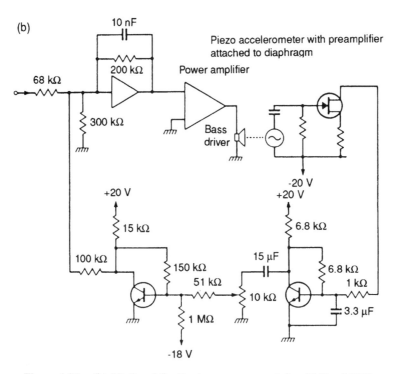

Figure 4.36. (b) Motional feedback arrangement (after Philips MFB)

can be an important consideration when a loudspeaker may be subject to occasional overload on loud programme transients.

Technically most feedback designs employ an accelerometer attached to the diaphragm, the former usually concealed beneath an enlarged dust-cap. Its output is electrically processed and fed in anti-phase to the accompanying power amplifier (Figure 4.36(a)).

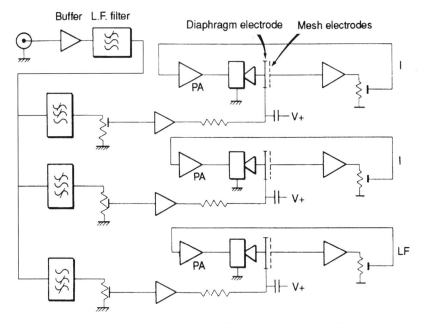

Figure 4.37. General system of 'Monitor 8'. Metallized diaphragms are polarized at $V_+ = 400\,\text{V}$. Capacitive electrostatic pickup via mesh electrodes spaced in front of diaphragms. Incoming filtered signal mixed with error signal from electrode and fed to power amplifiers as control signal for correction of frequency response and linearity (after Backes and Muller)

While other sensing transducers including magnetic and electrostatic types have also been proposed, a version of the latter has been employed in a system manufactured in West Germany. Motional feedback is applied to all the drivers over their respective piston ranges; the piston bandpasses are wide since the mid and treble are handled by dome units and the bass by an array of four closely coupled, 100 mm cone drivers (Figure 4.37). At large excursions feedback systems may become unstable and headroom limiters are common.

Acoustic Adjustability of Loudspeakers

So far equalization has been considered as an adjunct to a more or less completed classic design. Sixth-order, low frequency alignments hint at an additional flexibility through additional electronics while the 'Ace' bass principle goes further and seeks to electrically control and modify the electroacoustic constants of the driver itself.

This concept may be taken much further. Imagine the drive unit as a raw transducer developed to produce its maximum acoustic performance unfettered by the usual prime target of a maximally flat axial response over its operating range. Full system synthesis is well within the capability of modern electronics, whereby the natural poles and zeros or resonance properties may be cancelled in the circuitry feeding the power amplifier. Any desired response may then be readily imposed on the system, including all the classic low frequency alignments plus more recent room matched expositions.

The outcome would be versatile speaker systems possessing a high degree of acoustic adjustability. This can be used to good account in optimizing the performance in a variety of rooms and specific positions in those rooms. A further benefit would be a much reduced enclosure size for the same low frequency extension. Requiring auxiliary electronics, such adjustable systems are likely to be in the upper price group for some years.

With increased use of digital programme greater demands are being made in the low frequency range. Low phase shift at low frequencies demands an extended bandwidth, possibly to below 20 Hz. Power handling demands here are low, particularly in the absence of the troublesome rumble energy contained in analogue disc programmes. So a large radiating area and excursion is not necessarily required for such extended bandwidths.

KEF have built an experimental system with a 5 Hz bandwidth to explore the effect of low frequency cutoff on digitally recorded sounds. Initial results suggest a promising future for extended low frequency response loudspeakers (see Reference 39 in Chapter 9).

Comparison of Second-, Fourth, and Sixth-order Systems

Keele [31] published a fascinating paper on direct low frequency driver synthesis with many examples. One is reproduced in Figure 4.38 and Table 4.1. A 57 litre (2 ft^3) compact system is designed for a 30 Hz, − 3 dB point, maximally flat, capable of 0.2 W of acoustic power. In the suggested system implementations the wide disparity of driver

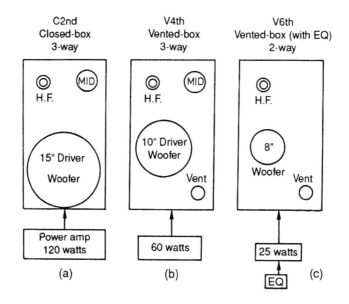

Figure 4.38. Line drawings of likely implementation of three systems. Each of the three systems has an internal volume of 0.057 m^3 and provide identical mid-band acoustic power of 0.2 W with response down to 30 Hz (−3 dB). (a) Three-way closed-box system with 15 in woofer driven by 120 W power amplifier. (b) Three-way fourth-order system with 10 in woofer driven by a 60 W amplifier. (c) Two-way sixth-order system with 8 in woofer driven by a 25 W amplifier with high-pass filter/equalizer. Relative sizes have been preserved (after Keele [31])

Table 4.1. System and driver parameters for the three system types. A bookshelf system designed for specific enclosure size, low-end limit, acoustic power output, and diaphragm excursion

	Parameter	System type		
		Closed-box C2ND	Vented-box V4TH	Vented-box V6TH
System parameters	f_3	30 Hz	30 Hz	30 Hz
	V_B	←	5.66×10^{-2} m³ (2 ft³)	→
	η_0	0.18%	0.36%	0.90%
	f_B	—	30	30
	P_{AR}	0.2 W	0.2 W	0.2 W
Driver parameters Thiele–Small	f_s	13.0 Hz	30 Hz	30 Hz
	V_{AS}	0.28 m³ (10 ft³)	0.060 m³ (2.12 ft³)	0.12 m³ (4.24 ft³)
	Q_{TS}	0.31	0.40	0.30
	Q_{MS}	3.9	5.0	2.2
	Q_{ES}	0.33	0.44	0.35
	V_D	539 cm³ (32.9 in³)	189 cm³ (11.5 in³)	124 cm³ (7.6 in³)
Electromechanical	M_{MS}	550 g	60 g	12.9 g
	C_{MS}	2.7×10^{-4} m/N	4.7×10^{-4} m/N	2.2×10^{-3} m/N
	Bl	29.6 T . m	13.0 T . m	6.7 T . m
	R_E	6.5 Ω	6.5 Ω	6.5 Ω
	R_{ME} (∼ cost)	134.4 N . s/m	26.0 N . s/m	7.0 N . s/m
	S_D	8.5×10^{-2} m²	3.0×10^{-2} m²	2.0×10^{-2} m²
	Effective diameter	0.33 m (12.9 in)	0.20 m (7.7 in)	0.16 m (6.2 in)
	x_{max}	6.35 mm	6.35 mm (0.25 in)	6.35 mm
	P_{ER}, $P_{E(max)}$	112 W	56 W	22.4 W

Note the very wide differences in driver size, input power, moving mass, and magnet requirements, with the vented-box drivers on the low side. If driver size can be allowed to decrease (as in this example), the vented-box drivers can be extremely cost effective (Keele [31])

size and required input power is most instructive. In Table 4.1, note also the range of moving mass (M_{MS}) and force factor (Bl) involved.

Effects of Parameter Variation, Particularly Climate

In a number of the references, sensitivity analysis has been applied to examine how the target responses are affected by variation in the parameters. It would appear that the more complex the order, the more sensitive it becomes to such variations.

Consider a system designed for a world-wide market where air conditioning and temperature control cannot be assumed in the place of use. Humidity can vary from almost 0 to 100%. A significant change in moving mass can result with pulp cones due to the hygroscopic nature of many diaphragms including, though to a lesser extent, the synthetic nylon-based materials such as polyamide.* Temperature can vary from 15°C to 30°C with 22°C as a general mean, but a speaker system with its back to a sunlit window can experience an overall temperature rise of 10°C or so above ambient, making a total rise to 40°C. These are serious effects since the surround of many modern low colouration bass units employ resistive materials such as nitrile rubber whose mechanical resistance and compliance vary strongly with temperature in the range concerned. As an example, an LF reflexed driver with a nominal 25 Hz resonance can exhibit a 17 Hz resonance at 27°C, but rising sharply to 35 Hz at 10°C. With the surround often representing the dominant driver compliance it would be worth choosing an alignment which exhibited reduced insensitivity to such variation. Over the specific range cited, a reflex design might misbehave in such a manner that reflex loading would be a risk that a designer might not consider worth taking.

As regards other factors, almost any driver parameter can and will vary due to production tolerance, probably the greatest of these being the Bl factor. Appearing 'squared' in the LF analysis relationships it is certainly the first parameter to tolerance closely, both for production spread and its effect on the resulting system alignment (see Chapter 3, Section 3.9).

4.6 LONGEVITY, RELIABILITY, TOLERANCES

Whereas many consumer products may be considered worn out or out of date after only a few years and are frequently discarded, loudspeakers tend to last a long time. Well-made examples are still operating after 40 years of use. Drivers built of traditional materials such as cured resin fabric for both surround and cone suspension can be very stable over temperature and time, whereas some more modern materials have shown early degradation, prematurely shortening the life of speakers.

After manufacture, the components of a loudspeaker are inevitably subject to ageing, both from use and the environmental factors. Many plastics used in loudspeakers are susceptible to the effect of light, particularly ultraviolet, to heat, and to degradation from oxidation and chemical pollutants. Nickel plated or tinned connector systems will tarnish and result in poorer electrical contact. Reticulated polyester foam was once very fashionable as a grille

---

*Even silk tweeter domes may vary with humidity.

material, but may crumble to dust after some years in adverse conditions. Nevertheless it is still favoured for its superior acoustic properties and its low cost means it is inexpensive to replace. When used inside a loudspeaker, e.g. for air flow damping purposes, it has a much longer life, probably tens of years thanks to its protected location.

For the bass driver many designers still favour a foam surround for a moving coil unit chosen on grounds of lightness and good cone termination. Historically, a very large number, typically made of a urethane polymer, have disintegrated in service over periods of 5 to 10 years. The protective addition of carbon black can extend the lifetime substantially and polymer purity also plays some part.

Surrounds made of plasticized PVC and other similar materials also may change over a period of a few years due to migration or loss of the plasticizer. In one case a famous speaker system, produced in many thousands, had a bass driver whose surround compliance fell by a factor of four after a few years, causing an almost complete loss of bass.

The phenomenon of 'creep' in plastic suspensions also needs to be controlled. Sag at this point can result in a driver voice coil rubbing in the gap after a period of years. Normally a loudspeaker is mounted with its axis horizontal. In many applications, notably bandpass and sub-woofer systems, the mounting axis may be set vertically. The constant application of gravity often causes a driver cone to sag over a period of time causing the voice coil centre to drop with respect to the optimum gap. If known in advance a driver may be built with a suspension bias that helps to offset the long-term effects of gravity.

Adhesives may not be as stable as the materials they are joining. Sometimes the materials themselves degrade the adhesives and vice versa, while adhesives that appear mechanically strong on first use may become brittle with time, or worse, soften and exhibit creep. In manufacture, adhesives need to be selected not only for their correct mechanical properties but also for manufacturing cycle times and for their long-term reliability.

Crossover networks are generally reliable unless the power resistors fitted are of inadequate power rating and hence may over-heat. Before failing, power resistors can char printed circuit boards, corrode track and melt the insulation on nearby cables. It is easy for the designer to under-estimate how hard a speaker may be driven when operated on modern high power amplifiers. A careful analysis is required, ideally backed by running tests with equivalent dummy loads substituted for the drivers where damage is likely.

The Running-in Phenomenon

It is observed that loudspeakers show subtle changes, generally improvements, over the initial tens of hours of use, a phenomenon reviewers call 'running in'. Listeners have noted improvements in treble clarity and sweetness, gains in definition and transparency as well as superior low frequency extension and uniformity. Helpful explanations are available for many of these phenomena.

A good designer is well aware of the freeing up of a loudspeaker suspension in the early hours of use and will take care to design prototypes with drive units that have been run in and whose parameters are stabilized. The free air resonance of a bass driver may fall between 10% and 12% from first manufacture after normal conditioning, and this will materially affect the whole set of low frequency parameters, including the optimized low frequency tuning or loading.

Many parts of a loudspeaker consist of materials of a fibrous content bonded with resins. These may be unduly rigid and exhibit lower Q factors when new. After use some of the initial stiffness is broken down and the fibres in the material, whether cone pulp, woven carbon, Kevlar fibre or the woven cloth used for the centring suspension, begin to move more freely with respect to each other. An increase in damping generally accompanies these changes. Running in can provide a smooth, more relaxed sound together with the attainment of the designed bass performance. Some critics and designers also note that there is an acclimatization process for the electrical components in a speaker, especially the cables and capacitors. Factors include a reduction in the dielectric stresses which result from the manufacturing and handling processes.

For tweeters the concept of running in is less obvious. Here one must consider the stresses arising in moulding and similar thermal forming processes. In use these materials are relieved of their initial excess stresses and may operate with lower Q values and a sweeter sound. This process may be compared with a car whose suspension is built with a complex mix of metal and polymer springs and dampers. For the first few thousand miles the suspension will seem stiff and unyielding and the optimum ride and handling qualities are only attained after a running-in period.

Once run in, after a period of a few hundred hours of use and settled at a stable temperature and climate, a high performance loudspeaker system can benefit from a judicious re-tightening of the driver and support fixings. In the critical area of comparative review this can make a significant difference to the outcome with resulting improvements judged as high as 10%–15%.

Driver Build, Gaps and Tolerances

Typical cones used for loudspeakers in the 120 mm to 250 mm frame size will be of variable thickness according to density and range from 1.5 to 3 mm. If of metal, e.g. aluminium alloy, a 1–2 mm thickness is common, often deeply anodized up to 30% of the depth to engineer hard skins of 'ceramic' (alumina) on the central alloy core. The cone mass on an alloy diaphragm for a 110 mm frame size will weigh from 5 to 7 g, rather more than a paper equivalent at 3–5 g.

As regards gap tolerance this size of driver may have an overall magnet gap of 0.9–1.1 mm taking a two-layer coil wound with 0.2 mm wire on a former of similar thickness.

For the high frequency unit the reduced movement allows for a 0.7–0.9 mm gap containing a two-layer coil wound in 0.09 mm wire, again on a thin former, often aluminium foil to aid heat conduction. The dome itself, depending on the metal or alloy chosen, will be drawn from foil of 30 to 100 μm thickness, and requires delicate handling.

Tolerancing

The control of sensitivity is vital for serious loudspeakers. Without this control a designer's good intentions count for little as quite large changes in sound quality occur with relatively small changes in relative driver level. In one example a stereo pair was found to have poor focus where the image was pulled laterally, depending on the frequency reproduced, and one channel sounded louder than the other. When individually measured the speakers fell within the tolerance defined for the overall specification namely ± 3 dB, 100 Hz to 15 kHz. Close examination showed errors of as little as 1 dB for each driver relative to the design centre defined for the system. A comparison of the systems showed that a worse case error scenario pertained. For one channel the tweeter was 1 dB high, the woofer 1 dB low; for the other the tweeter was 1 dB low, the woofer 1 dB high. On the criterion that for a consistent sound 0.5 dB tolerancing is worthwhile for the treble level relative to the mid range, this pair of speakers was delivering an unacceptable 2 dB error — one 2 dB too bright relative to the mid, the other 2 dB too dim. Production procedures are necessary which can trap such errors and provide data on the causes of these variations thus facilitating tighter quality control.

In the above example, swapping over the driver combinations would result in two tonally balanced speakers though with one measuring a noticeable, but not fatal, 2 dB louder than the other.

Where variations occur, pre-checking the driver sensitivity on a simple test fixture is really useful. For critical applications they may be sorted into colour-coded groups and matched with the other drivers of equivalent sensitivity. Good sample and pair matching does result in a significant performance gain, which may not always be clearly expressed by an observer. Controlled testing does show a link between better sound quality overall and close tolerance matching.

Virtually every component of a driver can affect performance and sensitivity. Key factors are the cone or diaphragm mass, the number of turns and the exact gauge of wire in the voice coil, coil concentration, the gap tolerance and the build quality of the magnet system. Small errrors can accumulate, resulting in serious 1 or 2 dB changes in loudness, and these must be controlled. In driver quality control there is also no substitute for a flux probe which can read directly the field strength attained by production parts and methods.

4.7   TRANSMISSION LINE ENCLOSURES

In theory a transmission line or labyrinth is capable of being extended to infinity, providing a perfectly resistive termination to the driver by absorbing all the rear directed energy. Hence in this respect it may be regarded as a special case of a very large sealed box, which is itself the practical realization of an 'infinite baffle'.

Physically the line resembles an acoustic pipe which, in its folded form, is also known as a labyrinth. The cross-section must be sufficiently large in order that the rear directed energy is not impeded, and yet it must also be long enough to terminate the energy down to the lowest desired frequencies. Since the path length must be comparable to the wavelength absorbed, if genuine low frequency operation is the aim, ideal transmission lines quickly become overlarge, even when folded. For a given closed line, the absorption reduces as the cutoff frequency is approached. This means that the energy is beginning to be reflected back to the driver and the system will then approximate to a sealed box of the same total volume. For example, a filled absorptive line intended to operate effectively down to 35 Hz must be at least 8 m long. If a cross-section of 900 cm^2 is adopted (roughly equivalent to that of a 36 cm diameter bass driver) then the enclosure volume will be inordinately large at 0.72 m^3.

Practical lines are of necessity smaller in cross-section to reduce the total volume, and pipe resonances and reflections at the higher frequencies may become a problem unless the absorption is very carefully arranged. It was discovered by Bailey that a filling composed of long-haired sheep's wool at a density of $8\,kg\,m^3$ possessed better acoustic properties compared with other materials, such as glass fibre. More specifically, at low frequencies (30–60 Hz) the speed of sound in the line appears to be reduced by 50%, which has important implications [18–20,23].

While in theory a damped line is a perfect absorber and the termination at the far end is of no consequence, it may be closed or open; in practice, commercial designs invariably leave the line open. Energy propagated below the line's cutoff frequency will emerge from the aperture and, depending on its phase, will augment or cancel the main contribution from the front of the driver. Over a limited range of frequencies such an enclosure will behave as a reflex, the propagation delay down the line providing the required phase inversion. This is where the wool filling comes into its own since, for a given length of line, the delay is effectively doubled, thus lowering the cutoff frequency of the system as a whole.

It has been suggested that the mechanism of this increased delay is at least partially due to the air velocity in the line reducing at low frequencies, due to the movement of the wool mass. As such, the latter represents a non-linear inertial component in the energy path down the line, thus accounting for the delay.

A further effect at low frequencies is produced by the air mass in the line moving with the driver, added to the driver's own moving mass. Whereas in conventional boxes the air mass is quite low, often less than 20%, in an open transmission line, the mass addition may be equal to that of the driver itself, thus reducing the latter's loaded resonant frequency by a factor of $\sqrt{2}$.

Two main resonant modes are possible in this conventional type of open pipe, occurring at half wavelengths and at odd (as opposed to even) multiples of the effective quarter wavelength. Taking the example of a 2 m filled pipe with moderate absorption (packing density below $5\,kg/m^3$) the quarter-wave mis-termination occurs at about 30 Hz and the half-wave mode at 73 Hz. If the pipe is overdamped, too much energy will be absorbed, and the sound output will possess a gradual rolloff from as high as 70 Hz. In commercially viable systems, i.e. those of reasonable size, it is difficult to find a satisfactory compromise between a uniformly extended LF response and an upper bass/lower mid-performance with minimal colouration-inducing line resonances (Figure 4.39).

Finally, the performance achieved with a transmission line is often no better than that of a properly designed reflex enclosure of similar dimensions. In fact for the same expenditure the results from the reflex sub-class, the ABR, may even be superior. Currently transmission line speakers are characterized by lower than average efficiency, notwithstanding their large size.

Quarter-wave 'Line' Loading

It is worth keeping in mind that for a useful 100 dB s.p.l. at a mid bass frequency of 50 Hz even a 210 mm (8 in) bass unit needs to move ±7.5 mm under sealed-box loading.

Considering the drive for a column of sufficient length, e.g. 1.5 m intending 'quarter-wave' operation, the system behaves as an infinite baffle above 100 Hz, but by 50 Hz the open-ended column is imparting a mass load on the driver, the additional larger radiating area maintaining the overall acoustic output while diaphragm excursion in this example is reduced to just ±2.5 mm at 50 Hz (as of course does reflex loading). The total moving mass is typically increased four-fold at low frequencies, reducing the f_s of a 50 Hz driver to 25 Hz.

Figure 4.39. A four-way, floor standing system with separate transmission line loading at mid and low frequencies. The larger line is folded and a thick foam lining helps absorb higher order pipe modes. (Courtesty TDL)

A one-quarter wavelength constitutes the low frequency limit; 30 Hz has a 12 m wavelength, so the required column will be 3 m long.

While columns differ in theory from transmission lines they also suffer from harmonic modes, and since damping is discouraged on the grounds of maintaining efficiency, the unwanted higher modes are often more troublesome. The most severe is the third harmonic, and this may be controlled by mounting the drive unit in the upper third of the column length on the node. In practice, further steps such as tapering the duct (which also is usually folded within the enclosure) and judicious placement of sections of fibre absorbent are required to control colouration due to secondary resonances.

In one example of a quarter-wave system the 'mouth' of the column was altered (successfully) to a quasi-reflex type by adding a slotted port which tuned the volume to 30 Hz — the lower system cut-off. In this instance a mix of the two resulted which was subjectively clean sounding in the bass.

4.8   SUB-WOOFERS

There is no theoretical distinction for low frequency design between full range systems and sub-woofers. A sub-woofer is designed to reproduce low frequencies only and may be part of, or an accessory to, a wider range system. Often, exploiting the ear's poor directional acuity at low frequencies, a single sub is used to carry a mono-based signal working with two-channel, stereo, upper range speaker systems. In this context these are called 'satellites'.

A specialized low frequency reproducer comes into its own when a given speaker system or installation can benefit from a more powerful and extended low frequency range. Properly matched, the sub fills in the band below the main frequency range and in the largest and best examples may extend the system response below 20 Hz into the infrasonic region. Here sound pressure waves are felt only; the ear does not distinguish pitch at such low frequencies unless it is distorted. There are cogent arguments for extending the low frequency range if the cost can be borne and if there is no impairment of sound quality; the low frequency definition and control, in what is termed the 'fundamental' bass range, i.e. 35 Hz to 150 Hz.

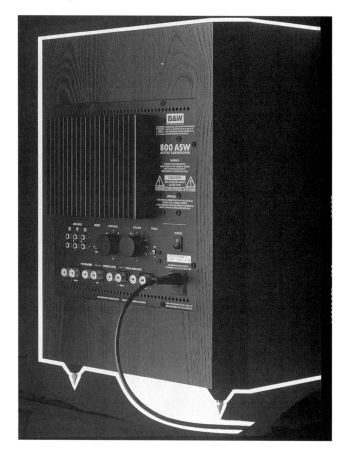

Figure 4.40.

Trials with a pair of active closed-box sub-woofers with 330 mm long throw drivers, optimally corrected for their corner boundary matching, showed significant acoustic power down to 7 Hz, sufficient for any likely programme power envelope (Figure 4.40). These test are revealing, their use improving the sound quality in the superior audio system used. In particular, the perception of bass colouration and damping control is shown to be related not only to the specific alignment of a given system, but also to bandwidth.

While it is true that the ear is prepared to accept a relatively limited low frequency response, and indeed can partially reconstruct the impression of a missing bass fundamental from those remaining higher frequency harmonics which are reproduced, bandwidth limitation must be considered a compromise dictated by considerations of size and cost. Experience has shown that it is worthwhile achieving a uniform room-matched low freqency bandwidth to 10 Hz, noting that the dynamic headroom required at very low frequencies is not great, perhaps only 20% of the required peak level at 50 Hz. Interestingly there are also gains in the perceived realism and timbre of sounds rather higher in the frequency range, e.g. female voice and solo violin.

Good low frequency extension lends a subtle sense of poise and scale and the high-pass effect that colours low frequency sounds is reduced with increasing extension. From the group delay point of view, it is well understood that the high-pass filter represented by a given cutoff frequency in the bass, which in some cases can be quite high order, fifth or sixth, and placed typically in the region between 35 Hz and 60 Hz, has a significant effect. Group delay theory suggests that these distortions do alter the perception of tonality — this tied up with perception of absolute phase. As a side issue, design for superior low frequency performance will result in improved characteristics — transient response and linearity higher in the bass range, where the subjective benefits are not subtle.

Pros and Cons of Sub-woofers

Pros

- Extension of low frequency.
- Reduced low frequency distortion.
- Potential improvements in distortion and headroom for partnering upper range systems/satellites.
- Possible flexibility in room placement.
- Active sub-designs may include user adjustments for boundary conditions, phase, contour and sensitivity.
- Lowered system colouration (if well designed and matched).

Cons

- Crossover design difficulties for passive subs.
- Difficulty in achieving a smooth transition to the partnering system.
- Not uncommonly, out-of-band colouration and resonances from an inferior sub will actually impair the performance of an existing, high quality system even when it has intrinsically limited low frequency bandwidth.

- Additional audio system complexity—extra wiring and in the case of active subs, further line level connections and power supplies.

- Need to locate and accommodate additional enclosures.

- Ideally the sub need to be installed aligned and calibrated for the system it is intended to partner.

- Significantly higher system cost.

- Tests have shown that sub-woofers perform better as two-channel stereo reproducers, the non-directional sensitivity of the ear being an approximation which breaks down in systems of very high quality.

Regardless of the additional complexity, sub-woofers or auxiliary bass systems are proving increasingly popular and very large numbers have been sold. The two areas of greatest growth have been in connection with (a) home theatre, where high power bass of great impact is now considered a vital part of the experience of a movie sound track, and (b) attractive stereo systems* using very small easily located satellite speakers for left and right channel and a disguised supporting sub-woofer. The potential customer is easily seduced by the visual unobtrusiveness of the stereo speakers perhaps only working down to 200 Hz, while the bass system may be surprisingly compact if of the active bandpass variety. Care needs to be taken with the upper range of the woofer system if leakage and colouration are not to extend into the mid range making the aural identification of its position more likely. If well filtered and placed in reasonable proximity to the stereo satellites the acoustic join between the two may be well disguised.

Design Details

Some designers attempt vertical mounting for the bass driver with the diaphragm facing towards the floor and the cabinet elevated on low feet. While there is some helpful filtering action due to the indirect path, such an arrangement may result in a resonant drumming sound on the floor owing to the magnet reaction vibration transmitted through the enclosure. Opposed push–pull driver systems avoid this problem, while for single drivers mounting on a side face of the enclosure is best.

With the increasing use of low-cost, custom-built amplifier and electronics packages, it is common for sub-woofers to be active, greatly easing the acoustic design and improving the sound quality thanks to the direct drive to the woofer and the matching filter and equalization electronics.

More sophisticated designs include circuits which monitor the power envelope of the programme and use controllable filters and/or limiters to tailor the output to best fit the programme demand without incurring audible distortion. The home theatre specifications, particularly those defined for THX approval, are demanding; e.g. 105 dB at 30 Hz, well beyond the compass of even some larger two-channel stereo speakers.

A number of sub-woofers incorporate feedback systems that sense cone excursion and provide linearization of both distortion and frequency response. If not applied with

*"Lifestyle Stereo".

caution, such arrangements may perform poorly in overload with a rapid onset of distortion as the control systems reach the limits of their operating range.

Interfacing

With active or passive sub-woofers the designer has to resolve the question of interfacing with the rest of the audio chain.

Passive

Taking the passive design first, usually the bass enclosure includes some form of high-pass filter to the satellite or stereo speakers operating at around 100 Hz. This is a difficult region for a crossover network because the target rolloffs are disturbed by the complex electrical impedance represented by both the satellites and the woofer itself. Most designers would admit that significant compromises are involved; the innate character of the satellites may be altered and ideally the low frequency alignment of the woofer is determined with the effects of its crossover section taken into account. Crossing over at low frequencies requires large values for the inductors and capacitors and, due to losses, their quality is not as high as the smaller values typically used for mid range and costs are increased.

With higher budget designs where the woofer is matched to the satellite a close view is taken of the overall system design which may include compensation networks for complex impedance to improve crossover accuracy.

For amplifier connection, generally the speaker leads go first to the sub. The low frequency energy is usually summed to mono, and after traversing the high-pass filter emerges at a pair of terminals taking the cable to the satellites. The summation has its own problems. Some designers use two woofers in one box, kept independent, relying on acoustic summation. Some have used two windings on a single voice coil and others use two inductors meeting at one bass driver. In home theatre systems the friendlier electronic control systems often include a dedicated sub-woofer channel either at power amplifier level or at line level making for easy implementation.

Active

While the active system benefits from a direct amplifier link to the woofer, there may be some impairment of the high-pass or satellite channels. In one option the line level signal is sent to the sub-woofer electronics, is high-pass filtered and returns via another cable to the main stereo amplifier. In top quality systems this loop may result in losses, particularly if the cable and sub-woofer electronics are not designed to the same standard as the carefully selected electronics of the main system. Where possible, critical enthusiasts try to leave the main stereo system intact and run an auxiliary line to the sub-woofer aiming to fill in the very lowest required bass frequencies. For general purpose use the active sub-woofer may offer considerable versatility, accepting inputs at line or speaker level and offering phase inversion or a smooth phase control, variable crossover point, sometimes independent for high- and low-pass channels, control of woofer level and of Q (for effective alignment). These facilities make it much easier to achieve a good blend in a variety of room acoustics and placements.

4.9 HORN LOADING

The specialized subject of horns can only be briefly covered in this book, and interested readers are again referred to the bibliography for more detailed information [21]. While the author is aware of the danger of generalization on this subject, in his view, horn loaded enclosures are not capable of top class subjective quality and most designs are inferior to typical direct radiator systems. The main reason for adopting horn loading, traditionally employed in public address situations, is to attain a high efficiency coupled with an improved control of directivity, vital considerations when large audiences are to be covered. Studio monitors also exploit high quality horns to deliver high sound levels up to 120 dB at 1 m.

Improved Acoustic Matching

The intrinsic low efficiency of direct radiator diaphragms is due to their poor matching to the acoustic impedance of the air load. Almost any value of acoustic impedance may be produced at a horn throat by suitable geometrical deisgn, and thus the match to the driver diaphragm may be optimized. Two benefits result: first, the efficiency is greatly increased and secondly, resistive termination presented to the diaphragm may greatly reduce the amplitude of any intrinsic response irregularities.

Efficiency

While optimum magnet design together with horn matching can result in efficiencies of nearly 50% in narrow band designs, if the frequency range is increased, great difficulty is experienced in maintaining both a smooth response and a high efficiency. A full multi-way horn system capable of 40 Hz to 20 kHz over a useful 60° forward polar response, potentially may have an efficiency in the 10%–20% range, but in practice this is often much less.

Bandwidth

The greatest problem is physical size. For a simple horn the lower cutoff frequency* is proportional to the effective diameter of the mouth when radiating into free space (i.e. 4π steradians). The mouth area $= \lambda_c^2/4\pi$, where λ_c is the cutoff wavelength. For example, for an f_c at 40 Hz, the mouth area should be 5.9 m^2. Such a design would need to be custom built into a location as part of a fixed structure, and is clearly impracticable for most domestic situations. Figure 4.41 shows the influence of horn proportion and size.

The response may be extended in two ways. A small closed box may be added which loads the rear face of the diaphragm and provides an inductive impedance component. This may be adjusted to offset the increasingly capacitive throat impedance below

---

*The frequency at which the acoustic impedance becomes reactive rather than resistive, i.e. the resistive component has fallen by 6 dB.

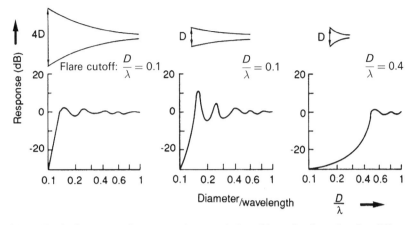

Figure 4.41. Typical response frequency characteristics of horn loudspeaker for different mouth openings and different flare cutoffs as a function of the ratio of the mouth diameter to the wavelength (after Olsen [22])

cutoff. Additionally, if the radiation space is reduced, e.g. by mounting the system in a wall (2π) or between a wall and floor (π) or in a corner ($\pi/2$), the impedance match improves proportionately. Effectively, the flare of the horn is extended by the adjacent wall surface thereby reducing the cutoff frequency. With corner mounting a bass horn may have a basic mouth area one-eighth that of the free space mounted version, since the effective radiation space is reduced from 4π to $\pi/2$.

Position

Certain problems will arise if a horn is mounted in a corner. Maximum excitation of the room LF modes is inevitable, these producing irregularities in the radiation impedance and consequently in the sound pressure. In addition there will be mid-band colouration as the reproducing assemblies cannot always be flush fitted to the adjacent walls; this being virtually impossible with three-way horns.

Propagation Delay

Propagation delay in a horn produces large phase displacement which complicates the crossover points in a multi-way system. In theory the mouths should be aligned so that the propagation time to the listener from all energizing diaphragms is the same, but the necessary physical displacement of the assemblies is impracticable. However, the recent development of electronic delay lines of satisfactory quality means that the mouths of multi-way horns may now be conventionally aligned to the designer's convenience, and the differential delay and resulting phase discrepancy may be electronically compensated. The resulting system must of course be powered by separate amplifiers with active filter crossovers.

Folded Horns

Theoretically a horn structure should be linear, and the smaller mid and treble types do adopt this format. However, large bass horns (15 m or more for a free field model) require some folding technique, unless the horn can be concealed; for example, built into the sub-floor structure. If the folds are too severe, reflections will occur, resulting in irregularities in the throat impedance and hence the frequency response.

Horn Shape

Basic horn flares include the exponential, hyperbolic and tractrix forms (Figure 4.42). The relationship between the cross-sectional area of an exponential horn and the axial distance x from the diaphragm is given by

$$S = S_0 c^{mx}$$

and the cutoff frequency by

$$f_c = \frac{mc}{4}$$

where S = cross-section; S_0 = throat cross-section = distance from throat; m = flare constant and c = velocity of sound.

The hyperbolic horn offers a short flare for a given cutoff combined with good driver termination, although the throat distortion is fairly high for a given output level. Horns

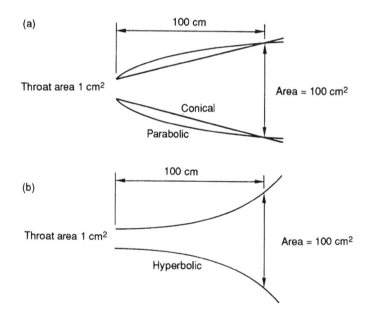

Figure 4.42. (a) Longitudinal sections of conical and parabolic horns. (b) Longitudinal sections of a hyperbolic horn (from *Acoustics* by L. Beranek, copyright 1954. Used with permission of McGraw-Hill Book Co.)

with combinations of exponential, conical and hyperbolic profiles have also been used, the results largely depending on the empirical skill of the designer and the method of construction. As cutoff is approached the response ripples cause colouration. Designers use their skill to optimally balance these effects.

Directivity

The variable geometry of horn shapes has been taken even further by recent commercial designs where in an effort to maintain a constant horizontal directivity over a wide frequency range, the horn may be 'flattened'. The resulting mouth resembles a radial flared slot. Alternatively the horn may be 'thinned' in the vertical plane to strengthen the diffraction radiation effect to give good dispersion in the lateral plane.

In addition to the well-known multi-cell methods, acoustic lenses and diffraction plate structures are also commonly employed to disperse the energy.

Upper Range Problems

The lower range cutoff has been discussed but a high range cutoff will also be present due to a combination of several factors. The driver diameter is of significance, as in a simple horn, where the throat is directly coupled, an upper frequency cutoff exists where the dimensions of the throat cavity are comparable with a sound wavelength. One solution involves reducing the throat cross-section considerably and, in addition, fitting a multi-channel coupling block in the remaining acoustic space. This places the throat-cavity resonance at a higher frequency, but eventually breakup will limit the maximum working frequency of a given diaphragm.

Figure 4.43 illustrates various throat structures which smooth and extend the response. An unfortunate by-product of reducing the throat section is a rise in the throat pressure change at lower frequencies, thus increasing the distortion. In part, this explains why it is impossible to design a single, wide range horn capable of high quality performance.

Comercial Horn Systems

Many high output commercial systems employ horn loaded mid and treble transducers where the resulting high efficiency endows the design with a considerable power handling capacity. The LF range is usually a compromise, and generally employs a parallel combination of two or three 300 mm–350 mm high power, coned bass drivers in an optimum reflex enclosure. The latter will utilize a high efficiency response alignment which incorporates bass lift equalization in the accompanying power amplifier.

Efficiencies of the order of 5% are possible using this technique and acoustic outputs of the order of 125 dB at 1 m are obtained with the energy maintained over a 60° forward arc up to at least 15 kHz.

The subjective performance is not on a par with direct radiator monitoring systems, but these horn designs are very satisfactory for their prime application, namely large audience sound coverage, and high-level monitoring. This class is also enjoying a revival with low

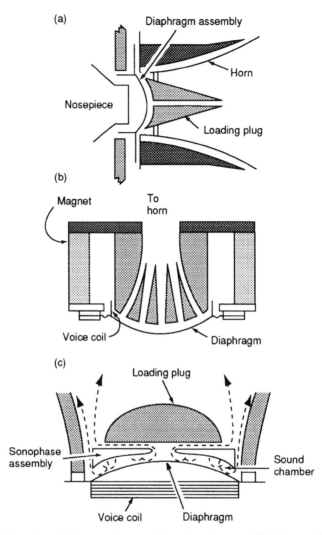

Figure 4.43. Examples of phase plugs used in the throat of HF horns (diagram (c) after Electrovoice)

powered amplifiers,* where better than 6 dB/W sensitivity is a great advantage. Subjectively good horns have an excellent resolution of dynamics, good low-level detail, and a fast sound, appreciated in some market sectors despite the known colourations.

4.10 LINE SOURCES

Much confusion exists over the theory and practical performance of line source radiation and 'line' speaker systems, the type vaguely described as designs where the

---

*Mainly single ended 'SE' triode types below 10 W.

effective radiating height is many times the width, forming a tall structure. Such column speakers are common in public address applications and are generally effective over a good speech bandwidth, 100 Hz to 8 kHz. Usually they are relatively remote to the listeners, i.e. the latter are in the reasonably well behaved far field and their directivity is augmented by the use of acoustic shading and absorptive baffling techniques. Tidy responses can be obtained where the directivity is intentionally poor in the vertical plane, so reducing roof and floor reflections, while the radiation is intentionally wide in the lateral plane to give good audience coverage.

In recent years a number of so-called line source speakers have appeared on the hi-fi market. While areas of excellence have been recognized in the performance of the better designs, on the debit side a significant degree of sound quality variation has been observed between different makes, sizes and models, this uncertainty adding to the normal variation due to different room acoustics.

Theoretical analysis suggests a fundamental but not necessarily disabling flaw in the line source principle when used at the relatively close distances typical under domestic conditions, i.e. 2–5 m. 'Slot' radiator heights of 1 or 2 m are common. When laboratory tested such speakers have tended to show greater than usual source response variations with microphone distance and measuring angle. Sensible correlation between subjective opinion and frequency response has only been achieved when averaged, multiple position response measurements have been undertaken in-room.

The objections to such finite length lines can be overcome by the special case of a full height, floor to ceiling line — if you have the right room — where the two plane boundaries form acoustic mirrors, increasing the apparent line height to infinity. However, in general, a practical line is significantly mismatched with its acoustic environment due to its inevitable truncation, i.e. its finite height.

A brief return to theory (Lipshitz and Vanderkooy [35]) will help clarify the position. Noting that a monopole is the equivalent of a small sealed back or IB radiating loudspeaker and that a dipole corresponds to an open back arrangement, two kinds of line are seen; a vertical stack or equivalent long 'ribbon' element built in a closed back system, or alternatively the more popular dipole or 'open back' type with a bi-directional radiation pattern.

The usual small box loudspeaker is regarded as a point source and hence obeys the inverse square law of intensity; sound pressure level reducing with the square of distance (twice the distance = one-quarter the sound pressure or a loss of 6 dB). In addition the pressure and particle velocity are considered to be in phase. As one approaches a given radiator, one enters the near field of that source. This is defined by the distance r and the wave number k, where $k = 2\pi/\lambda = \omega/c$ and the transition to 'near field' thus depends on frequency.

Near field is the region where k times r is rather less than unity, far field is where k times r is much greater than unity. Thus the 'near field' extends farther from the source at low frequencies and incidentally is the cause of the so-called bass boost 'proximity effect' noted with recording microphones. Proximity also influences the use of loudspeakers for close-up, near field monitoring, e.g. at 1 m located on the top of a studio mixing console, such a speaker pair positioned to flank the desk operator.

Thus, at 75 Hz and at the industry standard 1 m measuring distance, $kr = 1.4$, too close to unity to be regarded as far field. This results in an inflated microphone reading at lower frequencies compared with the higher frequency range. It explains why some

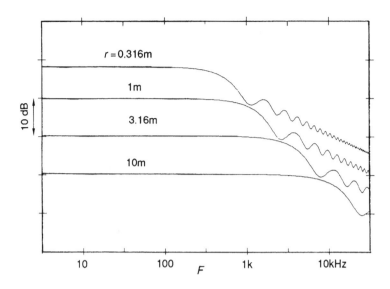

Figure 4.44. Variation in theoretical frequency response with measuring distance *r*, for a line
source monopole (closed back) (after Lipshitz and Vanderkooy [35])

manufacturers and this author prefer a 2 m (ideally 3–4 m) mic distance for assessing the
overall anechoic frequency response of wide-range loudspeaker systems.

Alternatively described as 'extended radiators', practicable line source loudspeakers
suffer from both diffractive and proximity effects at normal domestic listening distances
and the theoretical computation of the expected frequency responses this is not trivial.
An example of the computed results for a typical sealed-back theoretical model is given
in Figure 4.44.

A flat response is nowhere apparent except in the far field at some 10 m distant, and
even this shows a rolloff above 8 kHz. At a typical 3 m distance, 3 kHz now defines the
rolloff frequency. For the alternative dipole or open-back case (Figure 4.45), the front
to back cancellation effect is now superimposed on the line characteristic, the
cancellation being the familiar 6 dB/octave, 20 dB/decade uptilted slope. For narrow
lines or small sources this slope begins at a relatively high frequency, while for a 1 m size
of square panel or similar approximation, the cancellation effect does not occur until
below 150 Hz or so.

Of particular interest is the fundamental variation of response with distance; these
effects also present at different angles off-axis. Lipshitz and Vanderkooy conclude that
there is little to recommend such line source designs and comment that the amplitude
responses are non-flat in the diffractive near fields, and exhibit broad slope errors
exacerbated by response ripples due to line truncation. Response variation with
distance and poor vertical directivity are also cited weaknesses. Finally the presence of
waveform distortion is noted, a feature of the two-dimensional wave equation defined
for such lines, which denotes a dispersive impulse response with a 'wake' following the
main pulse.

Conversely the above authors also note that the line analysis quoted is theoretical
and only covers the restrictive case of a narrow, unbaffled source working over the

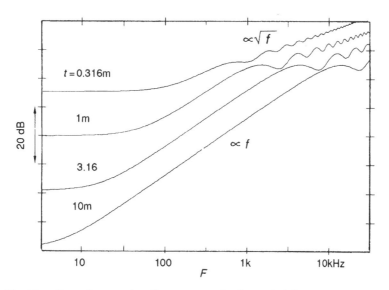

Figure 4.45. The effect of measuring distance r on the theoretical frequency response of a finite line source dipole element (unbaffled). Note proximity effects (after Lipshitz and Vanderkooy [35])

whole frequency range, under the 'significant mass', 'constant volume acceleration' regime. This is relevant for normal moving coil drivers working above their fundamental resonance. If the 'line' diaphragm is of low mass, such as an electrostatic, then the 'constant volume velocity' mode is operative, requiring a $j\omega$ factor in the equations, equivalent to a clockwise rotation of the responses by 20 dB/decade.

Practicable line loudspeakers can be made to work by the use of the following design features which are often derived empirically. Commercial 'line' speakers generally use a baffle for their sources, for example a 'line' dipole ribbon of 9 or 12 mm effective source width is generally mounted in a panel typically 160 to 220 mm wide. This extends the upper 'monopole' frequency range down to 750 Hz. By asymmetric placement of the ribbon element the roll-off 'corner' is softened and diffraction ripples are significantly reduced. Both the effects of line truncation and excessively narrow vertical directivity may be satisfactorily controlled by gently curving the radiating element in the vertical plane. In one design, nicknamed the 'banana' ribbon, lightweight pressure pads displace elements of the ribbon in the forward direction to widen the wavefront in the vertical plane, so improving the directivity.

Finally, the most effective design tool is multi-way operation. Generally the 'line' sections only work over a 500 Hz to 20 kHz range, the region below handled by an approximation to a rectangular, partially baffled panel radiator or, in some cases, a closed-box system. With such detailed design 'line' speakers can be made to work well and deliver their main sound quality advantage most effectively. Good examples have a sound entirely free of box, panel or cavity colourations. In this light the paper cited [35] may be considered to be a valuable theoretical contribution but is perhaps unnecessarily dismissive of the technique.

Supporting analysis of line sources [37] has helped sort out one fallacy, namely that as elements are added to increase the height of a line source the lateral directivity is widened. In fact, the lateral directivity is constant, defined by the source width, while adding elements to increase the height progressively reduces the vertical beam width and increases the loudness in the forward direction. Taking another viewpoint note that if one commences with a radiator of constant area, as it is progressively changed from a circular or square shape, to a narrow vertical strip, the directivity is so shaped that the effective beam width in the vertical plane narrows as the horizontal beam width widens.

4.11   THE MOVING COIL SPACED DIPOLE

Bank at Celestion has developed a compact, directional low frequency system for the 25 Hz to 100 Hz range which employs two 330 mm moving coil drivers per channel, one spaced several centimetres behind the other, each rigidly mounted on a small open baffle. The simple array constitutes a dipole with little radiation to the sides and with well-defined axial lobes. Used a metre or so from the rear wall the enclosure axis is ideally set at an angle to the rear boundary to set up a more even distribution of low frequency modes in the listening room. With a low conversion efficiency of under 1% the acoustic output is largely unaffected by local acoustic impedance effects, in contrast to low mass electrostatic dipole panels.

The axial output of an open baffle falls at 6 dB/octave at frequencies with wavelengths well below that of the baffle size — due to increasing front to back cancellation. A pre-equalization (6 dB/octave boost) imparts a flat response to the system down to the resonance cut off of the LF drivers (Figure 4.46).

Advantages of the technique include freedom from enclosed box or reflex port colourations, no panel resonance since the enclosure is vestigial, and finally a directional low frequency output which interfaces more smoothly in the listening room by reducing the incidence of side wall reflections. In a commercial example the system was used as a sub-woofer for a high quality miniature two-way design with an active crossover at 100 Hz. The dipole woofer also serves as a substitute stand making for an unusual looking three-way design.

4.12   BI-POLAR SPEAKERS

Bi-polar is the name of a sub-class of speaker with radiation directed at the front and at the back. In contrast to the di-polar types where commonly a single element is open at the back and where the rear output is 180° reversed compared with the front, the bi-polar employs separate driver elements in a box enclosure and front and rear outputs are nominally in phase.

Due to driver size and cabinet width the radiation pattern will have fore and aft lobes or maxima, but the output does not null at the sides as in the case of the di-polar. In practice, for a typical listening position the different path lengths between the front and rear facing drivers of a bi-polar will result in a set of cancellations due to the progressive phase shift with frequency between the drivers with respect to the listener.

The bi-polar intention is to drive the room with a more uniform energy spectrum since such a speaker is more omni-directional than the usual single axis type, and thus increases the power and improves the spectral balance of the reverberant energy

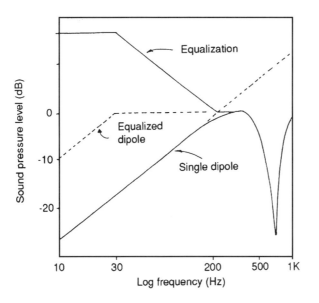

Figure 4.46. Response of a 300 mm driver in a vestigial open baffle operating as a simple dipole. Adding closely spaced dipoles in a line along the dipole axis increases the overall level while the shape is unchanged. The notch at 700 Hz is the first of a series of baffle cancellations and may be moderated in a larger baffle by asymmetric geometry and offset positioning for the driver. The equalization response required is a 6 dB/octave bass boost to 30 Hz, starting at 180 Hz, and resulting in the equalized response (dashed)

contribution. Subjectively such a speaker sounds more spacious, and the driven room acoustic sounds louder, more spacious and less coloured. Conversely, there may be losses in stereo focus and image presence due to dilution of the directed to reflected ratio. Placement well clear of boundaries is advised to avoid standing wave reflections from the nearest wall in the mid range.

REFERENCES

[1] Harwood, H. D., 'Speakers in corners', *Wireless World*, **162**, April (1970)
[2] Allinson, R. F., 'The influence of room boundaries on loudspeaker power output', *J. Audio Engng Soc.*, **22**, No. 5, 314–19 (1974)
[3] Thiele, A. N., 'Loudspeakers in vented-boxes', Pt. 1, *J. Audio Engng Soc.*, **19** (1971); Pt. 2, **19**, No. 6 (1971)
[4] Small, R. H., *Synthesis of Direct Radiator Loudspeaker Systems*, University of Sydney, Australia
[5] Small, R. H., 'Simplified loudspeaker measurements at low frequencies', *J. Audio Engng Soc.*, **20**, No. 1 (1972)
[6] Small, R. H., 'Direct radiator loudspeaker system analysis', *J. Audio Engng Soc.*, **20**, No. 5 (1972)
[7] Small, R. H., 'Closed-box loudspeaker systems', Pt. 1, 'Analysis', *J Audio Engng Soc.*, **20**, No. 10 (1972); Pt. 2, 'Synthesis', *J. Audio Engng Soc.*, **21**, No. 1 (1973)
[8] Small, R. H., 'Vented-box loudspeaker systems', Pt. 1, 'Small signal analysis', *J. Audio Engng Soc.*, **21**, No. 5 (1973); Pt. 2, 'Large signal analysis', *J. Audio Engng Soc.*, **21**, No. 6

(1973); Pt. 3, 'Synthesis', *J. Audio Engng Soc.*, **21**, No. 7 (1973); Pt. 4, 'Appendices', *J. Audio Engng Soc.*, **21**, No. 8 (1973)

[9] Small, R. H., 'Passive radiator loudspeaker systems', Pt. 1, 'Analysis', *J. Audio Engng Soc.*, **22**, No. 8 (1974); Pt. 2, 'Synthesis', *J. Audio Engng Soc.*, **22**, No. 9 (1974)

[10] Small, R. H., 'Direct radiator loudspeaker system analysis', *J. Audio Engng Soc.*, **20**, No. 5 (1972)

[11] Harwood, H. D., 'Loudspeaker distortion association with low frequency signals', *J. Audio Engng Soc.*, **20**, No. 9 (1972)

[12] Harwood, H. D., 'Non-linearity of air in loudspeaker cabinets', *Wireless World*, **80**, No. 1467 (1974)

[13] Small, R. H., 'Closed-box loudspeaker systems', Pt. 2, 'Synthesis', *J. Audio Engng Soc.*, **21**, No. 1 (1973)

[14] Gelow, W. J. and Rhodebeck, M. G., 'The scientific design of a three-driver loudspeaker system', *Proc. A.E.S. 46th Convention* (1973)

[15] Garner, A. V. and Jackson, P. M., 'Theoretical and practical aspects of loudspeaker bass unit design', *Proc. A.E.S. 50th Convention*, London, March (1975)

[16] Small, R. H., 'Vented-box loudspeaker systems', Pt. 3, 'Synthesis', *J. Audio Engng Soc.*, **21**, No. 7 (1973)

[17] Harwood, H. D., 'Motional feedback in loudspeakers', *Wireless World*, **80**, No. 1459, 51 (1974)

[18] Bradbury, L. J. S., 'The use of fibrous materials in loudspeaker enclosures', *J. Audio Engng Soc.*, **24**, No. 3 (1976)

[19] Bailey, A. R., 'Non-resonant loudspeaker enclosure', *Wireless World*, October (1965)

[20] Bailey, A. R., 'The transmission line loudspeaker enclosure', *Wireless World*, May (1972)

[21] Dinsdale, J., 'Horn loudspeaker design', *Wireless World*, **80**, No. 1459, 1461, 1462 (1974). These articles include a comprehensive bibliography on the subject

[22] Olson, H. F., *Modern Sound Reproduction*, Van Nostrand Reinhold, New York (1972). Copyright 1972 by Litton Educational Publishing Inc.

[23] Adams, G. J., 'Computer-aided loudspeaker system design', Pt. 1 and 2, *J. Audio Engng Soc.*, **26**, Nos. 11 and 12 (1978)

[24] Stahl, K. E., 'Synthesis of loudspeaker mechanical parameters by electrical means', Preprint 1381 (K-3), *61st Audio Engng Soc. Convention* (1978)

[25] Werner, R. E. and Carrell, R. M., 'Application of negative impedance amplifiers to loudspeaker systems', in *Loudspeaker Anthology* (Cooke, R. E., Ed.), *J. Audio Engng Soc.*, **1-25** (1979)

[26] Keele, D. B., 'A new set of sixth order vented-box loudspeaker alignments', *J. Audio Engng Soc.*, **23**, No. 5 (1975)

[27] Fincham, L. R., 'A bandpass loudspeaker enclosure', *63rd Audio Engng Soc. Convention* (1979)

[28] Holl, T., *Engineering the AR9*, Teledyne Acoustics Research (1978)

[29] KEF Electronics Ltd, 'Model 105', *Keftopics*, Issue 3, No. 1 (1978)

[30] Greiner, R. A. and Schoessow, M., 'Electronic equalisation of closed-box loudspeakers', *J. Audio Engng Soc.*, **31**, No. 3 (1983)

[31] Keele, D. B., 'Direct low frequency driver synthesis from system specifications', *J. Audio Engng Soc.*, **30**, No. 11 (1982)

[32] Margolis, G. and Small, R. H., 'Personal calculator programs for approximate vented-box and closed-box loudspeaker system design', *J. Audio Engng Soc.*, **29**, No. 6 (1981)

[33] Kerno, P. R., et al., UK Patent Application No. GB2045578A (Absorptive duct for vented box), 19 March (1979)

[34] Geddes, C. R., 'An introduction to band-pass loudspeaker systems', *J. Audio Engng Soc.*, **37**, No. 5 (1989)

[35] Lipshitz, S. P. and Vanderkooy, J., 'The acoustic radiation of line sources of finite length', *Proc. A.E.S., 81st Convention* (1986)

[36] Han, H. L., 'Frequency responses in acoustical enclosures (rooms)', *Proc. 82nd A.E.S. Conference*, March (1987)

[37] Gander, M. R., 'Dynamic linearity and power compression in moving coil loudspeakers', *J. Audio Engng Soc.*, **34**, No. 9 (1986)

[38] Adams, G., 'The room environment', Chapter 7, *Loudspeaker and Headphone Handbook*, (ed. Borwick, J.), Butterworth (1988)

[39] Colloms, M., 'Bass Profundo—bass perception and low frequency reproduction', *Stereophile*, Dec. (1991)

[40] Augspurger, G. L., 'New guidelines for vented box construction', *Speaker Builder*, **2**, 12 (1991)

BIBLIOGRAPHY

Ballagh, K. O., 'Optimum loudspeaker placement near reflecting planes', *J. Audio Engng Soc.*, December (1983), also 'Loudspeakers, Vol. 2', *A.E.S. Anthology*, Ch. 4.

Benson, J. E. 'An introduction to the design of filtered loudspeaker systems', *J. Audio Engng Soc.*, **23**, No. 7 (1975)

Beranek, L., *Acoustics*, McGraw-Hill, London (1954)

Cooke, R. E. (Ed.), *Loudspeaker Anthology, J. Audio Engng Soc.*, **1–25** (includes References 2, 3, 5, 7, 8, 18 and 25) (1979)

Fincham, L. R., 'A bandpass loudspeaker enclosure', *J. Audio Engng Soc.*, **27**, July/August (1979)

Gander, M. R., 'Dynamic linearity and power compression in moving coil loudspeakers', *J. Audio Engng Soc.*, **34**, No. 9 (1986)

Hanna, C. R. and Slepian, J., 'The function and design of horns for loudspeakers', *J. Audio Engng Soc.*, **25**, No. 9 (1977)

Harwood, H. D., 'New BBC monitoring loudspeaker', *Wireless World*, May (1968)

Harwood, H. D., 'Some aspects of loudspeaker quality', *Wireless World*, May (1976)

Henrikson, C. A. and Ureda, M. S., 'The manta ray horns', *J. Audio Engng Soc.*, **26**, No. 9 (1978)

Hilliard, J. K., 'A study of theatre loudspeakers and the resultant development of the Shearer 2-way horn system', *J. Audio Engng Soc.*, **26**, No. 11 (1971)

Hoge, W. J. J., 'A broadcast monitor speaker of small dimensions', *J. Audio Engng Soc.*, **26**, No. 6 (1978)

Keele, D. B., 'Sensitivity of Thiele's vented loudspeaker enclosure alignments to parameter variations', *J. Audio Engng Soc.*, **21** (1973)

Long, E. M., 'Design parameters of a dual woofer loudspeaker system', *Proc. A.E.S. 36th Convention* (1969)

Rank Leak Wharfedale, *Airdale SP Loudspeaker*, Press Release

Lord Rayleigh, *Theory of Sound*, Dover Publications

3A, Zone Industrielle 06600 Antibes *Loudspeaker Catalogue* (1976)

Walker, P. J., 'Wide-range electrostatic loudspeakers', *Wireless World*, May, June and August (1955)

Wente, E. C. and Thuras, A. L., 'Auditory perspective/loudspeakers and microphones', *J. Audio Engng Soc.*, **26**, No. 718 (1978)

Wright, J. R., 'An exact model of acoustic radiation in enclosed spaces', *J. Audio Engng Soc.*, **43**, No. 10 (1995)

Zacharia, K. P., 'On the synthesis of closed-box systems using available drivers', *J. Audio Engng Soc.*, **21**, No. 9 (1973)

5

Moving Coil Direct Radiator Drivers

In contrast to the theoretical analysis presented so far, this chapter covers the practical aspects of moving coil driver design. The techniques employed in high performance models are described, with critical review where appropriate. Inevitably some overlap of content with theoretical subjects — driver analysis, low frequency loading, etc. — will occur, but essentially the viewpoint taken here is a practical one.

As has been mentioned elsewhere, moving coil drivers are used in the vast majority of loudspeakers. There is a great variety of chassis, magnets and coils available, as well as numerous diaphragm shapes and materials to choose from, and thus several classes of specialized drivers covering individual frequency bands have been developed.

It has also been indicated that a single full range driver* cannot as yet satisfactorily meet the requirements of a high performance reproducer. At its simplest, the latter must be 'two or more way', i.e. incorporating at least two drivers, each responsible for separate adjacent frequency ranges, which together with an enclosure and crossover form the complete system. Several categories of drive unit may be broadly classified as in Table 5.1. After a short review of the essentials of the moving coil transducer, drivers with coverage approximating to these bands will be individually examined.

5.1 MOVING COIL MOTOR SYSTEM

The heart of a moving coil driver is the motor, which consists of a light, hollow coil immersed in a strong radial magnetic field, suspended in such a way as to allow free movement only in the axial direction. Some sort of diaphragm or radiating surface must be attached to the motor coil to couple the air to the forces generated by currents flowing in the coil, and hence to permit acoustic power (i.e. sound energy) to be radiated from the assembly.

Every component in a moving coil driver unit has some influence on the quality of the sound produced, whether it be in terms of pressure response, level, distortion, colouration, frequency response or directivity. Some may influence certain aspects only;

---

*i.e. pistonic. See also distributed mode radiator section 2.5.

Table 5.1.

Usable frequency range	Type of driver
30 Hz–1 kHz	Low frequency (LF)
30 Hz–5 kHz	Low to mid frequency (L/MF)
150 Hz–5 kHz	Mid frequency (MF)
700 Hz–10 kHz	Upper mid frequency
1–20 kHz	Upper mid/treble frequency
3–20 kHz	High frequency (HF)
8–30 kHz	Very high frequency (VHF)

for example, a lack of uniformity of the magnetic field may affect the distortion at a certain sound pressure level without significantly altering any other aspect of the performance. Changes in the strength of the magnetic field, however, will alter both output and frequency response, the latter due to the variation in electro-magnetic damping factor near the fundamental resonance (see Figure 3.38, Chapter 3).

The main components in a moving coil driver are illustrated in Figure 5.1, a cross-section of a cone model.

The Diaphragm or Sound Radiating Element

This may vary in diameter from 12 to 500 mm and may be formed in a variety of shapes and profiles, ranging from domes to cones, both convex and concave.

The Surround or Outer Support

Found in cone units and comprising the flexible structure joining the cone rim to the chassis. Most dome drive units do not have a separate surround, the edge suspension often being the only means of support for the main structure.

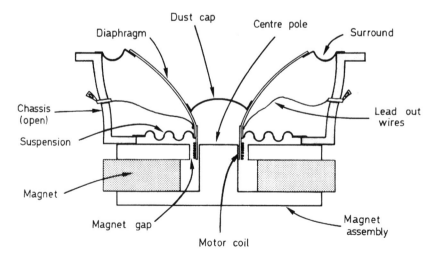

Figure 5.1. Moving coil driver components

The Suspension or Inner Support (Spider)

Usually found near the join between the motor coil and the diaphragm, this rear support member is a concentrically corrugated fabric disc joining the cone apex to the chassis. The function of both the suspension and the surround is to allow relatively free axial or piston movement of the diaphragm whilst offering great resistance to lateral displacement or rocking motion of the diaphragm.

The Motor Coil (or Voice Coil)

This is an assembly consisting of a shallow coil wound on a light, hollow tube firmly attached to the diaphragm. Flexible leadout wires convey power from the amplifier to this coil.

The Magnet

This usually refers to the complete magnet assembly which provides a strong uniform radial magnetic field through the motor coil. This is the polarizing field against which the alternating fields produced by the coil act, resulting in mechanical force on the moving assembly.

The Chassis (or Supporting Basket or Frame)

The chassis is usually a separate pressed steel or cast framework which provides a rigid foundation for all the other components, and a means by which the driver is fitted to an enclosure. In some dome high frequency units, the front magnet plate alone provides all the functions of the chassis.

The Dust Cap or Centre Dome

As the name suggests, this element does not usually fulfil any acoustic function and merely keeps dust out of the gap. However, it may be an integral part of the acoustic design in some units.

The Phase Plate

This may be a partial acoustic obstruction in the front of the diaphragm, or it may be a short horn-like structure adjacent to the dome. Its function is essentially that of equalization. Irregularities of sound pressure, directivity or phase with frequency may be corrected, particularly in dome diaphragm units.

The design of all these components is strongly related to the driver application (i.e. the frequency it is designed to cover) and as such the details will be included in the individual discussion of driver types which follows.

5.2 LOW FREQUENCY OR BASS UNITS

Low frequency design is intimately associated with the enclosure or loading method employed (see Chapter 4). A driver intended for high performance applications will possess a level reference region above the fundamental resonance. The total Q of such a driver is likely to be well below unity and its actual value is a vital factor in determining the design of the enclosure, whether sealed or reflex. For example, a sealed-box system Q greater than unity implies a significant rise at system resonance, a condition likely to be unacceptable for a high performance system.

As shown in Figure 3.30, for a given motor coil resistance and winding length the system Q is predominantly controlled by the flux density in the magnet gap. For any given driver there is an optimum value of magnet strength for the maximum level output. Greater efficiency may be achieved at the expense of low frequency extension and/or by increasing the volume of the enclosure.

Since bass drivers essentially operate as pistons, the acoustic power generated at a given frequency is proportional to the area of the cone and its peak excursion, i.e. the volume of air actually 'pumped' by the diaphragm. For each halving of frequency the required excursion for constant power is doubled and, for a given power and driver size, one can quickly reach a low frequency limit at which the driver ceases to operate linearly due to limits of available cone movement. The latter may be due to suspension constraint and/or the coil movement exceeding the region of constant magnetic flux.

In the case of the infinite baffle or, more precisely, 'sealed-box' enclosure, once limits are set for the cone size and matching enclosure volume, as well as for tolerable non-linearity, the resulting response must be a trade-off between low frequency extension and efficiency. This is why small, wide-range loudspeakers are inevitably inefficient.

Diaphragm Materials (LF)

While designers may hope that bass drivers operate as pistons, this is rarely the case over the working frequency range. Attempts have been made to reproduce highly rigid structures; Celestion, KEF Electronics, Yamaha and Leak have all employed reinforced polystyrene diaphragms in various forms. While Leak employed a cone, the 'Sandwich', the others utilized solid structures. An aluminium foil coating on both sides of the diaphragm provides reinforcement (Figure 5.2). Resonances in such rigid diaphragms are difficult to control, and when such a stiff structure does enter breakup, the effects may be severe. The more gentle breakup of an ordinary diaphragm is generally preferred in applications where the driver must be used in or near the breakup range.

An established example of a reinforced polystyrene bass driver is the B139, manufactured by KEF Electronics (Figure 5.3(a)). The unit operates as a pure piston up to approximately 800 Hz, where the first breakup mode occurs. It is used successfully in a number of current high performance systems, ideally with a high slope crossover at 200–300 Hz, although some designers have worked up to 450 Hz. The radiating area approximates to that of a 250 mm diameter circular unit and the model has been employed for both large reflex and compact sealed box designs. In the effort to combat the weakness of conventional paper/pulp cones, other formulations have also been tried. EMI developed an elliptical laminated structure comprising a reinforced glass fibre layer bonded to a conventional cone. A further development consisted of two stiff cones

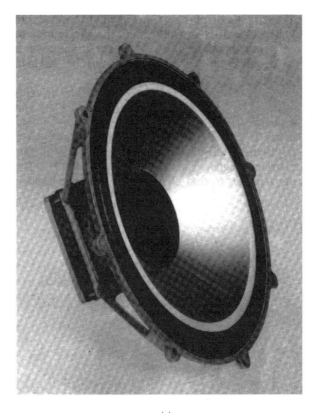

(a)

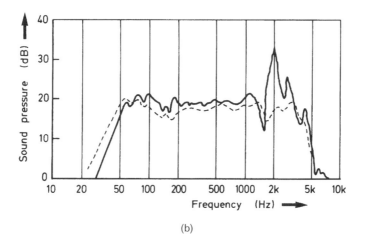

(b)

Figure 5.2. (a) A 'sandwich' cone 380 mm LF driver. (b) Typical response curves, with (– – –) and without (———) nodal mass loading control

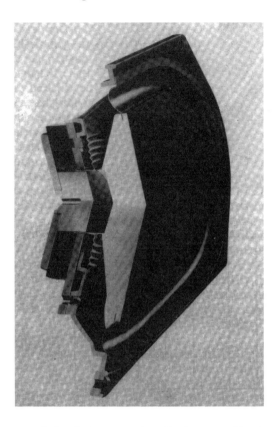

Figure 5.3. (a) An expanded polystyrene wedge diaphragm, with stressed aluminium skins
(courtesy KEF Electronics)

laminated with a flexible damping layer between the two. Thick pulp cones formed under
low compression are also favoured, since their high internal damping lowers colouration.
Technics have developed a three-layer, pulp cone material for their advanced bass drivers,
consisting of stiff outer layers with a soft centre layer of high damping.

More recently Technics have produced planar diaphragm drivers (see Figure 5.3(b))
using a foil-stressed aluminium honeycomb of special manufacture and of high radial
symmetry. Related research work* has also been carried out by engineers at Sony and
Pioneer [7]. The latter company's endeavours resulted in an unusual but costly design of
a three-way planar coaxial driver, remarkably employing rectangular form motor coils.
Other manufacturers have sought to increase the rigidity of conventional cones by the
addition of some high strength matrix such as carbon fibre. Sony, Pioneer and Son
Audax are all actively researching this aspect [8].

While the developments mentioned so far have enjoyed reasonable success, special
mention is due to a synthetic isotropic cone material marketed under the name of
'Bextrene', and consisting of a mixture of polystyrene and neoprene. Well-designed
cones made from this material, if suitably terminated by an appropriate surround and

*These planar drivers are intended to remain pistonic.

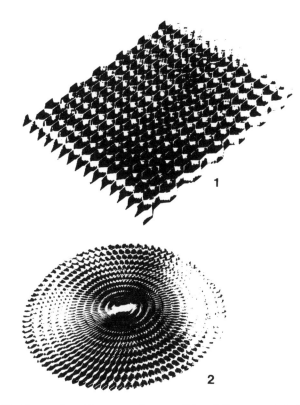

Figure 5.3. (b) (1) Conventional/honeycomb core, (2) axially symmetrical honeycomb core
(courtesy Technics)

coated with PVA,* such as Plastiflex, have offered a low/mid frequency performance
well in advance of other cone types. Not only is the fundamental 50–500 Hz range
covered in almost textbook fashion, but the favourable characteristics of low distortion
and colouration plus excellent response uniformity may continue for several octaves
above, making the choice of crossover points and slope less critical. The BBC engineers
who pioneered the use of this material in the 1960s succeeded in producing a two-way,
high performance monitor system with a 300 mm LF/MF driver crossing over at
1.7 kHz [1].

A successor to Bextrene emerged which, on listening and laboratory tests, appeared
to offer a superior performance. This plastic is based on polypropylene, with an added
mixture of polyethylene of molecular weights and mix chosen for optimum acoustic
properties. In addition, it possesses the additional advantage of requiring no extra
damping, as the internal losses are intrinsically sufficient. In similar thicknesses to those
employed for Bextrene drivers (25–37 μm for 200 mm drivers, and 37–65 μm for 300 mm
units), the cone mass is reduced resulting in a sensitivity improvement of the order of
3 dB for the same driver assembly. A further advantage concerns its excellent long-term
stability, a factor at times in doubt with Bextrene and certainly in doubt with paper.
Potentially, the material also offers great consistency from batch to batch, but remains

*Aqueous suspension of polyvinyl acetate.

difficult to glue. Talc fillings improve the stiffness and are popular, especially with a homopolymer formulation.

As with Bextrene, the performance of a well-designed 300 mm driver made from polypropylene is satisfactory to beyond 2 kHz, and has allowed the completion of a second generation two-way system based on the BBC example quoted above. The response curve of the new driver is smooth and well balanced up to 4 kHz (Figure 5.4). The commercial system in which it is employed uses a complex active equalizer, but the equalization is more concerned with compensation for enclosure diffraction effects than driver anomalies.

Metal cones have also been produced but usually by the time resonances have been controlled, the efficiency is reduced due to the high mass. In addition, as with the rigid polystyrene structures, the breakup modes in metal diaphragms can be severe. Nevertheless, aluminium diaphragms are in use by several manufacturers, one employing expanded foam metal for the LF unit and another a honeycomb structure.

The range of potential diaphragm materials and constructions is limited only by the imagination. As new plastics and polymer formulations are produced, drive unit designers eagerly try them hoping for a quick solution to the fundamental problems. In Table 5.2 this material diversity is well represented.

While some degree of internal damping is desirable in a material or construction, if the loss is too high compared with the E value or stiffness factor the diaphragm might present a smooth frequency characteristic but show higher distortion. An additional effect with high loss materials, hard to quantify, is a hysteresis phenomenon which subjectively appears to mask fine musical detail.

Despite much emphasis in commercial promotional terms concerning new diaphragm technology, there is still much life left in the paper/pulp diaphragms originally used in moving coil drivers developed by Rice and Kellog over half a century ago. Its sonic velocity ($\sqrt{E/\rho}$) is quite good, and its manufacture is well established and understood.

Table 5.2 is worthy of a closer inspection. The values for sonic velocity give a fair idea of the first bending mode resonance for various materials and might suggest the right choice for a piston mode dome hi-fi unit. Given that copper can be made to work up to 20 kHz as a 34 mm cap (with a sonic velocity of 4.1 units), its replacement by aluminium should lift the break point to around 27 kHz. Beryllium is undoubtedly an excellent choice but is very costly to process and poses a health hazard during manufacture. Boron is too brittle as a natural substance but can be successfully applied to a substrate such as titanium.

So far designers have tried to combine the surround with the diaphragm when using plastics. However, the figures for polyester (mylar) suggests that it could perform well as a piston dome if the surround function was separately served by the choice of an appropriate material. Likewise, paper/pulp domes should perform well here if used with matched surrounds. Their very low density will allow a relatively thick construction, durable and easily handled. Both aluminium and titanium foil have become popular choices for pure piston high frequency domes.

The values in Table 5.2 can only be a guide since they refer to the properties in sheet form. When made up into practical diaphragm shapes additional factors play their part, such as self-damping, and the resulting performance cannot be fully determined from the material properties alone. (With composites the concept of sonic velocity has limited meaning.)

(a)

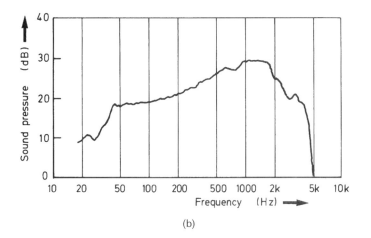

(b)

Figure 5.4. (a) A 305 mm polypropylene cone LF/MF driver. (b) Typical unequalized response curve showing extended, well controlled characteristics (courtesy Rogers (Chartwell) Electronics)

Diaphragm Shape

Some of the effects of diaphragm shape have been discussed in Section 3.1. These principles cannot be ignored even for bass drivers, and the determination of a suitable profile for a given material and driver size is to a large degree a matter of trial. The choice of surround is vital and, strictly speaking, the surround is best considered as a

Table 5.2. Material strengths and densities considered usable for loudspeaker diaphragms (and possibly enclosures)

	Density (ρ) kg m^{-3} × 10^3	Young modulus (E) N m^{-2} × 10^{10}	Specific modulus (E/ρ) m s^{-1} × 10^7	Sonic velocity ($\sqrt{E/\rho}$) m s^{-1} × 10^3	Q
Aluminium (sheet)	2.7	7.5	2.75	5.4	>200
Aluminium (honeycomb)	0.55	1.5	2.7	5.22	>50
Bextrene (p.v.a. doped)	1.3	0.19	0.15	1.2	10
Beryllium	1.8	25	15	12	>200
Boron	2.4	39	16.5	12.7	>200
Boronised titanium	4.2	25	6	7.7	>100
Carbon fibre composite	0.26	1.0	3.8	6.2	>100
Chipboard (600)	0.6	0.2	0.33	1.825	50
Copper	8.5	15	1.8	4.1	>200
Graphite polymer	1.8	7	3.9	6.2	12
Iron	7.9	20	2.5	5	>200
Magnesium	1.8	4.5	2.6	5.8	>200
Paper/pulp (typical)	0.15	0.05	0.33	1.83	10
Paper/phenolic	0.35	0.1	0.285	1.69	15
Plywood	0.78	0.86	1.1	3.3	>50
Polyester film	1.4	0.7	0.5	2.23	>50
Polyethylene	0.94	0.1	0.106	1.0	10
Polyamide film	1.4	0.3	0.22	1.45	>50
Polymethyl pentene	0.84	0.28	0.33	1.83	8
Polypropylene homopolymer	1.0	0.23	0.23	1.52	11
Polypropylene copolymer	0.91	0.14	0.153	1.24	10
Polypropylene (filled, talc)	1.3	0.30	0.23	1.52	10
Polystyrene*	0.95	0.19	0.2	1.45	31
Polystyrene foam	0.01	0.0003	0.030	0.548	8
Polystyrene (foam, alloy skinned)	0.027	0.20	7.5	8.6	12
Resin glass fibre (honeycomb)*	0.43	1.1	2.55	5.5	15
Titanium	4.5	11.6	2.6	5	>200

[Table augmented after Kershaw] *composite

continuation of the diaphragm even though it is a separate component generally made from a different material.

With vacuum formed thermoplastic cones, the best results to date have been given by flared, moderately shallow profiles. The transition between the neck and the motor-coil former should form a smooth curve, and to inhibit 'folding over' of the edge, the tangential angle between the cone rim and front plane should not be much less than 25°.

The flare rate, cone and motor-coil diameter, coil mass and cone thickness will all affect the linearity, slope, range and directivity of the pressure response. Even if a bass driver is not intended to operate beyond 400 Hz, it is worth designing the unit to give a well-behaved characteristic at least two octaves above, since this will generally reduce colouration and the crossover transition will be better defined.

Designers often ignore peaks in driver response if they appear in the crossover rolloff range of the final system, outside the driver's own bandpass. In high performance systems such peaks may present problems in terms of crossover slope integration and some ripples will still remain on the final axial response. Such cone resonances may also be excited by the succeeding upper range driver by acoustic coupling, and thus become audible.

Most synthetic cones are produced by some kind of suction or vacuum forming process where heated thermoplastic sheet is drawn over a porous cone mould. The process tends to form diaphragms where the thinnest, most stretched section is at the apex, the thickest least formed region is at the rim. Study of cone behaviour indicates that the reverse is preferable, e.g. a strong thick apex section tapering evenly out to a thinner rim. Injection moulding is an increasingly popular method and requires very close tolerances for the machine and tool parts in order to achieve the high degree of axi-symmetric uniformity required. If the cone is not well balanced axially, rocking and other dominant breakup modes may occur.

With more sophisticated vacuum moulding methods it should be possible to control the mould temperatures differentially for different sectors of the cone former and to heat the sheet at the moment prior to moulding such that the apex section does not experience the usual excessive thinning.

Surrounds

This component is important to all drivers but it is most critical in low frequency applications, where it must perform two roles. First, it must provide a large amplitude excursion at low frequencies with a resistance to the considerable attendant differential air pressure and, simultaneously, give effective absorption and termination of energy at the cone edge through to the mid range. It must also support the cone for years at a time without sagging. In the past, surrounds were generally formed by a series of concentric corrugations, either as an extension of the cone edge or an additional component. By contrast, almost all quality drivers today employ some form of half-roll surround. These are satisfactorily linear and can be made to the high compliance values required for low resonance sealed-cabinet drivers. In addition they may be formed from a variety of materials some of which offer excellent termination properties. The curved shape of a roll-surround endows it with a resistance to the high differential pressure between the front and rear sides of a bass driver under strong LF excursion. Materials which gradually overload give less distortion, and plastic foams and rubbers are greatly preferable to treated cambric or similar doped fabric. Although they are quite heavy, surrounds of moulded neoprene are extensively used with bass drivers (Figure 5.3), and recently other related substances such as nitrile rubbers, and the lighter material, highly plasticized pvc have also proved satisfactory. The latter was also found to be particularly suitable for mid-range applications due to its excellent termination properties. Hitachi have also patented a type of pleated surround of doped fabric offering improved excursion linearity and termination.

The preferred suspension materials such as pvc and nitrile have a density in the 1 to $1.1 \times 10^3 \, \text{kg} \, \text{m}^3$ range, comparing well with the density of polypropylene and Bextrene plastics. Butyl rubber has also proved popular with dense pulp cone drivers. For lighter grade cones, foamed polyurethane works well,* especially with an additional glossy coating. Adhesive problems have been noted with cone plastics, notably polypropylene and also surrounds, particularly plasticized pvc,** but in recent years these seem to have been overcome. Favoured methods now include RF bonding of surrounds and cones, this and compression heating are preferable to solvent-based adhesives, the use of which can result in serious physical distortion of the components.

Designers must take into account the divergent requirements, namely good dissipation of vibrational energy above 150 Hz or so; a profile which allows reasonably linear excursion up to 8 mm peak to peak, and a mechanical structure which inhibits self-resonance. In a particular prototype driver, a response dip occurred at about 600 Hz, whose source was eventually traced to the anti-resonance in the surround. A flat section had been designed between the half-roll and the code edge, but another was unintentionally present between the outer 'roll' edge and the clamping point on the chassis. The unwanted circuit comprised the 'roll' vibrating with the compliance of the two adjacent flat sections. Readjustment of the roll dimension to eliminate the outer flat portion removed the dip.

The surround must be regarded as an additional radiating element producing an out-of-phase component and should, in theory at least, be as narrow as practicable for the required excursion. Another factor is the significant temperature variability of compliance shown by many surround materials, particularly nitrile rubber and p.v.c.

Subjective Effect of Bass Driver Suspension Hysteresis; Pace and Rhythm

The author has undertaken an analysis to investigate a particular difference in low frequency sound quality noted between some 400 speaker models, and has identified a factor controlling bass sound quality which may be separated from the usual criteria such as low frequency bandwidth, Q factor, and the bass–mid range frequency balance.

With continuing improvements in amplification and source material, particularly with digitally derived programme, a more critical awareness is developing among reviewers and users concerning not just bass response uniformity but also bass dynamics, timing, rhythm and tunefulness. For good tune playing in the bass, the low frequency register must be reproduced with the kind of low colouration performance

*Can degrade prematurely in tropical conditions unless UV stabilized and treated.
**Rare now due to toxicity.

which was originally only sought and expected from the mid-range region of high quality speaker systems. As regards the engineering aspects of this criterion, obvious influential factors include the use of non-resonant bass diaphragms, rigid, inert enclosures, and the absence of internal standing wave modes within the enclosure and, if used, a ducted reflex loading which is itself free of secondary resonances higher up the frequency range. Obviously the low frequency amplitude–frequency response shape must also be tailored for the expected environmental loading in order to optimize the response smoothness in conjunction with the room acoustics.

Attention to these factors alone will not guarantee the additional subjective qualities of expression, dynamics and rhythm in the bass. These aspects appear to be associated more with the complex mechanical impedance of the suspension of a moving coil driver. Two basic kinds of suspension may be identified: those with the soft viscoelastic, vinyl related surrounds, and those with rubber or foam plastic surrounds. The former are often of low mechanical Q but more importantly suffer from a considerable degree of hysteresis; whereby return to the centre zero or rest position is considerably delayed (minutes or even hours) after a large signal bass excursion has ceased. The mechanical response of the second type is largely Hookean, i.e. following a simple spring characteristic and possessing negligible hysteresis. Interestingly, some unpublished engineering work done at KEF correlates well with the following subjective observation, namely that the hysteretic class of surrounds can perform well on simple sine-wave related sounds such as the 'open' bass waveforms of acoustic orchestral bass instruments, organ, bass viol, bass drum, but are clearly 'slower' and less accurate when reproducing the faster percussive rhythm bass generally encountered in modern jazz and popular music. Even a well damped pedal drum sound can be heard to be subjectively distorted. In such cases the bass is said to sound 'compressed', 'undynamic', 'slowed' and fails to 'time properly' with the mid-range beat.

Conversely the majority of low hysteresis bass units and, for that matter, large, open panel film type transducers, have a bass which is fluid and dynamic, holds in rhythmic time with the rest of the frequency range and sounds appropriately 'fast' with suitable programme input.

The KEF data on this subject emerged in connection with research investigating the lack of correlation between frequency responses in the low frequency range derived in the steady state and for a narrow impulse (see Figure 5.7).

In a particular case for an enclosure, the discrepancy was associated with a semi-random movement of the internal absorbing linings of the enclosure. Its movement was uncontrolled on impulse excitation but settled into regular patterns on continuous, slow swept sine wave stimulus. This behaviour is analogous to hysteresis in a driver suspension material and investigation of the fine detail of the impulse response of bass drivers is expected to reveal further information on this subject.

Certainly many speaker designers have been intuitively aware of the subjective qualities of surrounds and it appears to have resulted in two identifiable schools of thought. Where a high hysteresis vinyl based surround is acceptable, this is usually exploited to produce the finest termination for the diaphragms of reference quality bass mid-range drivers. These are fundamental to the production of medium sized two-way speakers considered to be of monitor standard judged in terms of frequency balance, accuracy and low colouration. Their customers generally value mid-range quality and accuracy above bass definition. On the other hand, the alternative school values bass dynamics highly and, in some cases, will go to almost any lengths to maximize them,* including the use of false tonal balances, e.g. mid-range lift, curtailment of the low frequency extension and, excessively low system Q, all of which may subjectively 'speed up' bass percussion sounds. However the primary sacrifice is often in the area of mid-range colouration, resulting from the deliberate use of low hysteresis, low-loss rubber or rubber equivalent driver surrounds, often with poorer cone termination in the mid range.

The obvious technical solution is to adopt three-way system design and to use the surround types optimized for the respective driver frequency bands. Another solution is to design self-terminating diaphragms for the bass–mid range, placing far less reliance on the surround terminating function. Another fairly successful solution was seen in a Celestion 170 mm bass–mid range driver where the half-roll surround function was split into two parts which were bonded together; it comprised an outer rim of butyl rubber for optimum bass and an inner section of hysteretic vinyl chosen for its diaphragm termination property. Clearly there is scope for more elaborate designs which integrate diaphragm and the surround termination; variable surround thickness achieved by precision moulding may also be a useful technique.

Suspensions (LF)

The compliance of synthetic surrounds is usually temperature dependent; a p.v.c. type may vary by two-to-one over a range of 15°C to 30°C. A low frequency driver will often need a reasonably stable fundamental resonance, e.g. for a reflex loaded application. The surround of such a unit, if temperature sensitive, clearly should not provide the bulk of the diaphragm restoring stiffness. The latter must then be dependent on the remaining supporting component, namely the suspension.

Good quality suspensions are manufactured from a polyamide fabric (nylon or polyester) impregnated with a cured, temperature stable epoxy resin. If carefully designed, the suspension can provide a stable restoring force over the required excursion. The suspension should also be engineered so as to limit gently at the peak displacement allowed by the magnet assembly. This will reduce the severity of coil jump-out effect noted in the section on magnet/coil systems in Chapter 3.

It is also possible to control the suspension non-linearity, by suitable choice of material and geometry, to compensate for the magnetic non-linearity produced by large excursions [2]. This is important if an optimal efficiency, short-coil short-gap assembly is used, where the coil length is only 5%–10% greater than the physical magnet gap. Some of the **BBC** Bextrene driver systems employ this technique, but owing to the difficulty of its execution, the newer generation of reflex loaded monitors have reverted to a more conventional linear suspension and employ a motor coil with an approximately 25% coil overhang.

Poorly designed suspensions, particularly those which are large and heavy, may possess self-resonances at even a few hundred Hz. Additionally, poor quality cotton or linen fabric suspensions fitted to heavy coned bass drivers can take a 'set' away from the nominal centre zero coil position in a matter of days, particularly under humid conditions and if the units are stored horizontally. Vertical positioning relieves the suspension of the axial bias due to gravity, but the long-term stability is still in doubt.

*High Q_M is a key factor for the LF unit, $\geqslant 4$.

In practice, any suspension may develop a degree of off-centre 'set' if stored in one direction long enough, which is why vertical mounting, particularly of heavy coned bass drivers, is strongly recommended. A bias may be specified in manufacture to balance such an effect.

Double Suspensions (LF)

While there is little evidence that moderate degrees of suspension non-linearity are audible at low frequencies, some manufacturers have attained very low orders of bass distortion by utilizing a double suspension. The corrugation geometry is arranged to be complementary, so making the excursion characteristic identical in both the forward and the reverse directions.

With large drivers intended for high power applications, the lateral forces at the voice coil may be sufficient to cause momentary decentring. A spaced double suspension may then be used to advantage, resulting in a greatly improved resistance to decentration.

Matsushita have produced a suspension geometry with a strong resistance to lateral decentring forces, comprising a four-piece box section structure with a motion akin to a pantograph. Rocking modes are strongly resisted, which is vitally important with shallow drivers such as the honeycomb planar diaphragm designs. An additional benefit conferred is the much improved linearity; for a given size the 'box' suspension provides twice the linear excursion as compared with the usual corrugated cloth type (see Figure 5.5).

Motor Coil and Magnet Assembly (LF): Bandwidth Examples

The moving diaphragm mass is virtually a fixed quantity for a given unit size. The motor coil diameter is generally proportional to its thermal power handling, and with these two factors decided, the efficiency of the resulting driver is proportional to the square of the magnetic field strength in the gap. The design of magnet structure will depend on the loading and maximum undistorted output requirements for the system. For example, take the case of a sealed box with a long throw 200 mm driver, which is required to develop 96 dB from 60 Hz upwards. Suppose it possesses a magnetic field strength acting on the coil length immersed in the gap sufficient to provide a system Q of about 0.7. Since the large cone excursion is needed to produce 96 dB at 60 Hz, about $\pm$ 6 mm peak, a typical 6 mm magnet gap will require a coil overhang of at least twice this, resulting in a total coil length of approximately 18 mm. An unavoidable loss in motor efficiency occurs, since only one-third of the available power in the coil is actually used in creating acoustic power.

It is worth noting that if the response of this system was required to extend one octave lower to 30 Hz at the same power level, then the excursion must be multiplied by four, i.e. to a ridiculous 50 mm peak to peak. Sound power is proportional to the square of the diaphragm amplitude, and if undistorted reproduction down to 30 Hz is required of this particular driver, the overall power level must be reduced by 12 dB.

If a similar sealed box capable of 96 dB at 30 Hz were required it could be achieved within a sensible excursion limit by increasing the diaphragm size, since the power

(a)

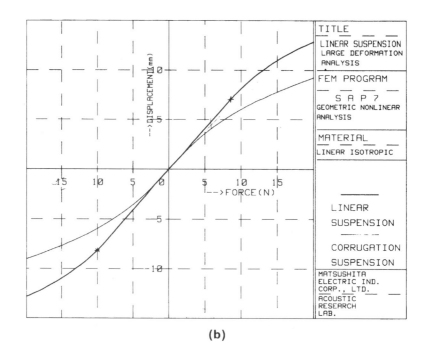

(b)

Figure 5.5. (a) Linear damper (courtesy Technics); (b) linear suspension, large deformation
analysis

radiated is proportional to the square of the product of the moving area and excursion. If the same ± 6 mm excursion is adopted, then the diaphragm area must be four times greater to radiate 96 dB at 30 Hz. Thus the required driver diameter needs to be increased to 380 mm.

Reflex loading may allow a given driver to produced more bass. The peak excursion of the LF driver at resonance is reduced by as much as four times with reflexing, which allows the motor coil to be reduced in length, in turn giving a great improvement in efficiency. The resulting increase in the *Bl* factor is in fact essential to control the working *Q* of the reflexed system. Typically a sound power increase of four times (6 dB) is possible with a given driver when optimized for reflex loading as compared with sealed box. This holds true even for the same low frequency cutoff, although the reflex enclosure volume is likely to be larger than the sealed box equivalent.

Taking into account the average spectral energy distribution of programme, a well-designed reflex enclosure with a 300 mm or 350 mm bass driver can produce upwards of 115 dB of wideband programme at 1 m (a factor of 6 dB or so is gained by the falling energy in most programmes below 70 Hz). Higher levels may be attained by multiple arrays of such enclosures or by horn loading, although the latter technique is unavoidably bulky at low frequencies, and its use is generally restricted to fixed installations such as cinemas, etc.

Power Dissipation (LF)

Motor coil diameter is roughly related to cone size and power rating, with 20–25 mm coils the rule for small 100–170 mm drivers; 25–37 mm coils in 200 and 250 mm units, and 44–100 mm diameters for 250–380 mm units. There is no obvious acoustic advantage for large diameter voice coils, except perhaps the argument that a large voice coil on a small cone means that no part of the cone is very far from the driving point, which is likely to result in a stiffer structure unlikely to break up until a higher frequency range is reached.

The large 75–100 mm coils will dissipate up to 200 W of continuous thermal power if constructed of suitable high temperature formers such as aluminium foil or Nomex, and utilizing matching heat-cured adhesives.

Since most loudspeakers are at best only a few percent efficient, the bulk of the input power is dissipated as heat in the motor coil. Surprisingly high temperatures (up to 200°C) may be developed under heavy drive, and even with models designed for domestic use, some manufacturers aim at short-term service temperatures of the order of 250°C. Several factors must be taken into consideration when temperature rises of this magnitude are to be accommodated. The increase in d.c. resistance is appreciable and will provide some degree of self-limiting with regard to the maximum power drawn from a given source. It will also add a degree of non-linearity, depending on the thermal time constant of the motor coil.

King describes an example of a 305 mm driver equipped with a 75 mm coil and designed for use with an electric guitar [3]. The coil reached a steady 270°C after 4 h running at a voltage level equivalent to 200 W into 6 Ω, the latter being the driver's nominal 'cold' specification. The 'hot' impedance was found to be double, indicating that the driver was in fact only drawing 100 W from the source to maintain this temperature (copper's resistivity rises by 0.35% per degree centigrade). The thermal time constant for the coil was 15 s, the time for a 34% drop in resistance.

Clearly operation which results in an average power much above 75 W would be inappropriate for this driver, and the example highlights an often neglected problem encountered with high power units. The best method of cooling a coil is via conduction to the magnet structure. This is aided by a narrow gap clearance, a condition difficult to achieve with LF drivers as the coil excursion is considerable. However, there are some ways round this particular problem. For example, the magnet gap may be made longer than the coil so that the latter is always in close proximity to a large mass of metal. Alternatively, the coil may be wound on a large heat-conducting former, such as anodized aluminium foil, helping to spread the heat over a larger area. Thirdly, the magnet/coil structure may be ventilated such that cold air is continuously pumped through the gap where the unit is under drive, thus providing forced cooling; this artifice is only effective at low frequencies, for example, below 200 Hz where the cone excursion is appreciable. The magnet structure itself may be fitted with blackened radial fins on the exterior surface to aid heat dissipation.

A further problem concerns the effect of conducted heat on the diaphragm. While the high temperature components of the motor–coil may be readily designed to withstand the stress, some thermoplastic cones readily soften at around 100°C and a hot coil will quickly become decentred or detached. Even pulp cones may char at the neck and suffer from premature ageing or disintegration. A solution is provided by the use of a non-conducting coil–former section adjacent to the cone, which isolates the hot section of the coil from the cone (Figure 5.6).

Some premature failures have been noted with Nomex formers due to its hygroscopic nature. After a period of disuse, a Nomex equipped driver started up at high power may produce rapid moisture outgassing from the former, usually resulting in structural distortion and even bubbles, often locking the coil in the gap. Kapton, a high

Figure 5.6.   Example of a high temperature working motor–coil assembly. The coil is wound on alloy foil, bonded to a ventilated 'Nomex' former (high temperature polyamide). Heat cured adhesives are employed (courtesy KEF Electronics)

temperature plastic, has been brought into use despite adhesive difficulties when bonding to cones. Hygroscopy is not a problem here.

Ferrofluids

A development from the United States is the use of a fluid which may be applied to the magnet gap of moving coil drivers. It consists of a stable, inert organic diester base containing a colloidal, and hence a non-settling, dispersion of ferromagnetic material. The liquid is sufficiently magnetic to remain firmly trapped in the regions of greatest field strength, i.e. the gap. It is thus self-locating upon injection into the gap on either side of the voice coil. It may be obtained in a range of viscosities from 3000 to 50 cp, with 100 cp suggested as suitable for low frequency drivers.

Interesting performance gains result from its application. In LF terms, the primary benefit is a greatly improved short-term power dissipation for the motor–coil, since the fluid exhibits good thermal conductivity, many times that of the air it replaces. While drivers in low power applications would derive little advantage from its use, the high power examples described might gain a short-term power handling increase of up to three or four times. If additional damping is a design advantage, then the fluid can provide this through choice of a suitable viscosity. Finally, the fluid provides a lowered reluctance path in the gap which reduces magnetic fringing and will marginally increase efficiency, by up to 0.75 dB, depending on the driver construction, and may also reduce distortion.

By maintaining lower coil temperatures the fluid can dramatically reduce the power compression effect noted earlier in connection with high temperature rise. Some extra centring force is also provided by a tendency for the fluid to form a uniform layer around the pole. This may eliminate the need for additional centring, for example, via a second suspension or may obviate the suspension altogether in a light mid-range diaphragm assembly. However, with heavy excursions the fluid may not flow laminarly resulting in an asymmetric distribution which may impede motion and reduce linearity. Another consideration is the catastophic failure of the fluid when overheated.

In one investigation [9], the temperature of an LF driver with a 25 mm motor coil of four layers and a 20 W input at 25°C was 185°C after 120 s, a rise of 160°C. With a gap intensity of 200 G, 0.6 cm^3 of ferrofluid of 2000 cp viscosity was injected into the gap. Over the same period the temperature reached 80°C, a rise of 55°C, one-third that of the untreated gap.

Where ferrofluid viscosity is used to provide additional mechanical resistive damping its own viscosity variation with temperature should be borne in mind. It was noted with one HF driver, of small thermal capacity, that its ferrofluid damped transient response varied markedly with level. At the 0.1 W level it worked as predicted, but at 3–10 W input the damping fell off considerably. Certain ferrofluid grades have also proved incompatible with some motor–coil adhesives, and formers can soak up some of the fluid. Finally, the fluid flow under large excursion can result in a back pressure build-up behind the coil necessitating a rear vent in the magnet, particularly in LF units.

Winding Techniques (LF)

Coils may be wound with a variety of conductors, but the most common is enamelled copper. Anodized aluminium is also used, more especially in HF units where the

Figure 5.7.   Edge wound motor–coil using ribbon conductors (courtesy Bose)

reduced coil mass and inductance (its higher resistivity allows fewer turns) is beneficial to response extension.

Ideally, the mass of a copper coil should fill the gap to provide maximum utilization of the available magnetic flux. In practice, a clearance must be provided to prevent rubbing as the coil moves and to allow for change in the coil former profile with age and thermal stress. A further loss occurs due to the wasted space in the winding and the thickness of the former on which it is wound.

If a rectangular cross-section wire is employed, then the space utilization factor may be improved, with a resulting 15%–20% increase in efficiency. The cost of these 'edge wound' coils is high and the technique is uneconomic except for a few expensive designs. Alternatively, the wire may be partially deformed either before or after winding to 'square' the profile and hence reduce the air space (Figure 5.7).

A further solution requires the use of conducting insulated foil, flat wound.

At present, the vast majority of motor coils are wound in insulated copper* which may be precoated with a thermosetting adhesive to aid bonding. Most high frequency coils are heat cured to bake out any solvents or moisture which might cause bubbling or mechanical distortion in service. The majority are two layers, though four is quite common and six have been used in some short coil, long throw LF drivers. The main consideration is mass. The long coils required for sealed-box systems, if wound in four layers, will be sufficiently heavy to curtail the response in the mid band to possibly 2 kHz or lower, and may also undesirably affect the coil mass/cone resonance. In addition, multi-layer coils have a higher inductance which may further restrict the response.

Linearity and Magnets (LF)

At low frequencies, below 150 Hz, distortion is a function of loading and available excursion, both mechanical and magnetic. In the mid band other factors also assume importance, for example, eddy currents in the poles. Lamination of the pole faces or the use of a special material of high permeability and low electrical conductivity around the gap will control eddy effects and in a typical driver may reduce the distortion in the 200 Hz to 1 kHz band from 1% or 2% to below 0.25% (see Figure 3.8).

*'CCAW', 'copper clad aluminium wire' may be used, lower mass and easy soldering are advantages.

Magnet poles are often operated in saturation since this gives good control of magnet strength variations, but by definition, it is also wasteful of flux. Saturation places a limit on the maximum flux density in the gap. The usual mild steel pole and top plate allows maximum values around 1.4 T for a 25 mm pole and 1.7 T for a 50 mm pole. The incorporation of a higher saturation material for the pole faces such as Permendur or alternatively laminated permalloy, will allow a gap flux increase. Using Permendur gap components, 1.9 T has been reported in a commercial HF unit having a 19 mm pole.

Using a cobalt steel alloy with a 6.25 mm top plate a gap flux density of 2.5 T is possible. Gap flux values are usually lower than the saturation value for the pole due to the leakage prematurely saturating the base of the pole. Undercut 'T' section poles used to improve flux linearity, have the disadvantage of earlier saturation owing to the thinned section. As discussed in Section 3.8 other factors can control linearity, including eddy current control rings.

Demagnetization (LF)

While the choice of magnet material (Alnico, Alcomax, Magnadur, Ceramag, etc.) is usually dictated by the cost for a given pole structure, one other point also deserves mention. If a permanent magnet is stressed, either mechanically or magnetically, it tends to lose some magnetic strength. While ceramic types are exceedingly difficult to demagnetize, older iron alloy based materials are less so. Drivers employing the latter type of magnet may be unsuited to high power applications where the peak coil flux is considerable. King cites the example of a driver fitted with a long four-layer coil. 37 mm in diameter, with an Alnico cup magnet structure. Twenty-five watts of drive at 50 Hz resulted in 2 dB mid-band efficiency drop. The worst demagnetization effect occurred when driven in the frequency range of greatest excursion, i.e. fundamental resonance, particularly with long multi-layer coils. A two-layer coil under the same conditions gave only a 0.5 dB loss [5].

The Dust or Centre Cap (LF)

The dust cap may influence the performance of LF units, though its effects are usually more noticeable at mid frequencies. Functionally it prevents the ingress of dust to the magnet gap and may allow the differential air pressure at the pole to equalize that in front of the diaphragm. Conventional caps are made from porous treated fabric and contribute little to the acoustic output. However, a few diaphragms are fitted with rigid pulp/paper or even aluminium caps where a small ventilation hole may be present at the apex. The function of such a rigid cap is twofold. The dome structure may be an additional radiating element to extend the frequency range; it may also serve as a stiffening structure helping to reduce cone breakup, particularly of those modes which tend to distort the motor coil from a circular to an elliptical shape.

In one recent design of LF–MF driver using a formed thermoplastic diaphragm the usual moulding apex/cap, normally cut away from the cone, is intentionally left intact. Shaped with a ledge it forms a rigid foundation for the attachment of the motor coil former. This construction seals the gap and also stiffens the cone apex.

Mass Control Ring (LF)

Occasionally reference is found to a mass control ring on a LF driver. This usually comprises a rigid metal ring attached to the diaphragm at the apex, which can be used to improve the performance of certain drivers in the following way. Suppose an ideal cone for a given design of driver is too light for its required purpose, for example, a wide range, low resonance application, and mass needs to be added. Increasing the cone mass by simply substituting a thicker material will alter its acoustic properties. Instead, mass is added as a ring weight at the neck. Additionally, the ring may provide some stiffening and in one example, the ring mass is attached via a compliant adhesive and is used to control a dominant concentric mode resonance (see Figure 5.2).

Chassis (LF)

With motor coil/magnet gap clearances of the order of $10\,\mu$m, and large LF driver magnets up to 10 kg in weight, a rigid and stable chassis is essential both for long-term stability and to prevent misalignment due to transport shock. If properly deployed, both die-cast alloy and pressed steel are suitable materials for chassis construction.

The designer must compromise between the maximum window area at the chassis rear and the quantity of material to provide the required structural strength. Too much material produces cavities behind the cone which may colour the output, too little will encourage chassis resonance. Long, thin-walled chassis sections are obviously weak structures and may resonate. A number of units have suffered from such resonances, usually in the 200–500 Hz range, caused by the magnet wobbling on a weak frame, and best results have only been produced when the enclosure is fitted with bracing to reinforce the chassis of such a driver.

The chassis may have a surprising effect on a driver. In one inexpensive system using an established 0.2 m pressed steel frame, some colouration and minor response irregularity was experienced in the 800 Hz to 2 kHz range. Flexing in the front fixing flange was detected and substitution of a new frame with a deeper rolled-over front flange was found to solve this problem. If the application can justify the expense, then well designed cast frames provide the best solution and lowest colouration and are usually more economic over long production runs.

5.3 LF/MF UNITS

The majority of smaller loudspeaker systems today are two way, and incorporate a combined bass/mid range driver plus a high frequency unit, with a crossover point at 2–4 kHz. The main driver must satisfy two possibly conflicting requirements, namely a clean, well dispersed and uniform mid band and a low distortion bass with adequate power capacity. Only in exceptional cases such as the 305 mm polypropylene driver mentioned earlier, can the larger LF units offer an adequate performance in the crucial mid band. In this instance there is still an inevitable sacrifice in terms of narrowed directivity near the crossover point (approximately 1.7 kHz).

The almost universal choice of chassis diameter for a wide range driver is 160–200 mm. Such a unit offers a unique combination of virtues, which accounts for its popularity in medium level domestic applications. If well designed, the bass power is sufficient for most domestic purposes, and an adequate bass output can be achieved in an acceptably small enclosure (15–35 litres). The frequency response and directivity may be satisfactorily maintained up to a usefully high crossover point, generally 3 kHz. The sensitivity and power handling are both sufficient for quite demanding use without excessive expenditure on the magnet structure and with careful control of the important design factors, a genuinely high performance may be attained (see Figure 9.46).

Diaphragms (LF/MF)

The extension of the working range to meet an HF unit requires that the diaphragm be particularly well controlled and consistent, since it will almost certainly be operating in breakup at the high working frequencies. To attain a satisfactory performance, consistent, uniform cone materials are essential.

Drivers for LF/MF duty range in size from 100 mm to 305 mm although the LF power handling is obviously much reduced with the smaller diaphragms. Mid-range drivers are in fact often designed to fulfil two purposes: to act either as a true mid-range unit in the more elaborate system designs, or as a bass/mid unit in the simpler enclosures. Similarly, the larger, wide range units (200 mm and above) may often be employed in sophisticated systems for LF duty only, where their inherent low colouration characteristics are beneficial to the quality of the system as a whole.

Metal Cones

Metal cones have been in production for many years and have been used in the UK since 1955 (GEC). The reasons for use include durability, resistance to adverse environmental conditions, e.g. humidity and temperature, and the exploitation of the inherent stiffness in order to widen the pistonic range of cone operation. Light metal, or metal alloys of aluminium and magnesium, are substantially stiffer than the usual pulp or plastic material — in fact between 10 and 20 times stiffer. Such greater stiffness places the first bending mode resonance of a typical cone shape into the upper range between 2 kHz and 8 kHz depending on diameter and material thickness, compared with the usual breakup frequency which can be as low as 500 Hz. The best seen with small conventional diaphragms is perhaps 2 kHz with superior materials.

The bending mode frequency is essentially proportional to the square root of stiffness, so substantial increases in stiffness are necessary to make a significant improvement. Working in the mid-1960s, Jordan extensively researched the design of a full-range 170 mm driver employing a flared aluminium alloy cone with a nominal range of 50 Hz to 15 kHz. It had heavily resistive edge termination while a large metal dome at the centre dominated the high frequency response. Many successful systems were marketed using this driver — the sound quality benefiting from the absence of any crossover network. This helped to make up for the imperfect frequency response. Metal cone drivers became popular again in the 1980s and designers had to re-learn how to build them. Typically cones are straight sided of thickness 0.8 mm, formed by drawing or spinning and are often deeply anodized to form a so-called ceramic (alumina) hard

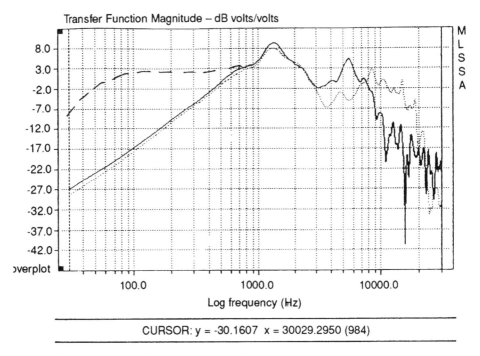

Transfer Function Magnitude – dB volts/volts

CURSOR: y = -30.1607 x = 30029.2950 (984)

Figure 5.8. Comparison of straight sided and flare profile metal cones. *Key*: ——— straight sided; - - - - - flared cone profile, 170 mm driver, aluminium alloy cone (- - - - nearfield, baffled correction)

layer on each side of the cone. Recently, flared curvatures have been rediscovered which ameliorate the inevitable and severe upper range resonance. Thin shell cones in magnesium alloy have also been successfully produced by die casting. For a 90 mm diameter cone in a 115 mm chassis the cone mass is typically 3–5 g. In the 140 mm cone size for a 160–170 mm chassis, it may weigh between 6 g and 12 g. Sensitivity is generally lower than for pulp or polymer cone equivalents due to the higher mass.

Characteristically the frequency response is uniform until the breakup region is approached. Prior to resonance there is a loss in output (as with metal dome tweeters). When breakup occurs, it is severe, peaking at 10 dB or 15 dB and may be followed by further harmonics. For a good 110 mm framed unit, the breakup may occur in the range 6–9 kHz, while for a 170 mm framed unit it is typically around 5 kHz. Figure 5.8 shows the improved control of resonance in a well-behaved metal cone driver when a flared profile is substituted for a straight-sided form.

For severe peaks above the crossover point a filter may be added to the network, thus providing good control over the overall response (see Figure 6.36).

Suspension and Surrounds (LF/MF)

Little needs to be said about these two components except that the suspension is less influential in the mid range, while the surround and its terminating efficiency become more critical as the cone enters the inevitable breakup modes.

Motor Coil (LF/MF)

Whereas a long heavy coil may be a necessary requirement for an LF unit, a satisfactory extension of response to cover the mid range may dictate a compromise whereby the coil is reduced in mass and inductance and hence shortened. Where this is impossible to effect due to the need to maintain bass performance, the costly* solution of a short, light coil immersed in a long magnet gap can be adopted.

5.4 MID FREQUENCY UNITS

True mid frequency units are characterized by a relatively high fundamental resonance, 100–500 Hz as opposed to the LF driver range of 15–60 Hz. The required excursion is small, and hence the motor coil and magnet may be optimized for maximum efficiency. Both dome and cone diaphragms are in common use, the former generally restricted to the upper frequency range (800 Hz to 6 kHz), while the latter may operate from 250 Hz to 5 kHz. The small diaphragms are generally less than 160 mm in diameter and typically range from 60 mm to 100 mm. Dome units smaller than 44 mm are more correctly classed as 'low treble' drivers.

The mid range, typically 250 Hz to 5 kHz, is undoubtedly a most critical band. This is the region where the ear's sensitivity and analytical ability are most acute, and it additionally includes the greatest concentration of information in normal programme material. Many brave attempts to produce a high performance system have been unsuccessful because the designer has failed to adequately appreciate these facts. Critical listeners may tolerate moderate problems in the LF or HF bands but cannot forgive inaccuracies in the mid range, whether they be in the form of spectral imbalances, response irregularities, distortion or colouration.

For this reason, MF units must be designed with care and used with considerable skill and judgement. It has been indicated in Section 6.1 on systems and crossovers that a reasonable response extension outside of the required bandpass is desirable to maintain smooth crossover transitions. Considering our ideal bandpass of 250 Hz to 5 kHz, we should add an octave to either extreme, resulting in an overall mid-driver ideal bandpass of 125 Hz to 10 kHz. In the author's view, only a cone diaphragm is capable of approaching this performance. Where a high quality, moderately sized LF unit is to be used in conjunction with a mid-range driver the lower crossover point may be lifted to the 600 Hz range or above, making possible the use of a dome unit. There remains, however, some reservation concerning the placement of a crossover point with its attendant polar and phase anomalies, since these occur almost in the centre of the mid range.

Diaphragm (MF)

As has been suggested, dome units give their best results in the upper mid range and cone units in the lower range. Figure 5.9 shows the theoretical effect of a cone front cavity compared with a plane disc mid-range diaphragm. The cone result is characteristic but will be modified by enclosure diffraction and the acoustics of the cone itself.

---

*This situation is now addressed by the neo-radial magnet (Aura Inc.).

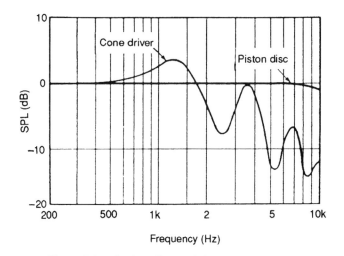

Figure 5.9. Cavity effect and frequency response

Phase Plug

It has been noted that the acoustic matching at the apex of a cone diaphragm is not well controlled and various schemes have been devised by designers, including a hard centre cap, various sizes of semi-soft dust cap, in p.v.c. or a coated cloth. These are varied to try to improve the smoothness of the upper range frequency response. Depending on price, these are variously successful (see Figure 5.10). Where an open pole centre is appropriate to the design, the cylindrical space down the motor-coil former to the pole constitutes a resonant cavity. Use of an absorbent plug controls the cavity resonance at the expense of upper range output. An open construction is often chosen to avoid the secondary modes present in many dust caps, which may also give rise to distortion at

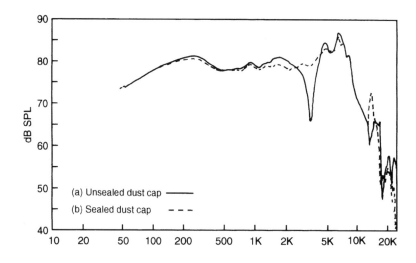

Figure 5.10. Effect on sealing an open fabric dust cap. (a) 1 m on-axis response (unsealed dust cap). (b) 1 m on-axis response (sealed dust cap) (after *Speaker Builder*)

low frequencies due to the effect of high pressures with the pole structure under high excursion conditions. The cavity effect may be well controlled by the addition of a short phase plug, fabricated or moulded in plastic or metal. This forms a short extension of the pole face, preferably tapered at the tip. The interfering cavity is thus removed, and the output is noticeably smoother.

Domes (MF)

Their use is fashionable at present, due to an illogical belief that a dome radiates sound over a wider angle than a cone of the same diameter. In fact, the converse may be true, since in the degenerate condition a dome approximates to an annular radiator whereas the cone approaches a point source, i.e. that formed by the smaller apex area.

The fact remains that dome units are popular among designers, and are also relatively easy to manufacture. A number of models employ press-formed fabric domes, impregnated with some suitable viscous damping material, usually synthetic rubber based. The combined suspension/surround is contiguous with the dome surface, with a small annular recess often provided for the direct attachment of the motor–coil assembly. A felt plug is usually fitted just beneath the dome to absorb most of the rear radiated energy. They range in size from a nominal 37 mm to 85 mm diameter, with a variety of surround and profile shapes. Most, however, approximate to a shallow spherical section.

Dome structural rigidity is poor and most of the so-called 'soft' domes are in fact in some form of breakup from an octave or so above the fundamental resonance which may range from 250 Hz to 500 Hz. The large surface damping component gives these drivers an almost resistance-controlled acoustic characteristic, but due to the unavoidable hysteresis effect of a rubber-based damping compound, they may not sound as transparent as their smooth frequency characteristics would indicate. The better examples utilize a rather tougher diaphragm and damping material in an effort to maintain piston operation to higher frequencies.

Some mid units often sound 'coloured' and two factors can be held responsible for this effect. The fundamental resonance is both relatively high and often inadequately damped as large dome units require large and costly magnet systems for optimum performance. Poor crossover design may result in a further weakening of the control of the main resonance, obtruding in the system sound as subjectively perceived colouration.

Adequate absorption behind the dome is difficult and the reflection and resonance behind the dome is often audible. In more costly units the centre pole may be hollowed or, ideally, tapered to reduce self-resonance, leading to a larger damped rear chamber.

A complete contrast is presented by the 'rigid' domes employing diaphragms of high strength-to-weight ratio, such as stiff card, titanium, beryllium, certain plastics and, more recently, boron-coated metal. With these diaphragms the structure is designed to be so light and rigid that breakup does not occur at all within the required bandpass. The most spectacular of these designs are undoubtedly those formed in beryllium which though difficult to manufacture, offer startling gains in rigidity versus mass [5]. The Yamaha 88 mm beryllium dome unit is an outstanding example of the art, and provides virtually pure piston operation from 400 Hz to nearly 12 kHz. This results in great clarity, though when the unit does enter breakup it does so rather more aggressively than soft domes, due to the intrinsically low damping of the diaphragm material. It should be noted that very high strength to weight ratios may result in off-axis response irregularities due to the coincidence effect caused by the high sound velocity in such a diaphragm compared with normal materials.

Other promising constructions include fabric impregnated with cured phenolic resin, and mylar, the latter improved by lamination with another plastic such as p.v.c., which has a high internal loss. Hard pulp/paper domes are also employed.

As has been explained in Section 3.1 on dome diaphragms, the profile and size is largely dependent on the chosen material. The more rigid domes (metal or strong plastic) are often quite shallow in profile, whereas the softer examples need to use a steeper, more conical shape to maintain sufficient structural rigidity and hence extend the working frequency range. However, even with a 'soft' dome, if the profile is too steep, the breakup mode may be sufficiently severe to produce a response peak up to 4 dB high.

Heavy well-damped domes may have a surprisingly restricted response, and one well-known example which is widely employed in European designs possesses a gently falling characteristic above 3.5 kHz. Others, by fortuitous choice of profile and material, may be extended by a further two octaves to approximately 10 kHz.

There is nevertheless one appealing characteristic possessed by good mid-range dome units. Their subjective sound may be entirely free of the particular 'edgy' hardness exhibited to some degree by most cone units when operating in the mid range and, in this respect, domes bear comparison with the film-type transducers such as the electrostatic. This largely accounts for the determined efforts on the part of many manufacturers to incorporate such units into their system designs.

Cones (MF)

A coned driver, if of moderate size, may be designed to cover the entire LF/MF range to a high standard. Such a unit may also prove a good choice for mid-range use only. A number of recent three-way high performance systems have utilized mid-range drivers which are also capable of bass reproduction in suitable enclosures. By definition, these units have a wide bandwidth, this being advantageous when selecting the optimum crossover point. Coned LF/MF drivers ranging in size from 100 mm to 200 mm have been used, with a separate sub-enclosure to isolate the driver from the back air pressure generated by the LF unit. Specialized mid units tend to be on the small side (typically 80 mm piston radius) since this provides sufficient output down to 250 Hz or so and yet presents a small enough source for satisfactory directivity at the high frequency end of the range. Models may be fitted with an integral enclosure or, more simply, supplied with a sealed chassis back, the space behind the diaphragm fitted with suitable acoustic absorbent. A typical response curve can be uniform from 250 Hz to 6 kHz.

The theory which concerns controlled smooth transitions from one vibrational mode to another is especially relevant in the case of mid-range diaphragms, since they generally operate in piston mode in the lower range and in controlled breakup in the upper band, with the broad region dividing the two generally appearing between 1 kHz and 2 kHz. It is thus essential that the diaphragm be well terminated at its boundaries, and that the diaphragm be well terminated at its boundaries, and that the shape and choice of material be conducive to good transient behaviour, free of significant delayed resonance. Shallow flared profiles have given quite good results especially in thermoplastic cones.

A new structure is being used by the British company of Bowers and Wilkins in the production of a mid-range unit that incorporates may of the above design principles. The cone is fairly rigid, formed from an open weave Kevlar fibre coated with resins which also serve to stiffen the structure.

There is a final point worthy of consideration when choosing a cone material. In a practical driver there is often some maximum sound pressure level above which an audible deterioration in quality occurs. This effect cannot be ascribed to the magnet, suspension or crossover non-linearity and is believed to result from a gross compression at the neck of the cone due to the applied force exceeding the elastic limit of the material. Operation in such regions means that the material fails to return on its original dimensions for some time (ranging from milliseconds to hours), and the resulting distortion, being of the hysteresis type, is particularly unpleasant.

It appears that some plastics chosen for suitable acoustic properties for cone manufacture suffer from this effect more than others, and it may well be an important consideration if high sound pressure levels are required.

Experimentation is continuing in the design of metal cones. Good results have been achieved with a 110 mm LF–MF unit with an anodized, spun aluminium alloy cone. The first breakup appears at 12 kHz; the diaphragm mass alone is 4 g. In another example a 170 mm cone was pressed out of ductile sheet between two formers. This shows a first 'break' at 6 kHz. These resonant modes are generally severe when they occur due to the very high Q of metals. The main reason for continuing perseverance with the metal technology lies in the particular clarity of the reproduced sound. Subjectively, this does match the fine quality of good metal piston HF units rather better that the usual pulp or soft polymer cones. Standard LF–MF diaphragms generally possess controlled breakup modes in their working range and inevitably add sounds characteristic of the particular material employed, a sonic signature of the chosen materials technology.

Suspensions (MF)

The suspension is usually integral with the diaphragm in dome units, a doped, half-roll surround being formed at the perimeter.

Rocking is one unwanted mode of vibration which may be troublesome with single suspension designs. Soft dome assemblies are noted for this defect which results in sub-harmonics and increased intermodulation distortion. The mass of the assembly plus the shape and physical constraints of the surround are critical factors here. One obvious solution is the adoption of a double suspension. However, even two moderately spaced suspensions take up considerable depth and necessitate a long motor-coil former, which may be disadvantageous in a dome radiator. The resulting assembly is difficult to produce and is generally restricted to the more expensive designs such as the German unit illustrated in Figure 5.11.

(a)

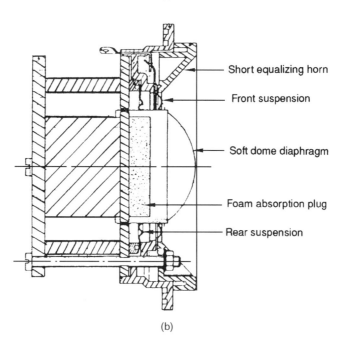

(b)

Figure 5.11. (a) 50 mm diameter double-suspension, fabric dome mid unit. (b) Cross-section of (a) (courtesy SEL/ITT)

Another solution could be to employ ferrofluid in the magnet gap, providing a centring contribution and helping to suppress the rocking mode.

In the case of cone diaphragms, the suspensions are usually manufactured from conventional corrugated fabric, but due regard must be paid to the possibility of any self-resonance which may be audible in the mid range.

Surround (MF)

The choice of a surround material presents a considerable problem for a designer of a unit covering mid and high frequencies. The commonly used material — neoprene, p.v.c., etc. — have mechanical properties which vary greatly with frequency. Ideally, the energy absorption property should be resistive and constant, but invariably the elastic rubber-type materials are hysteretic, and show a memory effect, i.e. a time lag exists between deformation and subsequent recovery. Consequently, with increasing frequency, these materials stiffen and become less absorptive. To date a lightly impregnated grade of foam plastic has proved most successful at the higher frequencies.

Since little excursion is required of a mid-range unit the surround profile may be a simple flat strip, no half-roll or similar device being strictly necessary.

Motor Systems (MF)

The magnet structures of cone and dome mid units present a great contrast in terms of their size; whereas a scaled-down assembly of typically 19 mm diameter is sufficient for the former, the dome requires a massive structure energizing a 50 mm or even larger pole.

Ideally the magnet diameter for a mid-range cone should be small, so that the minimum reflecting surface is present immediately behind the cone. This suggests the use of an Alnico or neodymium magnet rather than the ceramic 'pancake' type. Too many mid units have inadequate acoustic clearance behind the diaphragm, and the resulting reflections are clearly audible.

In an effort to lower the fundamental resonance of dome designs, the pole is often centre bored; ideally this assumes a tapered form to reduce pipe standing waves, with the aperture often leading to a sealed rear chamber filled with absorptive material or a line.

The use of ferrofluid may prove advantageous where the short-term power input is high and the cone excursion small. Large increases in power handling are possible with its use, up to two or three times in the short term (for example, 1.5 min).

Where very low distortion levels are required, secondary harmonic sources such as eddy currents in the poles must be eliminated. Suitable remedies include copper plating, lamination of the pole structures or the use of a high resistivity magnetic alloy for the poles themselves. Again the use of ferrofluid may offer some reduction in distortion.

The necessity for predictable and controlled breakup modes in mid-range diaphragm assemblies means that the location and method of attaching the lead out wires to the motor coil must be carefully considered. Ideally the exits should be symmetrical, preferably at the motor coil former rather than haphazardly stuck to the cone as is frequently the case. The latter order promotes asymmetrical breakup and high rocking modes, especially at the higher frequencies.

5.5   HIGH FREQUENCY UNITS

The general classification for HF units embraces a 1–30 kHz range, overlapping the upper mid band by an octave or so. The total range covers five octaves, which is virtually impossible for a single unit to achieve satisfactorily, and hence the classification is loosely split into low range (1–10 kHz), full range (3–18 kHz) and high range or super treble (8–30 kHz).

Conventional cone diaphragms are rarely used for this frequency band in high performance systems as this range operates a cone in the higher breakup region where the output is falling and is generally both irregular and unpredictable, although some cone/dome hybrids have enjoyed success in the lower cost sector.

Dome Diaphragms (HF)

Domes are undoubtedly the most common form of HF diaphragm and are available in a wide variety of shape, size and material. Additionally, the chassis and front plate structure may be acoustically tailored to provide equalization of both polar and/or axial response characteristics.

The 25 mm soft fabric dome first became popular in Europe and its use has since spread throughout the world, with more then 30 types now available from various manufacturers (Figure 5.12). (Soft dome HF units are also produced in the 34 mm and 19 mm sizes.)

Plastics are also used for the production of dome units, notably polyester, often in a laminated form bonded to a p.v.c. damping layer. Sizes 19 mm, 25 mm and 38 mm are all available with fundamental resonant frequencies ranging from 600 Hz to 2 kHz, and upper cutoff frequencies from 25 kHz down to 15 kHz.

Yamaha have produced a beryllium HF unit of 30 mm nominal diameter, whose first breakup mode is beyond 30 kHz, ensuring virtual piston operation over the entire 2–18 kHz usable band. The diaphragm thickness is 30 μm with a mass of 30 mg which compares with a soft dome counterpart at 100 mg and similar thickness. The rigidity of beryllium is too high to employ an integral suspension and instead a separate, tangentially pleated cloth surround is used, with a damping coating composed of two resins to dissipate energy at the rim. In this latter design the surround unfortunately forms a significant part of the radiating area, disturbing the upper band output (see Figure 5.13, surround moving in antiphase at 20 kHz). In the case of a more recent 37 mm metal dome driver (Celestion) care was taken to minimize the surround contribution. In this unit the diaphragm was electroformed in pure copper and the structure continued from the same section to provide an integral motor-coil former. This one-piece construction has the advantage of allowing the whole dome to act as a heat dissipator, a short-term rating of 50 W has been quoted. The unit exhibits a rather low sensitivity, of typically 82 dB/W, due to circulating eddy currents. Even with a material so unfavourable as copper (chosen mainly for its good electroforming

Figure 5.12.   25 mm 'soft dome' HF unit (courtesy Son Audax)

properties) the final breakup mode is held to just above 20 kHz. The Q is rather high, between 20 and 40 and in the commercial system a notch filter was fitted to the crossover network.

From the materials (Table 5.2), it is worth noting that amongst the more common metals of good conductivity, aluminium is a better contender than copper, and a one-piece diaphragm can be produced by precision deep drawing. Hard anodizing provides a final improvement, a good finish and an electrically insulating layer.

In recent years, metal dome high frequency units have become increasingly popular, with the 25 mm size as the norm. 32 mm and 19 mm types are also in production, but the larger sizes are quite expensive. Sensitivities of 88 dB W are typical for an 8 Ω unit. The dome is generally of hard anodized aluminium alloy foil, often with a 5% magnesium content, while titanium is also a popular choice. For the 25 mm size a first resonance in the 22–27 kHz range is typical, rising by some 10–20 dB. This first mode is often associated with the stiffness seen at the rim and the quality of bonding to the suspension and coil former in this area, rather than to the expected 'oil can' reflexing of the dome centre. With a resonant Q of this degree, the main peak is preceded by a dip of 4–6 dB in

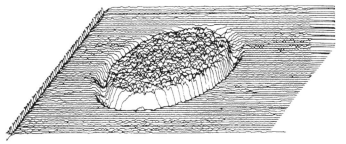

Figure 5.13.   25 mm ultra-hard dome treble unit; $f = 20$ kHz, surround in antiphase (after Bank and Hathaway)

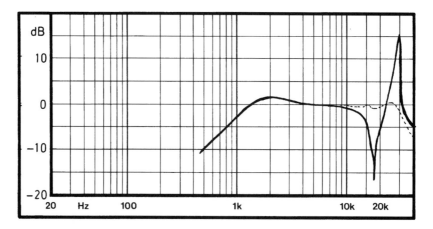

Figure 5.14. Natural response of a metal dome (25 mm); note the downwards drift, the in-band notch preceding the normally ultrasonic high Q resonance. The dotted response shown may be achieved by adding a contribution from the surround–suspension annulus and by the use of a correcting phase plate or a small reflector placed in proximity to the dome apex

the 18–22 kHz range making it difficult to deliver a flat response to 20 kHz. In any case the natural response of a dome unit in the piston range is a falling one (Figure 5.14 and Figure 3.2).

Some degree of equalization is often required in the matching crossover network while the transducer designer might resort to some acoustic tricks to level the response. Often a central baffle disc of 5 mm diameter can be used* to suppress the main peak and help fill in the notch. Meanwhile the usual half-roll surround, commonly of polyamide, may also be allowed to act as a weak annular radiator at high frequencies also helping to fill in the notch. The latter artifice is fairly harmless, since pure piston operation is still available over the fundamental 2–15 kHz range. However, the use of phase discs* or plates often induces cavity 'ringing' effects which may well be audible as a 'tizzy' grain, particularly off-axis. The main virtue of a pistonic transducer is its good clarity and harmonic purity, with an absence of the 'tizz' or 'grain' which is often audible in the upper range of soft dome drivers.

While a fabric dome often possesses a quite well-damped fundamental resonance due to the applied tacky compound, the plastic domes may show quite a high Q, as much as 10 at resonance. Various means may be used to control the Q ranging from applied damping to acoustic anti-resonant circuits formed by venting the rear cavity behind the dome into various resistively damped chambers present in the magnet structure.

In the upper range above 8 kHz or so, the plastic domes may suffer from breakup problems, either due to lack of rigidity or resonances in the surround. One example possesses a surround resonance near 10 kHz, which causes a marked rise in second harmonic distortion. This may be cured by appropriate damping on the inside of the suspension. Where the dome itself is in resonance a doping layer can be applied to the centre, or alternatively a small plug of polyurethane foam placed in contact with the offending area. Other techniques include the juxtaposition of a so-called phase

*Just in front of the dome.

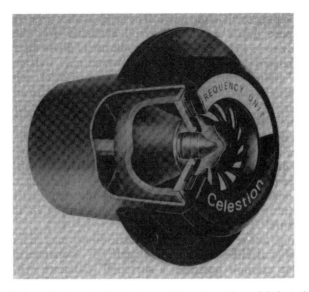

Figure 5.15. Rigid phenolic doped fabric dome HF unit, with sophisticated acoustic loading (from cavity resonator, rear energy-coupled resonator, diaphragm damping)

correction plate which may block, direct or delay the output. This works over limited frequency bands, controlling the radiation from specific areas of the dome to smooth the integrated far-field response.

An unusual and highly successful example of a hard dome HF unit is undoubtedly the HF1300 family, manufactured by Celestion (UK) (Figure 5.15). Variations of this design have been in production since the mid 1950s, and have been used in a number of systems. Almost all of the details of its construction have proved to be critical, these range from the particular grade of cured phenolic impregnated fabric used for the diaphragm, to the spacing of the phase correction plate. The centre dome is conical, about 19 mm in diameter, and has a shallow, rather broad surround actually larger in area than the dome itself. The unit as a whole has an overall diameter of 38 mm. The surround is in fact the main radiating element; piston operation holds to beyond 15 kHz and the diaphragm is particularly free of hysteresis effects. This, in conjunction with its relatively uniform axial frequency response, accounts for its unusually favourable subjective qualities.

The diaphragm's intrinsic pressure characteristic shows a peak at fundamental resonance (1.7 kHz) which is damped by resistively controlled air vents to the cavity within the enclosed pot magnet. The centre pole has a conical profile to closely follow the contour of the underside of the dome, this placing the first rear standing wave mode at a very high frequency. The output of the naked diaphragm falls above 7 kHz or so, and to correct this a perforated front plate is fitted, formed so as to follow a similar contour to that of the diaphragm. This results in a damped resonant cavity loading the diaphragm, and also provides a delay path between the dome and surround radiation. The output is thus uniformly maintained on axis to 14 kHz. Good examples may demonstrate a ±2 dB characteristic from 2 kHz to 14 kHz, with a range 4–12 kHz held within ±1 dB limits.

Motor Systems (HF)

With suitable coil winding techniques, the larger treble units with 30 mm centre poles and over will have power ratings of 8–15 W corresponding to a system rating of the order of 100 W on continuous programme. The smaller 19 mm motor coils may have ratings below 8 W, and these are often provided with fuse protection in the accompanying speaker system to guard against high-level, high frequency drive which may occur during tape spooling or amplifier instability.

As has been mentioned in connection with the other drivers, the application of ferrofluid to the magnet gap will offer considerable protection against thermal overload and a viscosity grade may be chosen which will also provide damping of the fundamental resonance and control of rocking, the latter proving a problem with HF as well as MF dome units. At the design stage the significant variation in ferrofluid viscosity with gap temperature, often up to 40°C, must not be forgotten if such damping forms part of a crossover analysis.

When high acoustic levels are required of dome mid and LF units their relatively low efficiency necessitates very large magnets and these can prove costly as a result. Where a moderate quality deterioration is tolerable, horn loaded treble or mid domes are employed. The horn offers control of directivity and gives a better acoustic match between the air load and the diaphragm. Using phenolic or aluminium domes with horn loading, efficiencies of up to 20% are possible, contrasting with the 1% to 2% typical of simple direct radiator dome units.

5.6 FULL-RANGE UNITS

A few examples of full-range moving coil drivers are in manufacture, and although at present they do not completely satisfy the 'high performance' criteria, certain of the techniques involved are worthy of note.

Full-range Cone Driver

Jordan designed an interesting example of an aluminium cone unit which was first produced in the mid 1960s. The diaphragm possesses a hyperbolic flare with a centre dome stiffener and a matched, viscous treated plastic foam surround. The cone diameter is 100 mm which was considered to be an optimum compromise between LF radiating area (the unit was intended for use in domestic applications) and adequate dispersion at the higher frequencies. It certainly achieved most of the designer's aims and although not produced in very large quantities was undoubtedly a commercial success. It demonstrated at a fairly early stage the value of a flared cone profile and good termination, plus the effectiveness of a high linearity magnet system and non-resonant chassis.

In Germany (Manger Systems*) another development concerns the design of a full-range driver (Figure 5.16(a), (b)). This employs a flexible diaphragm engineered so that

*Development by Professor Manger.

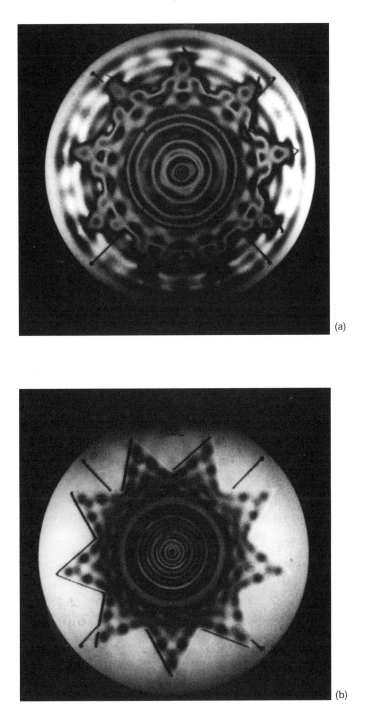

Figure 5.16. Manger diaphragm modal vibration at (a) 3 kHz, medium source size, and (b) at 7 kHz showing reducing radiating area

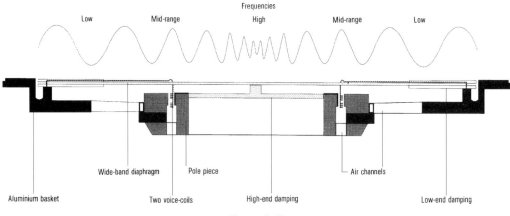

Figure 5.17.

the mechanical load presented to the motor coil is almost wholly resistive and independent of frequency. The moving structure consists of a pre-loaded flat web of synthetic fibre (polyamide/nylon group) impregnated with an air drying visco-elastic coating, probably of the PVA type (Figure 5.17).

A split motor coil, using differential drive, is employed with optional the injection of an offset current to allow centring of the coil in the magnet gap. A star-shaped plastic foam section acts as a supporting and stiffening component. With a fundamental resonance partly dependent on the drive amplitude (20–40 Hz), the unit is nominally flat to 15 kHz. The maximum displacement was quoted at 3 mm and the sensitivity as 3.2 W for 96 dB s.p.l. at 1 m. At low frequencies the full diaphragm area moves, but with increasing frequency the radiating area contracts smoothly toward the centre with resistive control, resulting in the small source required for good high-range performance [6].

In another example the multiple driver Bose 900 series loudspeakers use a number of small full-range 'modular' drive units. Here a special electronic equalizer is employed before the power amplifier to provide a uniform frequency characteristic.

A Full-range Moving Coil Panel Transducer

The isodynamic principle as used by Magneplanar, Apogee and others, centres on the use of a planar film diaphragm driven over its surface by a distributed motor. This comprises an open array of fixed magnets and a zig-zag or similar current path. The latter may be realized as a deposited or laminated metallic pattern or even by bonded round wire conductors (Magneplanar). The idea is to drive the film relatively uniformly over its surface to reduce the majority of resonance effects.

In the case of the Aria speaker by Sumo, designer Paul Burton [11], he has chosen to drive a large rectangular mylar polyester film panel, quite tightly stretched like a drumskin, from a single, near central point. This is defined by a low-mass moving coil motor. Intentionally, the system operates as a moving plate at low frequencies, constrained by diaphragm tension. With rising frequency the outer areas become progressively decoupled relative to the central region. Thus, with careful design the axial output can be made fairly linear with frequency and shows a progressive desirable reduction in radiating area with increasing frequency. Thus, a good directivity is maintained over a wide frequency range. In the final octave the residual acoustic source comprises the motor coil and its piston-like metal cone centre cap. Low frequency modes are partly controlled by acoustic resistance panels on the semi-open back of this floor standing dipole design, while the necessity to avoid a corrugated type of rear suspension or spider has generated a design solution in the form of a long magnet gap, short coil and a selected ferrofluid filling which performs three functions: a fluid bearing for coil centration, mechanical damping and cooling.

The magnet system is very powerful, mounted on strong rear crossbeams to help maintain good alignment. A sensitivity of around 85 dB/W is achieved. The motor details are interesting; to achieve a wide electrical bandwidth and a good electromechanical response to the higher frequencies the motor coil inductance is held to 0.1 mH (out of the gap) with just 45 turns on a 33 mm former, coil height 15 mm. The excursion limit is ±10 mm and a copper capped pole and aluminium eddy current damping ring are used in the magnet. A 200 W peak programme capacity is claimed while the complete coil assembly including centre cap weighs only 2.2 g. Whatever residual problems such design may have, for example centration, and rocking modes at some frequencies, the design has the fundamental qualities of simplicity, a nine octave bandwidth achieved without the need for multiple sources and without crossovers or equalizers.

The single motor coil connected directly to the amplifier and the system thus has a potential for good clarity and dynamic performance.

ICT — The Inductively Coupled Transducer

During the design of a one-piece metal dome tweeter with integral coil former, transducer engineer Elei Boaz investigated the complex effects of the shorted turn, represented by the continuous conductive element. Some electrical power is lost in the shorted turn, while some of the benefits conferred include control of damping at fundamental resonance and a reduction in magnetically induced distortion. He conceived the idea of a full-range concentric driver where the HF element was simply a one-piece metal dome with integral former, placed over the centre pole and sharing the magnetic gap of an LF–MF unit (see Figure 5.18 (a) and (b)). The theory describes the inductive coupling between the motor coil of the main driver and the single or shorted turn represented by the HF element with the potential for designing a full range, two-element transducer operating with direct coupling to the amplifier and without the need for any crossover or filter network [10]. As with other concentrics, problems arise at higher frequencies due to the acoustic mismatch between the dome radiator and the immediate surroundings, resulting in variations in output in the upper frequency range. A simple phase ring is helpful, but a more complete solution can be found in the form of a phase correcting plug placed over the dome radiator, similar to that used in high frequency horns, thus providing the appropriate delay paths.

Simple elastic sleeve suspensions are effective for the dome which is built as a close fit over the centre pole. The properties of the suspension, its damping and stiffness, define

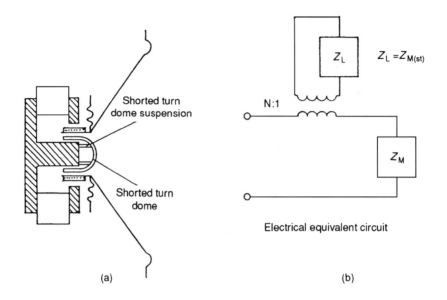

Figure 5.18. (a) The ICT or inductively coupled transducer as a two-way coaxial unit. (b) The construction and simplified equivalent circuit (after Boaz [10])

the mechanical crossover point for this radiating element. Established techniques, i.e. cone flare, mass, and attachment to the motor coil are exploited for the lower range diaphragm in order to design an appropriate low-pass rolloff for the bass mid unit.

Among the known advantages of the ICT design are the absence of any power handling limit for the HF section which is often the weakest link in a multi-unit arrangement, and reduced system complexity. High quality drivers for automobile use are now available, but additional development will be required for high fidelity application (see Figure 5.19).

Dual Concentrics

Variations of dual concentric drivers have been produced for a very long time and for many reasons, not least the convenience of a single frame, full-range device. A high frequency driver mounted concentrically with a larger low frequency section confers many benefits, especially a symmetrical off-axis directivity (Figure 5.20). Those awkward off-axis lobes present in the crossover region which are common with spaced multi-driver system can be avoided. Easing crossover design and enhancing performance, steps can be taken to align the effective radiating planes for virtual time coincidence. In the case of a centre horn type such as the Tannoy, this may be achieved by electronic delay to the high frequency section (see Figure 5.21).

Thanks to the development of a high performance magnet material using neodymium it has proved possible to miniaturize the magnet assembly for a fairly standard 25 or 32 mm diameter dome HF unit to the point where it may be placed on the centre pole of the conventional larger cone driver (Figure 5.22). Provided care is taken over the gap

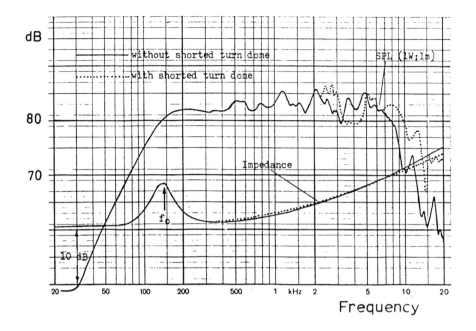

Figure 5.19. The practical performance of an ICT unit comparing the results with and without the dome in place (after Boaz [10])

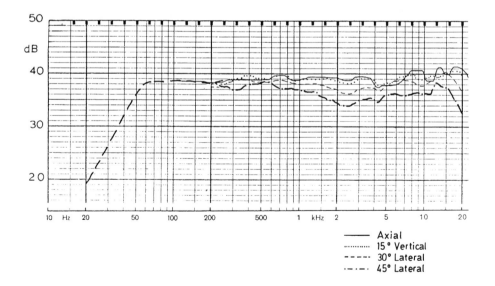

Figure 5.20. A example of the beneficially uniform off-axis responses achieved by a concentric system where a miniaturized HF unit is located on the open pole face of the LF–MF unit (UNI-Q, model C95) (see Figure 5.21)

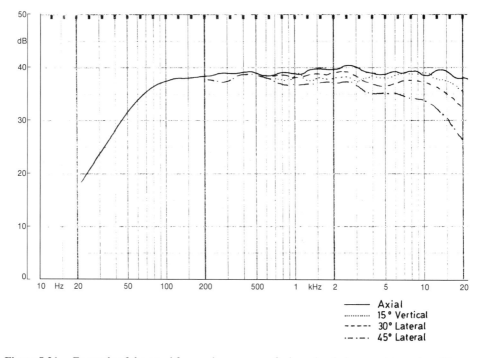

Figure 5.21. Example of the good forward responses of a horn loaded concentric system (the rear mounted piston HF unit employs the main diaphragm as a horn driven via apertures in the centre pole). The responses are for 2 m measuring distance with a crossover at 1 kHz (Tannoy DC 1000)

tolerances and the adjacency to the start of the cone flare, fine results can be achieved despite some axi-symmetric response ripple above 10 kHz (significantly ameliorated by the time the listener is 10° off-axis in the preferred direction). With careful design the time alignment is very close, significantly aiding in the design of the crossover; the off-axis frequency characteristics are predictably well behaved in consequence.

To date the finest exposition of the principle has been in a four-way speaker design where the concentric driver operated above 800 Hz and is not called upon to provide significant cone displacement. Some doubt exists concerning the viability of 200 mm drivers and also smaller concentric types for full-range duty since significant intermodulation products can be produced between the LF and HF ranges.

Substantial LF power input generates a displacement of the main diaphragm in a dual concentric which in turn modulates the acoustic load presented to the HF dome. Frequency and amplitude modulated components can be generated at audibly higher levels than with a conventional, equivalent, separated-driver system.

One claimed benefit of the coincident driver approach is the high degree of control and uniformity of on- and off-axis amplitude–frequency responses achieved by virtue of a good crossover performance as an acoustic target function. A floor mounted system using the UNI-Q (KEF ®) principle was measured in one-third octave bands at 2 m in an anechoic chamber for the axial, 15° vertical, 30° lateral and 45° lateral axes. The resulting graphs (Figure 5.23) illustrate the remarkable correspondence between the on- and off-axis responses and their overall conformity. Such forward lobe consistency

Figure 5.22. A UNI-Q driver, a full-range coaxial with a miniature dome HF unit mounted on the centre pole. The effective sound sources are coincident

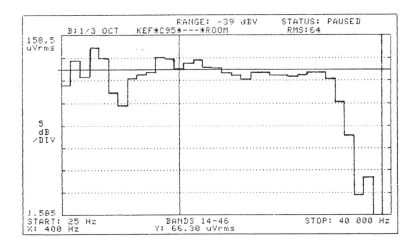

Figure 5.23. In-room averaged response for the floor-mounted system in Figure 5.19. Note the good correlation shown between the superior off-axis performance and the uniform in-room characteristic, particularly in the 3 kHz crossover region

must aid perceived stereo focus by markedly reducing the incidence of amplitude and phase differences normally present for a stereo pair of speakers as heard by a normally seated listener.

Examining the RAR or multiple room averaged response, the typical mid-bass anomaly is due to interference from the prime, floor reflected mode, while at higher frequencies the smooth frequency characteristic confirms the uniform nature of the output in the forward direction of a well-designed concentric system.

5.7   DYNAMICS AND ENGINEERING

The term 'dynamics' is used in a general sense to characterize certain aspects of sound quality and relates to a feeling of liveliness in the reproduced sound. The exciting attack and speed of natural acoustic instruments, percussive transients, the feeling that the loudest parts of a programme section are transduced without audible limiting, distortion or compression, are pertinent. The term 'micro dynamics' is also used where lower, quieter transient and percussive sounds still retain that sense of attack and presence.

All reproducers show some loss in dynamic power and expression compared with the original sound, and frequently renewed acquaintance with live sound is always a salutatory experience for a sound engineer, clearly illustrating the gap that remains between live and reproduced music. A number of aspects of loudspeaker engineering, both obvious and subtle, affect dynamics and there are also more logical associations with other aspects of sound quality such as frequency response and energy balance. For example, a false emphasis in the lower treble may well improve the subjective rating for dynamics but this is likely to be achieved at the expense of false timbre for broader range sounds, and increased listener fatigue is also likely.

Amplifier design principles indicate that dynamics may be comprised when one quantity controls or modulates another. For example, heavy bass transients may cause compression and or loss of detail in the mid and treble registers. Loudspeakers also suffer from equivalent effects and these are examined in more detail.

Limitation of Dynamics at High Power and/or with High Cone Excursion

Modulation of motor coil inductance with position in the magnetic gap

In addition to the inherent non-linearities present in the inductance characteristic of the motor coil, its overall value is also subject to variation according to the length of the coil immersed, and thus flux linked, to the magnetically permeable pole system. With large cone excursions there is a substantial variation in inductance; bass energy thus modulates the mid range by modulation of the inductance and consequently the termination the driver presents to the crossover network. Solutions include the use of an overlong, high inductance coil where the proportion changing with bass excursion is reduced. The theoretically favoured solution is a short coil operating in a long magnetic gap, but this has been a disproportionately expensive procedure. Even at small excursions a component of the non-linearity in coil inductance results from the eddy currents induced in the iron parts of the magnetic circuit. Remembering that the moving coil motor principle includes the equivalent of an audio frequency transformer, the latter would normally be built with a finely laminated core to minimize eddy currents — quite a contrast to the thick components of a magnet system. It is possible to minimize the eddy currents by placing elements of high conductivity, made from aluminium or copper, in or near to the gap. These preferentially draw in the eddy current fields producing a linear back e.m.f. In addition, the effective inductance of the coil is beneficially reduced. The reduction in distortion is audible both as improved timbre and superior dynamics, the greatest benefit is in the third harmonic which can improve by a factor of two or three.

The passive crossover benefits from a reduced and more constant motor coil inductance, stabilizing the performance of the filter and thus minimizing the modulation of the mid-band filter alignment by the low frequency variations.

Modulation effects due to common baffle

Another source of modulation between two frequency ranges is a result of the local acoustic load for one driver altering owing to the proximity of another driver. For example, a tweeter placed in proximity to a woofer sees the woofer surface as part of its acoustic load. When the woofer is under heavy excursion, part of the acoustic boundary from which the treble energy is launched is now moving, thus modulating the output of the tweeter. This is most serious in the case of concentric drivers where the lower frequency diaphragm defines a very high proportion of the baffle or local acoustic load for the treble unit. Specifically, the acoustic impedance of the radiation load on the treble unit is modulated in both amplitude and frequency. With moderate size, full-range concentrics this is an audible factor in reducing dynamic accuracy, and we can also include the secondary problem at the highest frequency where the physical blend from the edge of the tweeter to the cone flare may result in a step of variable height under bass excursion, resulting in short-term irregularities of typically 10 dB peak-to-peak. Thus the most successful examples of coincident concentric drivers [Uni-Q ™] are used in three-way or more systems where cone excursion is minimized for the upper range section of the design.

Effect of flux density on modulation and dynamics

As a general rule higher strength magnet systems and higher efficiency tend to maximize dynamics. For moderate excursion signals the voice coil non-linearities are reduced in proportion to the ratio of fixed to varying flux. The first is defined by the magnet, the second by the audio current in the coil. Viewed from an alternative viewpoint, it can be shown that the distortion resulting under small coil excursion is strongly dependent on current. This explains why low sensitivity speaker systems, which draw relatively high currents, suffer greater distortion than high sensitivity systems which require much less current to produce an equivalent sound level. Associating linearity with dynamics, it is obvious that great care is required in the design of small, full-range systems if they are to retain good dynamics. In general, larger systems do have better perceived dynamics even at moderate sound levels.

Effect of bass alignment, bass mid-range drivers

Low frequency alignments which tend to reduce cone excursion provide a payback in improved dynamics provided that they are executed without significant loss in transient response control or an increase in colouration. For example, bass reflex loading can be tailored to minimize cone movement in this region of maximum bass energy for a typical programme, considering both level frequency and the typical time history of the programme. If the cone generally moves less, and excepting an occasional lapse due to unusual momentary peaks perhaps at very low frequencies, then from a statistical viewpoint modulation effects are minimized and the dynamic quality of the mid range is enhanced.

Doppler modulation

Even if a driver has a perfectly linear motor system large excursions may increase Doppler or frequency modulation to significant levels. Historic analyses have been restricted to fairly modest sound levels (see [12]) or have relied on pure tone analysis for single-channel or monaural conditions. A more critical view is required when stereo sources are taken into account at more typical higher sound levels. The aural sensitivity to several of the more subtle distortion parameters is certainly substantially higher when using high quality stereo programme and experienced listeners.

Low frequency misalignment

The consequences of position-dependent inductance for the motor coil has been noted above. Of similar importance is the variation of fundamental driver parameters with power, e.g. Thiele–Small.* These are not constant with power. T-S frequency parameters are generally measured and specified at low power levels of 1 W or less. Typically the classic low frequency design of a system is referenced to these parameters.

   Unfortunately, these values may vary dramatically with power and frequency. In consequence, under heavy drive, and depending on the thermal history of the proceeding programme section, the low frequency alignment may be severely altered. Direct consequences are an uneven and ill-controlled low frequency range and a change in tonal balance between bass and mid range. For a given driver, increasing excursion results in non-linear stretching of the suspension and surround thereby increasing the effective fundamental resonance. At the same time the *Bl* factor is reduced as less of the coil is fully coupled to the available flux and this increases the Q. For the box itself, the smaller domestic enclosures, if reflexed, have comparatively small ports which are only linear at relatively low power levels of up to 90 dB. Increasing power beyond this point causes turbulence which reduces the active area and increases damping and, *in extremis*, the port output reduces near to zero. Considering the low frequency dynamics, these mechanisms result in a modulation of damping, of response uniformity and of extension with level. It could be argued that a designer has to consider a trade-off between the sealed box, and its greater tolerance of low frequency misalignment, and

*T-S.

the reflex which has a greater sensitivity to misalignment but has the potential for superior mid-range dynamics. This discussion logically leads to the design of larger speakers of three-way design with a sealed-box low frequency section and an independent mid range.

Temperature effects

After prolonged use at high power motor coils will reach high enough temperatures to materially increase their electrical resistance. In a multi-way system in particular, the result can be an imbalance between the frequency ranges of each driver due to differential sensitivity changes. There will also be some attendant mistermination of the crossover filters.

Note also that motor-coil resistance is a major factor in the value of driver Q and the low frequency alignment of a speaker will change significantly when the bass driver has reached a higher coil temperature. These changes constitute longer term dynamic effects.

Mechanical vibration

Moving coil drivers generate substantial accelerations at their mounting points through magnet reaction. Vibrational energy can be coupled to the enclosure and also cause the driver frames to resonate. Where several drivers share a baffle the vibration may be coupled from one to another, this constituting a modulation of position of one frequency range by another. Subjectively there is a loss of clarity and definition; transients are blurred and there is a loss in dynamic quality. Where a tweeter is affected by the mid-range unit the treble sound may be roughened, invoking the word 'grain' and with an effect similar to excessive jitter in a digital audio interface.

Remedies include local reinforcement of the baffle, while rigid non-resonant driver chassis or frames are helpful, particularly with more generous provision of mounting points, e.g. from three or four, to six or eight points. Some designers have exploited separate baffle or enclosure designs to combat this problem (see SPL (sound pressure level) and see SBL in Figure 4.17).

REFERENCES

[1] Harwood, H. D., 'New BBC monitoring loudspeaker', *Wireless World*, March, April and May (1968)
[2] Harwood, H. D., 'Loudspeak distortion associated with LF signals', *J. Audio Engng Soc.*, **20**, No. 9 (1972)
[3] King, J., 'Loudspeaker voice coils', *J Audio Engng Soc.*, **18**, No. 1, 34–43 (1970)
[4] Ferrofluidics Corporation, Massachusetts, USA, Leaflet, *Ferrofluidics*
[5] Yuasa, Y. and Greenberg, S., 'The beryllium dome diaphragm', *Proc. Audio Engrs. Soc. 52nd Convention*, October–November (1975)
[6] Pfau, E., 'Ein Neuer Dynamischer Lautsprecher mit extrem nachgeibiger Membran', *Funkshau*, March (1974) (Also Manger — inventor)
[7] Nakazono, J. *et al.*, 'Coaxial flat plane loudspeaker with polymer graphite honeycomb sandwich plate diaphragm', *J. Audio Engng Soc.*, **29**, No. 11 (1981)

[8]   Tsukagosh, T. *et al.*, 'Polymer graphite composite loudspeaker diaphragms', *J. Audio Engng Soc.*, **29**, No. 10 (1981)

[9]   Mellilo, L. and Raj, K., 'Ferro-fluid as a means of controlling woofer design parameters', *J. Audio Engng Soc.*, **29**, No. 3 (1981)

[10]   Boaz, E., 'The application of an inductively coupled shorted turn and the dual coil loudspeaker system', *AES Reprint*, No. 2548 (G-2), *83rd AES Convention* (1987)

[11]   Butler, T., 'Tailored by Burton', *Hi Fi News*, **34**, No. 6 (1989)

[12]   Allison, R. and Villchur, E. 'On the magnitude and credibility of FM distortion in loudspeakers' *JAES*, **30**, No. 10 (1982)

BIBLIOGRAPHY

Beranek, L., *Acoustics*, McGraw-Hill, London   (1954)

Briggs, G. A., *More About Loudspeakers*, Wharfedale Wireless Works, Idle, Yorkshire (1963)

Cohen, A. B., *Hi Fi Loudspeakers and Enclosures*, Newnes-Butterworth, London (1975)

Gilliom, J. R., Boliver, P. and Boliver, L., 'Design problems of high level cone loudspeakers', *J. Audio Engng Soc.*, **25**, No. 5 (1977)

Ishiwatari, K., Sakamoto, N., Kawabata, H., Takeuchi, H. and Shimuzu, T., 'Use of boron for HF dome loudspeakers', *J. Audio Engng Soc.*, **26**, No. 4 (1978)

Jordan, E. J., *Loudspeakers*, London (1963)

KEF Electronics Ltd., *You and Your Loudspeaker*, KEF Electronics Ltd. (c. 1970)

National Panasonic, *The Technics SB1000 High Linearity Loudspeaker*, Technics Promotional Leaflet

Rice, C. W. and Kellog, E. W., 'Notes on the development of a new type of hornless loudspeaker' (1924). Reprinted in *J. Audio Engng Soc.*, **30**, Nos. 7/8 (1982)

Yamamoto, T. *et al.*, 'High fidelity loudspeakers with boronised titanium diaphragms', *Audio Engng Soc. 63rd Convention* (1979). Also *Loudspeaker*, Vol. 2, Audio Engineering Society

6

Systems and Crossovers

Previous chapters have shown that a single diaphragm driver generally cannot easily meet the standard implied by the phrase 'high performance'. Unfortunately, the need for a large area to give effective low frequency reproduction conflicts with the very small diaphragm necessary for satisfactory HF performance. In consequence, the high performance loudspeaker is invariably a 'system' which, in its simplest form, consists of an enclosure of defined acoustic properties plus two or more specialized drivers and an electrical filter. The latter directs the correct frequency range into the appropriate drivers and is termed the crossover network (Figure 6.1). In advanced systems it may also be responsible for other functions such as attentuation and equalization.

Certain benefits result from this division of the working frequency range. All modulation distortions are considerably reduced, particularly FM,* the latter produced by the physical movement of a low frequency diaphragm whilst simultaneously reproducing a higher frequency. Some residual FM will remain in all systems except for those special cases where bass horns or considerably spaced drivers are employed, as the low frequency unit will still occupy a proportion of the enclosure surface. Some energy from the higher frequency units will be incident on the LF diaphragm and will undergo modulating excursion as a result or, viewed differently, a proportion of the acoustic load on the higher frequency drivers will be modulated by the LF diaphragm movement. However, it must be said that crossovers do introduce additional loss and distortion and often result in a more difficult amplifier load.

With appropriate choice of matching drivers, and their working frequency range, acceptably close control of directivity may also be achieved. The accompanying uniformity of off-axis response improves stereo imaging and contributes to a neutral reverberant sound field in the listening room. The presence of discontinuities in the off-axis responses may often be heard in a resulting colouration of the reverberant sound field.

Skilful crossover design will result in good integration between the outputs of adjacent drivers to ensure a uniform forward output through the crossover frequencies. The optimum listening axis may be adjusted to that pertaining under actual conditions of use, as this may differ from a standard axial test position (Figure 6.2). Some speakers have been designed with the driver baffle tilted back with respect to the listener; the tilt

*Frequency modulation or Doppler distortion.

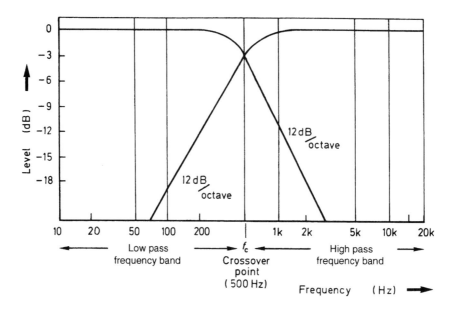

Figure 6.1. Two-way crossover filter response (second order 12 dB/octave). (High frequency power fed to HF unit, low frequency power fed to LF unit)

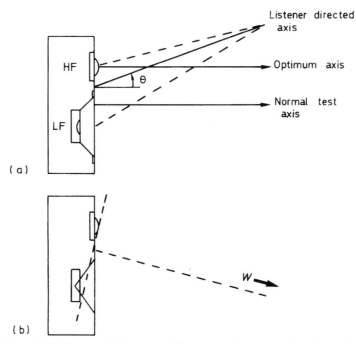

Figure 6.2. (a) Optimum radiation axes. (Crossover phase control produces listener directed axis at angle θ to normal axis.) (b) Combined in-phase wave front W directed off-axis by mounting method. Advancement of LF driver radiation centre to plane of HF driver will allow correction of inter-unit time delay/phase shift on axis. For (b), some designers simply invert the enclosure

may be adjusted in conjunction with the crossover alignment to compensate for driver delay.

With a loudspeaker designed for stereo use, the directivity in the vertical plane can be fairly narrow, as the relative heights of speaker and subject fall within reasonable limits. A suggested standard for vertical directivity might be 'the deviation from the axial response shall be less than 2 dB over a $\pm 10°$ vertical angle, up to 12 kHz'.

In the horizontal plane, an angle of twice this figure is desirable. While some designers aim for maximum spread at all frequencies, there is no real evidence that the stereo image quality improves as a result. Harwood suggests, on the evidence of BBC tests, that there may be an optimum for stereo. A very small, wide directivity BBC speaker gave poorer imaging results than a much larger model which was consistently closer to a $\pm 30°$ standard [1]. With wide angle designs the reverberant sound intensity will be higher relative to the direct sound than with narrower types, and perhaps this may partially account for the degraded imaging.

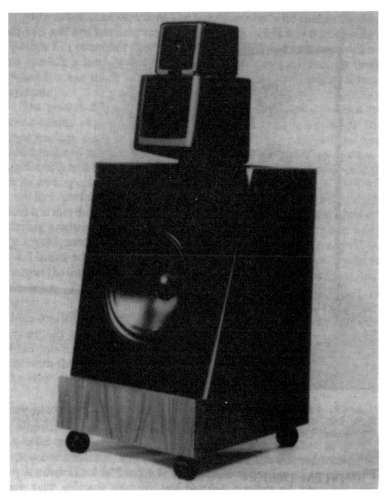

Figure 6.2. (c) A three-way system with time aligned drivers, 24 dB/octave, crossover (acoustic) and low diffraction enclosure forms (courtesy KEF Electronics)

Figure 6.3.   A three-way vertical in-line system. This monitor design may be tri-wired. Model S100, 300 mm Bextrene bass, 160 mm polypropylene mid, 19 mm doped fabric high frequency. (Courtesy Spendor Ltd)

With psycho-acoustic research confirming the importance of the 'early' direct sounds in the location of acoustic sources, within a 10–20 ms transient period, if near enough to the source local reflective boundaries are heard as a component of the source. In the case of a loudspeaker placed in its room environment, the strength of local secondary reflections in the mid-treble range will obviously depend on system directivity. A controlled forward directivity helps to improve the ratio of focused direct sound to reflected sound and thus sharpens the stereo image.

Within the desired solid angle of forward radiation, good uniformity is desirable to minimize differential phase and amplitude variations which occur with respect to a listener when listening to a stereo pair of speakers. Small differences in relative angle, head height or attitude should not result in noticeable phase and amplitude differences. Such defects confuse and defocus the sensation of stereo image precision and 'sharpness'.

The need for a uniform, symmetrical directivity in the horizontal plane dictates that the main drivers be mounted in a vertical in-line formation. There is no doubt that this particular configuration is important to stereo image stability and, if further evidence were needed, data collected during a consumer test of thirty pairs of loudspeakers provided strong indications that vertical in-line systems (Figure 6.3) gave superior stereo results [2]. (See also section on 'Phase' in Chapter 9.)

6.1 SYSTEM DESIGN

There are no hard and fast rules for the design of a successful system. Two distinct classes exist: first, systems designed in every detail by a loudspeaker manufacturer who makes his own drive units and secondly, systems assembled from o.e.m. (original equipment manufacturer) drivers from outside sources. In the case of the former, the engineer has complete control of the design process and can produce a system more or less precisely to meet a given standard of performance. From the desired specification he may establish the optimum driver and enclosure characteristics, design the units to meet these requirements and thus complete the system with the addition of a matching enclosure and crossover.

More often than not, even the manufacturer who makes his own drivers will not have the resources to design individual units for each system. For economic reasons he is forced to rationalize and is likely to produce a fairly well ordered range of drivers each capable (with possibly some small detail variation) of meeting the needs of several systems.

The perfect driver has yet to be developed, and each presents its own unique engineering and performance compromise. While the independent designer has a vast number of permutations and combinations to choose from, he has less control over driver uniformity and the very existence of such a wide selection brings its own special problems.

A high performance system is by definition a consistent product, and it is therefore vital that the driver characteristics remain constant during production. The possibility of any significant variation must be investigated at the design stage, a 'centre' specification should be established with allowable deviation limits, and provision made to correct the inevitable remaining tolerances, e.g. in the system crossover. The variations themselves may concern any aspect of performance, but once a unit is chosen and its manufacture kept under close control, the criteria of frequency response and sensitivity alone are usually sufficient to completely quantify the unit for production. With critical systems it may prove necessary to carry out the response test with the crossover in circuit to account for interaction between the two.

4 Ω Versus 8 Ω Speaker Impedance

Competition between speaker manufacturers is extremely keen. So far as sales are concerned it has long been known that as in the automobile market, where quicker is better, the corresponding law for loudspeakers is 'louder is better'. Good loudspeakers are often not very efficient due to the mutual conflict of physical laws governing frequency range, colouration, speaker size and sensitivity.

Traditionally most U.K., U.S., French, and Japanese speakers have been 8 Ω, based on a definition where the minimum impedance modulus is 6.4 Ω or higher, while countries following the German DIN specification have selected 4 Ω, largely based on the desire to extract maximum power from the less expensive solid state amplifiers. When first introduced, the latter provided ample current but a more restricted voltage swing.

Given that the effective description of a modern power amplifier is that of a 'voltage source of relatively unlimited current delivery', some manufacturers have recently adopted the $4\,\Omega$ specification. For a given volume setting or a voltage drive normally referenced to a 1 W, 2.83 V, $8\,\Omega$ level, the quoted 'sensitivity' is 3 dB higher than for the equivalent $8\,\Omega$ speaker since the $4\,\Omega$ example draws twice the current from the amplifier.

With complex multi-way speaker systems of normal design the instantaneous load value may fall well below the minimum indicated by a steady-state measurement of the modulus of impedance. For example, a three-way $4\,\Omega$ speaker with a nominal $3.2\,\Omega$ minimum modulus was shown to reach the equivalent of $2\,\Omega$ on music related transients. A $2\,\Omega$ peak loading is unwelcome, either for an amplifier, or when considering the effects of speaker cable impedance. Consequently, many $4\,\Omega$ speaker manufacturers are now producing models with a nearly constant resistive load impedance. This is achieved via conjugate compensation networks, justifying the contention that the 3 dB extra sensitivity is achieved without imposing undue frequency related stress on the cable or amplifier. With certain exceptions, notably some low current amplifier models and those few valve or tube designs, this is a fair assumption in practice (see Figures 6.25 and 6.29).

With very high quality speaker designs, some doubt remains concerning the subjective effects of the additional passive components required to effect the conjugate compensation. Some listeners have noted some loss of subjective dynamics and transparency associated with this technique.

Factors Affecting the Choice of Drive Units

Many system engineers rely on intuition when selecting a driver line up. Familiarity with a large number of units is a prerequisite, and the choice is generally made on the basis of the limitations of performance discussed in Chapter 5, but occasionally there are exceptions.

For example, the better class of 160–200 mm cone driver currently manufactured in the U.K. can produce a good performance that extends to 5 kHz, well beyond that which is attainable using older cones. This allows the crossover point to be placed between 3 kHz and 4 kHz, high enough to consider transfer to a small 25 mm or 19 mm high frequency dome unit for the upper range. Some excellent systems of necessarily limited maximum acoustic output have been produced along these lines from KEF Electronics (R103, R104, etc.), Spendor Audio Systems (BC1, SP1, SP2) and Rogers (LS7, Studio One).

Cone theory indicates that a 200 mm driver will be operating in breakup above 600 Hz or so and this is indeed often the case. However, the particularly consistent properties of the synthetic cone material employed allow the designer to adequately control the breakup modes such that the range may be smoothly extended by a further three octaves. A normal paper/pulp cone cannot often achieve this level of performance and in consequence the crossover point for a high performance application may need to be placed at around 1 kHz. This would entail the use of an additional mid-range driver to meet the bass unit at 1 kHz, and a further HF unit would probably be required to complete the frequency range.

Pulp cone designers are now meeting the challenge of synthetic materials and are producing new generations of better controlled, better sounding paper cones of usefully higher sensitivity than Bextrene. A number of these pulp cone bass–mid-range units are now well behaved to 4 kHz or so.

High Performance Systems

The need for adequate acoustic output from a true high performance system virtually dictates that the minimum size of bass unit should be 250 or 300 mm in diameter (the 200 mm based systems are generally insufficiently loud for professional monitoring).

Low frequency drivers

The choice of LF driver will depend on the level of colouration tolerable, together with considerations of bandwidth (LF excursion), efficiency, power handling and enclosure size.

Depending on the volume of the enclosure, the LF unit may be loaded in several ways, the electro-acoustic properties of the individual driver being an important factor in this context. In domestic locations compact systems are strongly favoured for aesthetic reasons and, in general, no loudspeaker system should be larger than is strictly necessary.

Few LF drivers will function adequately beyond 1 kHz and the maximum safe crossover frequency is generally 500–700 Hz for the better 250 mm units and 350–500 Hz for the 300 mm sizes. With these larger drivers a further reduction to 250 Hz will help to reduce subjective colouration, particularly as regards voice reproduction.

Mid frequency drivers

Taking 350 Hz as the basic lower limit, it is desirable that the mid-range unit gives an adequate performance from at least two octaves below, i.e. 100 Hz. It almost goes without saying that its range should possess a reasonably uniform pressure response free of significant resonances or colourations. Such a driver is usually a cone unit 80–200 mm in diameter, often capable of bass/mid range coverage in its own right.

The better examples of mid frequency driver will have a clean output up to 5 kHz or 6 kHz allowing the transition to an HF unit to occur at 3 kHz (ideally the upper range of the mid unit should extend further but this is rarely achieved).

To date, the dome mid-range units which have been produced range in diameter from 35 mm to 85 mm and considerable problems have been experienced in the attempt to provide a sufficiently wide response. To cover the basic 350 Hz to 5 kHz mid band, an overall 100 Hz to 15 kHz response is desirable, amounting to seven octaves. The best examples so far cope reasonably well from 350 Hz to 5 kHz, but necessitate crossover points around 600 Hz and 4 kHz with accompanying high slopes of perhaps third-order, 18 dB/octave rolloff. A worthwhile target is 90 dB/W for sensitivity, but this is rarely achieved. One 75 mm dome has attained this level but at huge magnet expense and cone units are generally far more economical.

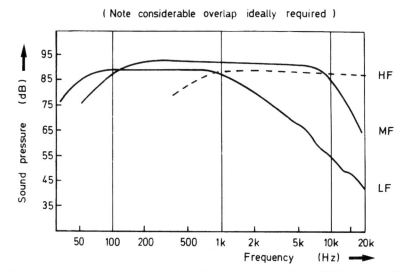

Figure 6.4. Ideal driver responses for three-way system (f_c at 500 Hz and 4 kHz)

High frequency drivers

Sensitivity will be an important factor in the selection of an HF driver since few high quality designs are efficient. Some recent examples of 25 mm dome units have managed to combine qualities of smoothness, high linearity and adequate sensitivity and these are now widely employed since at this quality level, cone units are virtually ruled out. If higher levels of acoustic output are required, then this may only be obtained via horn loading.

With a crossover point probably in the range 3–6 kHz, the HF unit should ideally have a response extension below 1 kHz for optimum crossover performance. Most 25 mm dome units fulfil this condition, although the magnitude of the Q at fundamental resonance must be taken into consideration. The broad overlap of driver response is best shown in terms of a system (Figure 6.4) illustrating the responses of an ideal 'three-way' set of drivers. Focal have succeeded in reaching 96 dB/W for a 1 in dome with very large magnets.

Sensitivity matching

Ideally, the in-band working sensitivities of the drivers making up a system should be equal when measured on the system axis, after equalization, compensation and crossover loss have been included. The latter effect can be difficult to predict and it is best to initially establish the low frequency sensitivity. The mid and treble drivers should be selected so that sufficient sensitivity remains in hand. Since even high quality drivers can vary in sensitivity by up to ±1.5 dB in production, often some selection or final balance adjustment may be required for the completed system. Moderate resistive attenuation does not unduly affect the driver and crossover performance for the mid and high frequencies, but is likely to significantly disturb an LF unit especially near the bass resonance, and sensitivity control must be ruled out for the bass unless an auto transformer is employed. It is therefore usual to direct couple the LF unit via the crossover, and to reserve any attenuation for the other drivers.

If the crossover is an active electronic design, separate power amplifiers feed each driver and sensitivity matching is easily accomplished via gain control. Nevertheless, the sensitivities and power handling must still be chosen with due regard to the maximum acoustic output of the system as a whole and the power spectrum of the programme to be used. Active crossovers are increasingly being used for professional monitors.

$2\frac{1}{2}$-way System Design

When designers expand a given range of speakers, it has become popular to take the basic design of a stand-mount two-way loudspeaker, enlarge it to make a floor standing model and add a further bass driver. Traditionally, a three-way speaker would be the result — for bass mid and treble — but these days with the use of smaller bass units, typically 5 in (110 mm) or 6.5 in (170 mm) bass units, there is a need for greater bass output and power handling. This can be achieved by having both bass drivers share the work load. With the tweeter mounted in its preferred, top-most position in the enclosure the drive to the lower bass unit must be rolled off to prevent its output interfering with the crossover between the upper bass driver and the tweeter. Because a full crossover is not present, this class of system has been classed as $2\frac{1}{2}$ way. In design there are some interesting aspects.

For example, the crossover equalizer to the bass–mid driver of the original two-way may be realigned to a higher sensitivity typically 2–3 dB by adjustment of the crossover, assuming that the treble unit can follow suit. The second, bass-only driver has a crossover that shapes its output to sum constructively with the upper driver to restore the overall mid-range level. Often a single inductor is sufficient, of value in the range 4–7 mH. As regards enclosure design, there are benefits in separating the two drivers into two enclosure sections thus allowing them to be individually loaded and tuned. This provides better bass control and extension. Note that the two bass driver sections are wired in parallel with essentially a halving of system impedance, unless corresponding adjustments are made to the drivers. Where an existing two-way system has a naturally high impedance, say 8 Ω, certain alignments of $2\frac{1}{2}$ way, particularly with a more efficient lower bass unit, may allow the latter driver to operate with a higher impedance coil of 12–16 Ω. This, when added to the main system, results in a satisfactory load impedance.

With such systems and related parallel bass systems, if defined as part of the design, it is possible to adjust the driver characteristics and their cabinet tuning such that the dips and peaks in their individual impedance characteristics overlap in the low frequency range. The object is a flat impedance curve for the system, with the objective of better matching to amplifiers, particularly SE types with their typically high output impedances, these measured in the range 1.5 Ω to as high as 4 Ω.

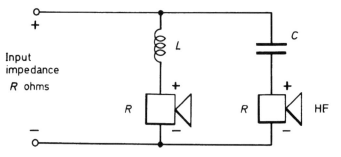

Figure 6.5. First-order circuit and equations, $L = R/(2\pi f_c)$, $C = 1/(2\pi f_c R)$; e.g. $R = 8\,\Omega$, $f_c = 3\,\text{kHz}$, $L = 0.42\,\text{mH}$ and $C = 6.6\,\mu\text{F}$ (6 dB/octave rolloff)

6.2 THE CROSSOVER NETWORK

Classically, a crossover network consists of a passive, high-power filter circuit designed using standard filter theory, commonly to a Butterworth response characteristic, which provides maximum response flatness with a well-defined rolloff.*

The simplest form of crossover is a two-way network of first order which consists of a single inductor to direct bass power to the low frequency unit and a single capacitor, which passes treble power to the high frequency unit. Such networks are inexpensive and are often found in low-priced systems (Figures 6.5 and 6.6).

While this form has been found inadequate for high performance applications, except in those exceptional cases where extremely wide range drivers of defined spacing (including a common radiation plane) and dispersion are employed, it is now enjoying increasing favour. In the latter case, it provides the most accurate method of achieving minimum phase shift between the drivers and hence allows the 'minimum phase' or, less accurately, 'linear phase' class of speakers to be designed. However, such systems, usually with displaced drivers, suffer from poorer vertical response due to the broad driver overlap and the resulting interference patterns.

Classical crossover theory tries to follow the idea of power matched filters of maximum transfer efficiency, the complex variants concerning 'm' derived networks where the concept of iterative impedance is employed. Here the termination impedance is supposed to be reflected via the filter to the input terminals. In practice, simple crossover filters, when used in their normal cutoff modes, cannot reflect such impedances and modern practical networks are based on constant resistance theory, i.e. for the complete network, the input resistance presented to the amplifier should be more or less constant. Constant resistance networks are less critical of impedance variations and are well suited to modern low output resistance amplifiers.

With single-order networks often proving inadequate due to overlap between driver outputs and consequent polar irregularities in the vertical plane, higher order networks are widely used. Second-order 12 dB/octave circuits are an effective compromise on cost grounds. With this network the driver outputs are out of phase by 180° at crossover. Assuming the units to be in the same time plane, a well integrated output will only be obtained with one driver, usually the HF unit, phase inverted (wire driver '+' to '−'

*Assuming uniform driver impedance, flat responses and no time delay!

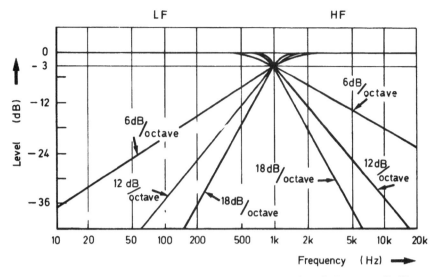

Figure 6.6. Two-way crossover responses (6, 12 and 18 dB/octave rolloffs)

terminal on crossover, i.e. reversing the connections). The rest of its frequency range then remains phase inverted, a possible cause of contention concerning system phase linearity. The second-order network has a parallel inductor across the HF unit, which helps to damp its fundamental resonance at frequencies below the crossover point.

Figure 6.7 illustrates the second-order configuration. The following classical 'm' equations:

$$L_1 = \frac{R}{\sqrt{2}\,2\pi f} \qquad L_2 = \frac{R}{\sqrt{2}\,\pi f} \qquad (L_2 = 2L_1)$$

$$C_1 = \frac{1}{\sqrt{2}\pi\,fR} \qquad C_2 = \frac{1}{\sqrt{2}\,2\pi fR} \qquad (C_1 = 2C_2)$$

may be compared to the modern constant resistance standard

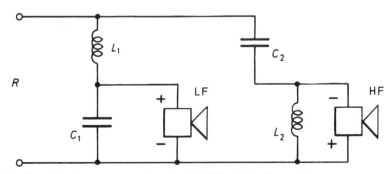

Figure 6.7. Second-order crossover circuit with HF polarity inverted, a common connection

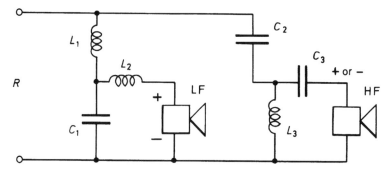

Figure 6.8.   Third-order crossover circuit

$$C_1 = C_2 = \frac{1}{\sqrt{2}\,2\pi f R}$$

$$L_1 = L_2 = \frac{\sqrt{2}R}{2\pi f}$$

For a lower Q_f the Linkwitz Riley form $(-6\,\text{dB})$ at crossover is designed using the formulae

$$C_1 = C_2 = \frac{0.8}{Rf} \quad \text{and} \quad L_1 = L_2 = \frac{0.32}{f}\,R$$

Because of the vagaries of driver acoustic output and the time delays present between drivers due to geometry and mounting, strict rules concerning crossover order and systems can rarely be applied. The crossover design rules must be interpreted with the aim of producing a smooth, well integrated acoustic output from the drivers concerned, this termed the acoustic target function.

One successful system using a low-order filter employs a transitional first-order low-pass with a second-order high-pass. In theory the phase shift at crossover should be between 90° and 180° but at the crossover frequency chosen the drivers show good integration with the HF unit when connected phase reversed, taking account of the final acoustic delays involved (25 mm dome, 200 mm cone LF-MF).

Higher Order Networks

Higher-order networks are also commonly used, with third order providing a 90° shift between drivers and steeper slopes, improving driver isolation. However, the final series capacitor in the HF circuit can cause difficulties at that driver's resonance (see Figure 6.22).

Butterworth networks generate up to 3 dB of peaking at crossover on the optimum axis. In practical designs the engineer can adopt a different f_c for the complementary high and low pass sections, sliding f_{c1} and f_{c2} apart from the original f_c and thus controlling the amplitude through the transition.

For the third-order network in Figure 6.8, generally

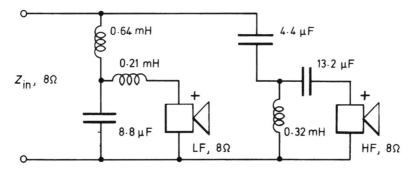

Figure 6.9. Theoretical 8 Ω, 3 kHz, third-order crossover circuit (based on Figure 6.8)

$$L_1 = \frac{3R}{4\pi f} = 3L_2 = 2L_3$$

$$C_1 = \frac{2}{3\pi fR} = \tfrac{2}{3}C_3 = 2C_2$$

A practical circuit is shown in Figure 6.9.

Fourth order is probably as high as is economically practicable or necessary and has several virtues. That helpful shunt inductor reappears across the HF driver and the rolloff slopes are so steep at 24 dB/octave that out-of-band driver problems are well attenuated. While fourth order reverts to an in-phase connection of drivers, in theory a 6 dB notch is present at the crossover point. In practice this is so narrow that any dip in the measured axial response of a system is hard to find and may be filled by a fractional overlap.

Due to its in-phase symmetry, fourth order, with delay spaced drivers allows a well-defined, minimum phase system to be designed. As proposed by Linkwitz, a cascaded fourth-order Butterworth design will provide the best axial polar symmetry. Interestingly this network assumes the properties of an even order all-pass network (see Figure 6.10).

Energy plots of the in-phase (Figure 6.11(e), (f)) fourth-order system show a desirably broad control obtained with symmetric minor side lobes. By comparison, the second-order type (a, b) has broader side lobes and a narrower central focus. The plot of third

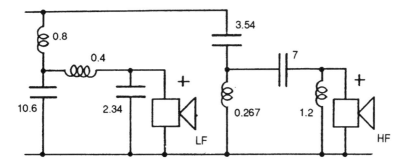

Figure 6.10. 24 dB/octave, 4th order passive crossover; 8 Ω, 3 kHz, scale to any frequency (cascaded Butterworth — after Linkwitz [4])

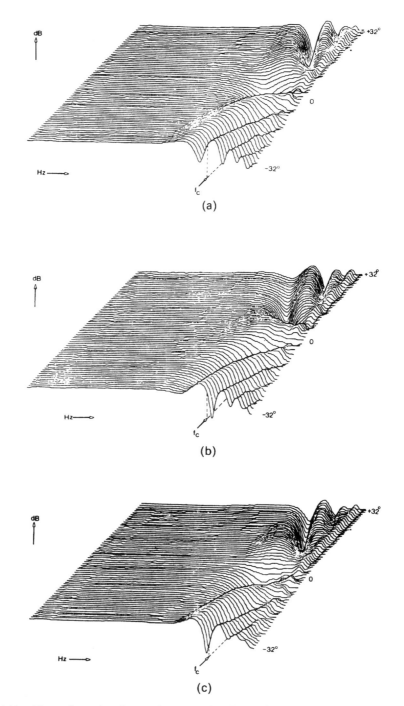

Figure 6.11. Three-dimensional second energy plots in the frequency domain (after Bank and Hathaway). (a) and (b) are second-order filter measured from below axis, (a) in phase (b) antiphase. (c) and (d) are third-order filter from below, (c) in phase, driver connection, (d) out of phase. (e) and (f) are fourth-order filter, from below, (e) in phase, (f) antiphase

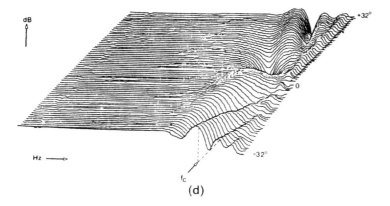

(d)

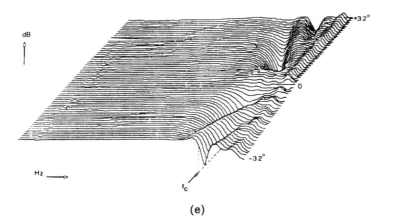

(e)

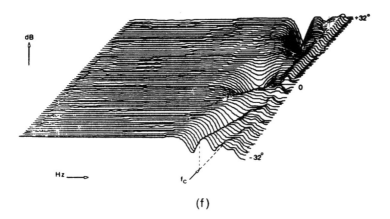

(f)

Figure 6.11.   *(cont.)*

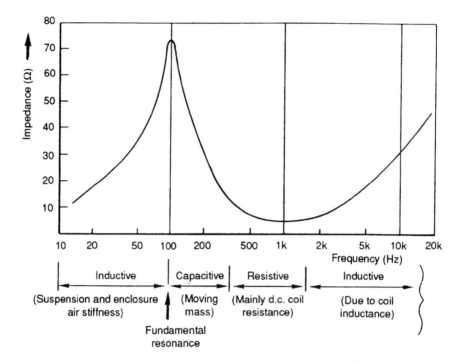

Figure 6.12. Moving coil driver input impedance

order with antiphase driver connection shows an asymmetry predicted by Linkwitz, with an upwards directed main lobe. This may be exploited both to shift the desired listening axis or conversely it may be used to balance the differential delays resulting from the greater depth of a bass driver (c, d).

Both theory and computed energy plots assume that the drivers are perfectly flat in response, have no differential time delay and present resistive (8 Ω) loads. Real systems do not conform to this and the designer needs to make many allowances and compromises. His goal is an acoustic target function whereby the final crossover generates the required crossover points and slopes in the acoustic output of the required drivers, on and off the required axes.

Mis-termination

In practice, crossover networks may be further complicated by matching problems. The constant crossover input resistance proposed by Butterworth crossover theory assumes that the input impedance of the drivers to have a constant resistance also, nominally 8 Ω for the network illustrated in Figure 6.9.

Region C of Figure 2.18 shows that at only one part of the frequency range is the terminal impedance of a moving coil driver a fairly constant resistance. Figure 6.12 shows that at lower frequencies a more or less well-damped fundamental resonance occurs with the motional impedance passing through the inductive, resistive and capacitive regions [3]. At higher frequencies the inductance of the motor coil becomes significant, producing a rise in impedance. When such a complex load is connected to a standard filter, mis-termination occurs, with resulting irregularities in both the amplitude and phase response of the filter.

An active crossover is essentially immune to these problems, since perfect filter network termination is provided in the circuitry, with power amplifiers voltage-driving the motor coil in an ideal fashion. The advantages of this approach are discussed more fully in Section 6.3 on active crossovers.

Figure 6.9 is a theoretical realization of a two-way 8 Ω network operating nominally at 3.0 kHz. The network from a successful commercial system is reproduced in Figure 6.13. Because it is difficult to see any resemblance between these two, how then may crossovers be designed if the basic theory appears to have so little relevance? The answer lies in an extension of the filter network theory to include the other relevant factors present in the total system. This involves viewing the crossover response in terms of the combined acoustic output of the drivers instead of theoretical voltages on the driver terminals. By moving outside and in front of the enclosure it becomes obvious that the crossover requirement is for the overall response to include the acoustic output of the drivers. Hence the driver parameters—motor coil inductance and resistance, frequency and phase response and the motional impedance near resonance—must be accounted for in the network theory while ensuring that the system as a whole still offers a sensible load for the matching amplifier. The characteristic loading generated in Figure 6.13 is shown in the system impedance curve (Figure 6.14), and is clearly not a constant 8 Ω.

The lack of correspondence between the measured acoustic output and the calculated crossover filter response is shown in Figure 6.15. The first-generation network of Figure 6.13 was further developed to incorporate compensation for both the motional impedance variation due to the fundamental resonance, and the intrinsic motor and coil characteristic of the HF unit. The equivalent circuit for these components was included in a computer program together with information concerning its amplitude response and a new network was synthesized which was designed to give the intended 18 dB/ octave Butterworth response in the driver's acoustic output. The success of the new

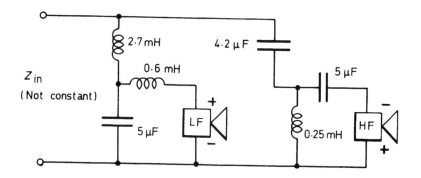

Figure 6.13. Successful commercial 3 kHz, third-order crossover (KEF R104 system)

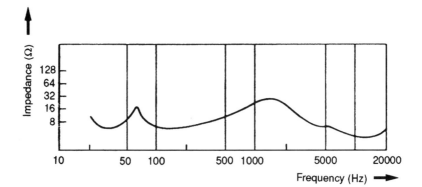

Figure 6.14. KEF Model 104, impedance frequency

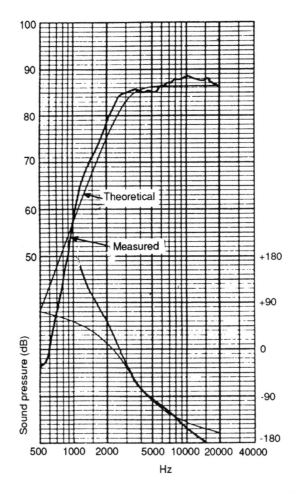

Figure 6.15. HF unit output; measured and theoretical responses (courtesy KEF Electronics).
Circuit of Figure 6.13

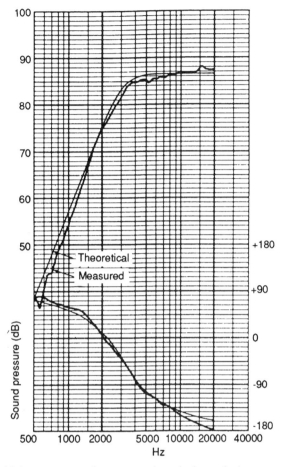

Figure 6.16. New high-pass network; measured and theoretical responses (courtesy KEF Electronics). Circuit of Figure 6.15

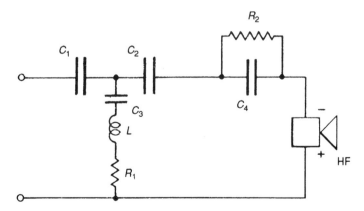

Figure 6.17. Theoretical circuit of new third-order compensated high-pass section (courtesy KEF Electronics

network is shown in the close agreement between the measured and calculated responses shown in Figure 6.16. Figure 6.17 illustrates the theoretical form of this new compensated high-pass network with Figure 6.18 illustrating a practical realization, and Figure 6.19 carrying the numerical values used in a commercial system.

Figure 6.18 was derived by applying the Star-Delta transform to Figure 6.17. It is sufficient to view the circuit as a lumped approximation and it can be readily seen that the controlled Q of the shunt indicator provides a first slope of 6 dB/octave, summing with the driver's natural 12 dB/octave slope.

The second capacitor C_4 is equivalent to the moving mass of the diaphragm and is part of the complete synthesis. C_4 causes the derivative of the current waveform in the driver to follow the Butterworth curve, resulting in constant motor coil acceleration. Interestingly the values of the basic 'T' section network are quite close to those established theoretically for a perfect driver.

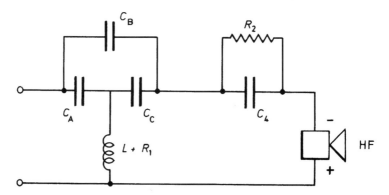

Figure 6.18. Practical realization of new network:

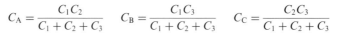

$$C_A = \frac{C_1 C_2}{C_1 + C_2 + C_3} \qquad C_B = \frac{C_1 C_3}{C_1 + C_2 + C_3} \qquad C_C = \frac{C_2 C_3}{C_1 + C_2 + C_3}$$

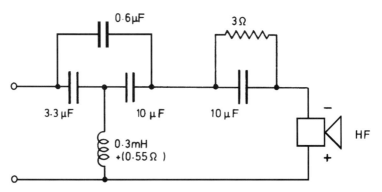

Figure 6.19. Values used in commercial system KEF R104AB (courtesy KEF Electronics)

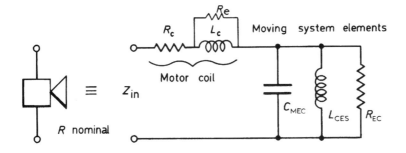

Figure 6.20. Moving coil driver equivalent circuit

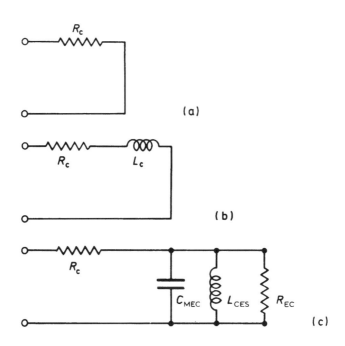

Figure 6.21. Simplified forms over specific frequency bands (from Figure 6.20). (a) At mid frequencies, (b) at high frequencies and (c) at near resonance; low frequencies

Driver Impedance Compensation

A moving coil driver may be represented by an electrical equivalent circuit consisting of the coil components R_c and L_c and the transformed mechanical components L_{CES}, C_{MEC} and R_{EC} (see Figure 6.20).

With an LF unit the fundamental resonance is deliberately exploited and does not require compensation in this context.* However, with mid and HF drivers the fundamental resonance can be sufficiently near the crossover region to cause mis-

*Mid-treble. For a low crossover frequency compensation is often useful.

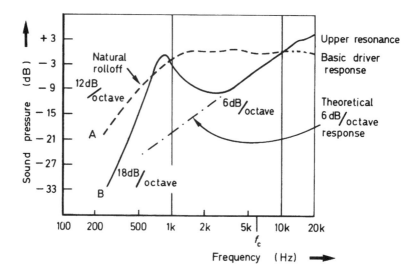

Figure 6.22. Moving coil driver response; A, alone; B, with first-order series capacitor,
$f_c = 7\,\text{kHz}$

termination and is worth neutralizing. Figure 6.21 can be simplified if specific frequency ranges are separately considered (Figure 6.21).

The consequences of crossover filter mis-termination are clearly shown in Figure 6.22. A crossover at 7 kHz consisted of a single capacitor and was to be used with an HF unit having the response curve 'A'. While the rolloff in the resulting curve 'B' approximates to the desired 6 dB/octave slope down to 1.5 kHz, the increase in motional impedance at resonance (850 Hz) causes the output to rise sharply, almost peaking to the driver output level in the absence of the crossover. Below resonance the slope now follows an 18 dB/octave rolloff due to the addition of the intrinsic driver rolloff. At the highest frequencies the crossover capacitor begins to weakly resonate with the motor coil inductance, producing a rise in output above 10 kHz. The overall curve bears little resemblance to the required smooth transition with a 6 dB/octave slope.

The motor coil inductance may largely be compensated by a series R and C combination connected in parallel with the driver terminals (Figure 6.23). If the coil values are R_c and L_c the equalization components to a first approximation are as follows:

$$R_{eq} = R_c \quad \text{and} \quad C_{eq} = \frac{L_c}{R_c^2}$$

where R_c is the a.c. resistance at the required frequency. A typical 25 mm diameter motor coil for an HF unit might have a 6.4 Ω resistance and a 0.15 mH inductance, giving respective values for R_{eq} and C_{eq} of 7.5 Ω and 0.366 μF. The fact that this compensation is possible gives the designer some freedom in adjusting the response at extreme frequencies should the driver appear to need this.

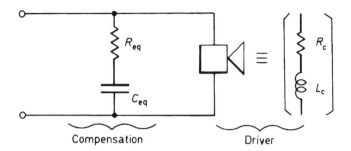

Figure 6.23. Circuit and equations for motor-coil inductance operation; $R_{eq} = R_C$, $C_{eq} = L_C/R_c^2$

Since the driver impedance rises at resonance, a simple compensation would consist of a series resonant circuit connected in parallel with the terminals, this largely accounting for the electrical equivalent components of the mechanical fundamental resonance, imparting a uniform impedance curve to the combination; the latter is shown before and after connection in Figure 6.24. The correction circuit used is shown in Figure 6.25.

The resonant frequency and its Q may be determined from the impedance curve and an equalization circuit synthesized to match this. Alternatively, the values may be found by experiment which in practice proves relatively simple to achieve.

Additionally, the appropriate viscosity of a magnetic fluid may serve to fully damp the fundamental resonance thus allowing a series capacitor circuit.

Low Order Crossovers

Designers may argue at length about preferred crossover orders and complexity. High order crossovers provide opportunities for fine tuning both the amplitude and phase response of the intended driver frequency ranges, and the steep roll-out slopes which

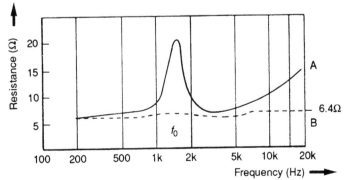

Figure 6.24. Curve A, impedance curve of 10 mm plastic dome HF unit. Curve B, as for curve A but with compensation circuit of Figure 6.25

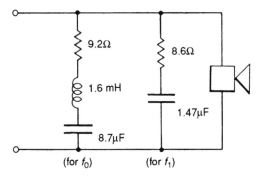

Figure 6.25. Compensation circuit for non-uniform impedance of HF driver in Figure 6.24
(values determined by experiment)

can be realized minimize the acoustic interaction between the drive units. This looks good on measurement, particularly in respect of a defined axial response, but it may still be argued that frequently a poorer transient response due to the sharper corners of a high order crossover is audible in the final sound. This may be because it is considered necessary to achieve a good axial response and a smooth power response, the latter reflected in a sensible assessment of the off-axis results. For this reason a number of engineers are moving towards low order crossovers, taking care to control those aspects of driver performance so that smooth results may still be achieved, despite the reduced number of variables under the control of the system designer.

It could be argued that higher order crossovers are the province of the designer working with available drive units, while the designer blessed with the resources to define custom units will have the freedom to use simpler crossovers with well matched design. When well executed low order crossovers, in particular first order, often sound better blended. The transition to the tweeter may be rendered undetectable while the power response of the speaker is more uniform thanks to the broader region of blend and overlap between the drivers. Benefits include a reduced number of crossover components, which may result in lower cost as well as greater transparency and improved subjective dynamics.

Control of phase is vital, otherwise broad suck-outs will occur in the frequency response. Phase control must be achieved via the characteristics of the driver diaphragms, the phase shifts due to the intended crossover, the relative phase of the drivers (polarity) and any helpful compensation for delay afforded by enclosure design and driver mounting.

Ideally the low frequency driver, placed in the required cabinet and located as intended with respect to local boundaries, will have a naturally flat frequency response terminating in a well formed rolloff at the intended crossover frequency. Such a characteristic may be achieved by control of the voice coil winding, its mass and inductance, for example using a four-layer winding. The variables of cone design may be iterated to generate the intended response shape by the control of material thickness, flare curvature and termination. Via the magnet strength, Q may be adjusted to give an optimum bass to mid range balance in a given enclosure design. If necessary a remaining degree of response rise may be easily controlled by a small series inductor, perhaps with a simple compensatory termination.

For the high frequency unit it is necessary to account for that essential series capacitor which keeps low frequencies out of this delicate driver. A series capacitor reacts unfavourably with the moving coil driver's motional impedance (see Figure 6.22). The response curve is strongly modified by the complex impedance loading presented by the driver. With more complex networks the interaction may be terminated and compensated, or high order networks may side-step the issue by virtue of their fast rolloff. With a first-order design the treble driver must be designed so that its impedance characteristic is largely resistive due to mechanical damping (the grade of the ferrofluid, suspension characteristics, acoustic chambers and associated acoustic damping). In effect the mechanical properties and frequency response of the tweeter, defined *in situ* in the enclosure, are tuned with the crossover capacitor for the desired response. In these designs the tweeter may well be accepting more power in its low frequency band and its power handling and linearity may need to be greater than for units intended for high order crossovers.

An interesting characteristic concerns the effect of varying the series capacitor (of the usual value between 2.2 μF and 4.4 μF) which in addition to moving the crossover point also has the result of acting as an attenuator over the treble range. Because of the broad overlap, small changes in capacitor value or in series inductance to the bass unit substantially alter the sound of the system; the effects are heard over several octaves. It is clearly a great advantage to use drivers with smoothly extended frequency ranges which also have low colouration outside of their intended working range. Conversely, the softer transition between the drivers may help to smooth dips in the frequency response of one driver thanks to the overlap from the other. If well executed, a more linear power response often results, with a more natural sound in typical listening rooms. Small errors in relative driver sensitivity change the sound quality less than with high slope crossovers, thus avoiding the 'step' effect that occurs between the driver pass bands when mismatched in level.

On Using a First-order Filter with an HF Unit

Generally the use of a single capacitor for a crossover 'network' to a tweeter is inadvisable (see Figure 6.22). However, it is possible to control the tweeter parameters to align with a single capacitor, e.g. 3.5 μF at 3.7 kHz (for a nominal 6 Ω unit). Via the right surround of suitably high suspension stiffness, taking care that the distortion performance is not compromised and ensuring a careful match to the viscosity of a gap damping fluid such as ferrofluid, the notional impedance at resonance may be firmly controlled; the resulting acoustic response is then matched to the choice of series capacitor.

Minor response variations that affect the overall trend may be fine tuned by adjusting the geometric shape of the front plate of the unit, whether via a shallow residual horn (see Figures 7.20, 7.21) or some form of 'phase correcting' assembly in front of the dome.

Simple crossovers of this type generally reward the designer with greater transparency and more natural treble sounds for the final system.

External Crossovers

Where the price point allows, there are good arguments for leaving the crossover outside the enclosure. Noting the possible increase in cable and connections, depending on whether a captive harness or detachable set of cables is chosen, a remote location for the crossover may improve sound quality by its removal from the acoustic, vibrational, electromagnetic and magnetic fields within the enclosure.

External crossover design requires an additional pair of small boxes. It does allow for easy changeover from passive to active operation, where the passive crossover is simply omitted from the cable run and an active filter is placed ahead of the required number of power amplifiers.

Aspect of the D'Appolito Configuration

In the D'Appolito configuration (also found in earlier loudspeaker designs, including Meridian) two moderate sized bass–mid range or mid-range drivers appear above and below the tweeter. The larger drivers carry identical signals which confer symmetry in the vertical axis. One could view the arrangement as a vertical section of a concentric driver and some of the directivity advantages of such a driver are attributable to the array. The crossover design aims to integrate the drivers over as accurately as possible with respect to phase in order to attain a sufficiently wide lobe of output in the vertical plane. It is advantageous to choose a lower crossover frequency than usual in order to widen the lobe, this still narrower than for a conventional mid-treble arrangement due to the increased height of the three-driver array.

The directivity in the horizontal plane is controlled by cabinet width; but note that in the crossover region the phase control required for some optimized axial responses may produce unexpected variations in off-axis power.

In a particular example, an enclosure 22 cm wide will allow the installation of a three-driver array using two 17 cm frame mid or bass–mid units and a 25 mm dome tweeter (typically 10 cm chassis). The overall source height at crossover is 45 cm. Note that a wavelength at the usual 3 kHz crossover point is around 10 cm and thus the directivity in the vertical plane will be narrowed. Some irregularities may be present in the tweeter response compared with the more usual flat baffle position due to the proximity of a pair of cavities formed by the mid-range cones. If of sufficient quality, a designer might wish to consider flat diaphragms in this application (Figure 6.26).

Comparing with the sound quality of a conventional two-driver array with the tweeter uppermost near the top of the cabinet you have to consider a trade-off between the potential for a symmetric but narrowed crossover lobe for the triple, compared with a potentially asymmetric but wider angle of vertical response for the double arrangement.

Making the assumption that in both types the designer has managed to achieve an accurate forward response and good integration, there will still be a significant difference in sound quality due to the different way in which the room acoustic is driven. For the double unit type with the tweeter in the uppermost position, in the lower octaves of the treble range the tweeter is radiating into a wide angle of acoustic load; it is effectively placed at the edge of a prism and significant energy is directed in the angles

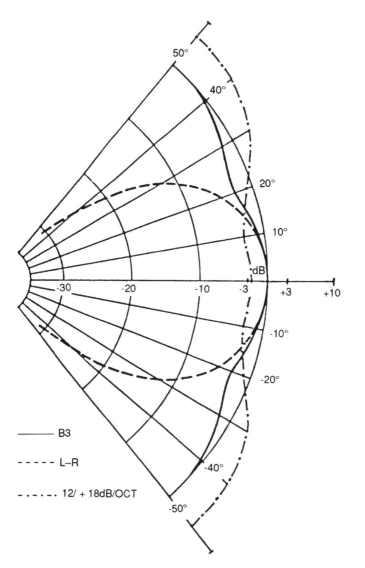

Figure 6.26. Vertical polar; three-driver, central tweeter for a 2 kHz crossover point (2 × 110 mm + 28 mm dome) (after d'Appolito). Key: − − − even-order Linkwitz type (in phase); ——— odd-order (third) Butterworth; − − · − · − − even and odd mix (12/18 dB, d'Appolito seeks 90° shift, constant through crossover region to maintain vertical lobe accuracy and uniformity

above the loudspeaker. As regards the room sound, it will contain a sample of that energy reflected from the ceiling whose brighter sounding acoustic helps give such systems their characteristic sense of life and air. For the triple array two factors conspire to reduce the amount of energy directed off-axis in the vertical plane which would contribute to that ceiling reflection. First, there is the defined narrower directivity due to the increased size of the array in the crossover region, and secondly, the placement of the tweeter farthest from the top edge of the cabinet, acoustically shadows the region above the cabinet. This consideration also applies for the equivalent and often popular orientation for a two-driver arrangement where the tweeter is placed below the mid or bass–mid unit, usually for reasons of phase and/or delay control. Consequently, the sound of these systems may be duller and less 'airy' and tweeter level matching, octave by octave alignment may need sensitive adjustment.

Open or Top Location for an HF Unit

This is a convenient point to consider the other extreme for tweeter placement, namely that of a baffleless 'bullet' style of tweeter mounting where the complete high frequency unit is barely larger than the dome itself and is presented on the top surface of the cabinet either alone or as an equivalent small baffle or narrowed tapered cabinet section. The effect of a vestigial or complete absence of baffle or mounting surface for the tweeter confers an unusually wide directivity in its lower operating range. This must be taken into account in the system alignment since it may cause problems when attempting to design for a smooth energy variation with frequency; e.g. the transition from the mid range (with its narrowing directivity) to the tweeter (a sudden transition to a much wider directivity). Without skilful control of the energy balance these systems have a tendency to leave the tweeter sounding 'exposed', thus lacking the best subjective integration achievable with the full baffle types.

Comparing this case with the D'Appolito central tweeter, or the two-driver, 'low mounted' tweeter case, the narrower directivity of these tweeters will require consideration with regard to the target energy response. A good balance between the energy response and the axial response is the designer's goal. Those balances are subtle but importantly different for these configurations, and may not be clearly elucidated in computer-aided syntheses. Listening work is essential to confirm the validity of computer analyses.

High Order Crossover Considerations

High order crossovers have their advocates. If executed with top quality components, with low losses and close tolerances, they can provide reassuringly consistent transfer functions for the whole speaker; variations in driver performance out of band are prevented from affecting the final result. Additionally, the narrow region of crossover overlap minimizes the effect of this troublesome region on the designed response.

In practice losses of both power and fidelity accrue with increasing order and fourth is as high as most designers venture; more precisely this is an acoustic target and not necessarily the sole responsibility of the crossover network.

Borrowing from the complex theory of mutual inductance coupling, long used in high slope bandpass filters in radio engineering, one designer (Richard Modafferi of 'Infinite Slope' USP 4771466) has patented its application to loudspeakers and shows how usefully high slopes may be obtained with minimal overlap while using a basic network, augmented by deceptively simple mutual coupling elements, essentially defined by the critical placement of existing inductors on the circuit board. In practice slopes of 100 dB/octave are possible while circuit analysis using the pole zero technique allows

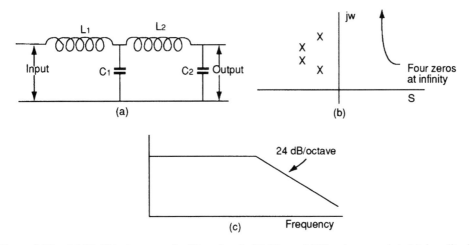

Figure 6.27.   (a) 24 dB/octave woofer filter circuit. (b) Upper LHP pole–zero plot. (c) Amplitude
response

the completed system to have a smooth phase response thanks to the clean join
achieved at the crossover point where the summation is most precise.

If near brick wall shapes are attained for the high- and low-pass functions at
crossover, then the negligible overlap of driver energy ensures tight control. Figure 6.28
shows a two-way design, essentially fourth order, where the steep initial slope is
generated by the additional components is derived from the small 'stray' mutual
coupling arranged between the inductors.

Briefly, poles and zeros are respectively the roots of the denominator and numerator
polynomials for the transfer function equation which characterizes the amplitude and
phase response of the filter network; for example, on the complex plane poles indicate
the level at or near their frequencies, while at or near zeros, the level is zero.

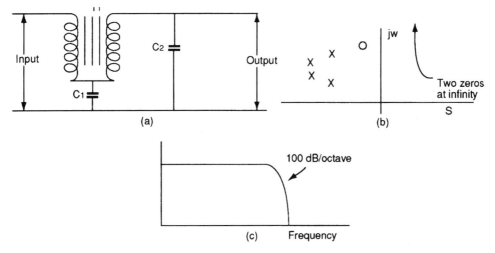

Figure 6.28.   (a) Infinite slope woofer filter circuit. (b) Upper LHP pole–zero plot. (c) Amplitude
response

In an example given by Modafferi (Figure 6.27) a conventional fourth-order low-pass crossover is compared with the mutually coupled alternative Figure 6.28. Greatly simplifying the electrical circuit theory which governs the operation, by altering the circuit such that an allowed pair of the four zeros, which in the standard form lie at infinity, are brought near to the poles (the diagram in Figure 6.28 shows one pole, the other is in the negative $j\omega$ quadrant and is not shown).

The magnetic coupling has an implied transformer which features in the underlying equations and results in the high slope shown in the final response.

Amplitude Response Equalization

Equalization of the driver impedance has been discussed, but to some degree this is rather academic unless the driver frequency response is already uniform, which is rarely the case. Thus in most high performance systems some form of response equalization is almost invariably included.

Equalization is only possible where the response irregularities are in the form of gentle trends rather than severe narrow band discontinuities with associated non-minimum phase characteristics. Clearly, equalization will also alter the input impedance of the entire system. A rise in impedance due to response correction is of little consequence, but a significant fall in input impedance as a result of compensation for response droop is undesirable due to amplifier matching problems. In practice, maximum sound pressure dips of around 2 dB may be equalized, and no limit appears necessary for the correction of excesses, provided that they are gentle.

A typical 170 mm plastic cone bass/mid range driver, when mounted in a 20 litre enclosure designed for stand mounting, may exhibit a well-behaved axial response continuing to rise from approximately 400 Hz to 2 kHz and possibly beyond. If the criteria of a balanced and uniform axial sound pressure response is the objective, then equalization must be applied. With active crossovers an equalization stage may simply be added to the filter system, while in the case of the passive crossover filter, the equalization is usually integrated with the crossover itself.

In the case of the 200 mm driver (Figure 6.29), the upwards response slope may be seen to approximate to 4 dB/octave, which characteristic could be compensated by a suitable series combination of inductance and resistance. When setting the values, the motor coil inductance must be taken into account, the latter typically of the order of 0.35 mH for a long-throw design. Approximately 2.2 mH of additional series inductance is required for the equalization, and the inductor is split into 1.6 and 0.6 mH sections, the latter with a parallel 22 Ω resistor (Figure 6.30). The latter reduces the slope to match the rate of rise at the higher frequencies.

The third-order low-pass 'T' configuration crossover may be successfully combined with such equalization by adding the necessary capacitor at the junction, and dimensioning it so that the desired turnover frequency is obtained (Figure 6.31). The Butterworth relation now of course no longer holds for the crossover and due note should be taken of the overall response of the filter/equalizer. With the configuration described, a peak may develop at the crossover point, which is usually adequately damped by an additional resistance; fortunately this is already present in the example given above (R in Figure 6.30(a)).

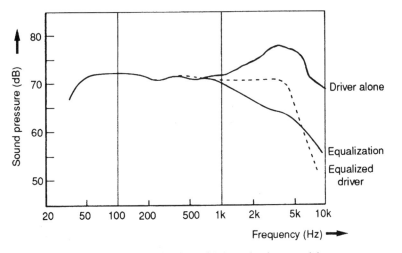

Figure 6.29.   Equalization of 0.2 m plastic-cone driver

Where large series inductance is required for equalization, the ability to readily shape and, if necessary, peak the response in the range close to crossover can be very helpful in fine tuning the amplitude and phase response of the system, particularly on the required listening axis. A wide range of responses may be readily obtained.

In Figure 6.30(b) an alternative shunt form of equalization is shown. This is useful for controlling a rise in output existing over a limited frequency range, perhaps half to one octave.

Finally, a useful circuit often employed to equalize a peak in response in dome HF units is given (in Figure 6.30(c)). This is a fairly low Q circuit and where a narrow band compensation is required more elaborate steps are necessary.

In a particular example using a dome HF unit the network was nominally 12 dB/octave but the acoustic output was modified by a significant 5–6 dB rise at fundamental resonance. Here the shunt inductor was set rather lower than usual (by 30%) to aid compensation. The driver also exhibited a first mode of rather high Q at 21 kHz, considered undesirable, and it was subjected to equalization. A series trap was insufficient and could not be produced to a sufficient notch depth. It required terminating with a second shunt network across the driver terminals. Above 21 kHz the system then inevitably presents a rather low and capacitative input impedance (Figure 6.31(b)).

When several filter sections are cascaded, as in the bandpass arm for a mid-range driver, further unwanted interaction may occur and hence modification of the calculated values is to be expected. Commercial crossovers are shown in Figure 6.32a,b.

Acoustic Centre and Delay

Advanced calculation of crossover networks including predictable phase control is aided by knowledge of the acoustic driving point relative to the mounting plane, in order to assess the relative delay between a mid and a high frequency unit.

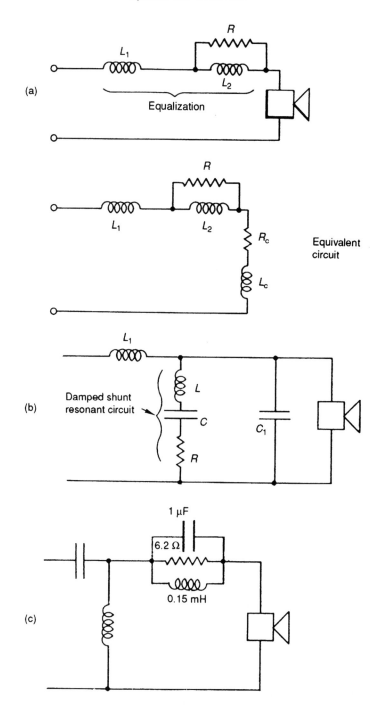

Figure 6.30. (a) Passive equalization; (b) damped shunt resonant circuit to provide equalizing dip in crossover characteristic (L is often combined with R by suitable choice of wire gauge); (c) series resonant circuit to provide equalizing dip in response to correct for a peak in an HF unit response (typically 12–15 kHz for a dome unit — example values are for $f_d = 13$ kHz)

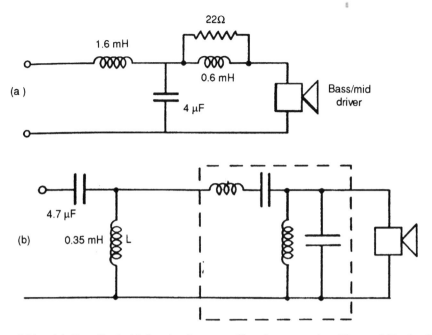

Figure 6.31. (a) Equalized third-order low-pass filter incorporating Figure 6.30. $f_c \simeq 3\,\text{kHz}$, 18 dB/octave; (b) terminated series/shunt notch filter for high Q dome resonance

For dome tweeters, the driving point is very close to the rim of the dome. For a cone driver, the apex, at or just below the dust cap, is a good starting point. For a typical 170 mm driver 2 in or 51 mm delay is used; perhaps 1.5 in (38 mm) for a shallow, large voice coil type. Relative to the mounting plane (flush) the tweeter delay will be in the range of 2–4 mm, which is almost negligible in phase terms.

The 170 mm example cited would show a practical delay of 180 μs. When a crossover frequency is decided, note that additional factors will affect the phase shift present due to the relative delay, e.g. the electrical inductance of the driver voice coil, together with a contribution from its natural, low-pass, characteristic frequency response.

For example, if a 3 kHz crossover point is selected and the driver has a common 12 dB/octave second-order filter together with a natural rolloff at 6 kHz, then this intrinsic low-pass characteristic will result in an additional phase shift at 3 kHz:

$$\text{delay } (\mu s) = \frac{\text{acoustic distance (in)}}{13\,500}$$

or

$$\text{delay } (\mu s) = \frac{\text{acoustic distance (mm)}}{343\,000}$$

For a given frequency f_c, the phase shift (in degrees) $= \text{delay} \times 360 \times f_c$.

Note that the effective driver delay is somewhat dependent on frequency. For example, the 25 mm tweeter delay does not settle down until above 4 kHz, somewhat

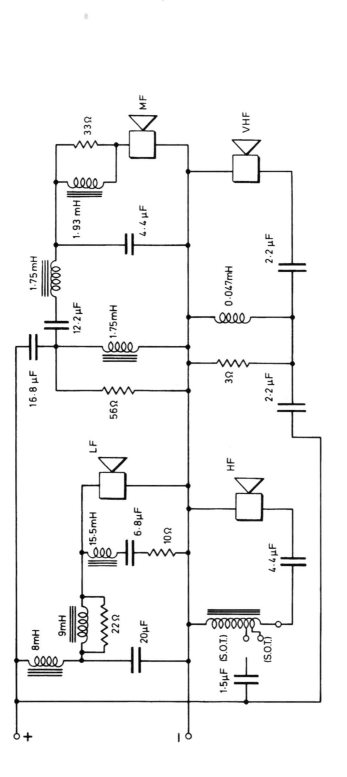

Figure 6.32. (a) Commercial high performance four-way third-order crossover (Spendor BC3). Crossover points at 500 Hz, 3 kHz, 13 kHz. Adjusted on test to match HF sensitivity. HF unit rolls off naturally above 14 kHz. LF is a 305 mm Bextrene cone, MF a 200 mm Bextrene cone, HF a 38 mm dome and VHF a 19 mm dome (see Figure 6.3) (courtesy Spendor Ltd)

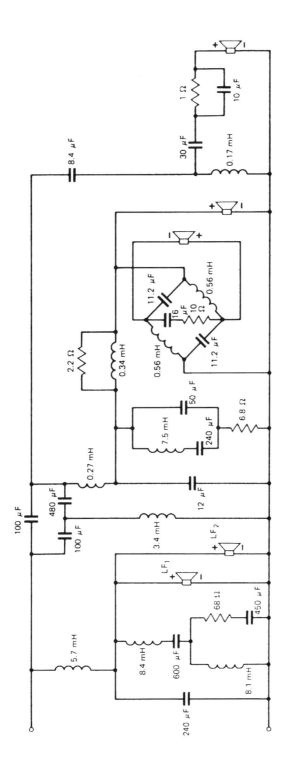

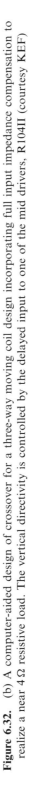

Figure 6.32. (b) A computer-aided design of crossover for a three-way moving coil design incorporating full input impedance compensation to realize a near 4 Ω resistive load. The vertical directivity is controlled by the delayed input to one of the mid drivers, R104II (courtesy KEF)

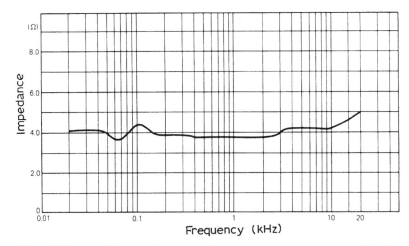

Figure 6.32. (c) The resulting load impedance for the compensated system of (b). Between 20 Hz and 12 kHz, it lies within $\pm 0.5\Omega - 4\Omega$

clear of the fundamental resonance. By 1.6 kHz the effective delay may be halved by the additional effect of the approaching main resonance. These changes are precisely in the region where phase control is important in order to predict crossover design and significantly complicate the issue.

This also helps to explain why, during the development of a speaker design, it is worth trying total phase inversion to the tweeter. That 180° of phase shift may well provide sufficient freedom to realign the networks to achieve the desired phase integration through the crossover region (see Figure 6.33).

Passive Delay

Several designers have tried passive delay networks to bring dissimilar drivers into the same time envelope. Multiple half-section networks have been used and have proved costly as the cumulative loss introduced must be held to sensible levels. Recently an all-pass delay network has been tried which gives an attractive theoretical solution. In practice, the basic network may be confounded in production by response ripple in and out of band, plus unwanted phase shift, partly resulting from mis-termination. It is suspected of causing as many problems as it solves (Figure 6.34).

Crossover Components

Modern high performance loudspeakers often need to handle considerable power, up to 200 W programme, equivalent to the rated continuous output of an amplifier when driven to clip point on programme of average energy distribution. Clearly the crossover network should exhibit low losses and possess an adequately high voltage and current capacity.

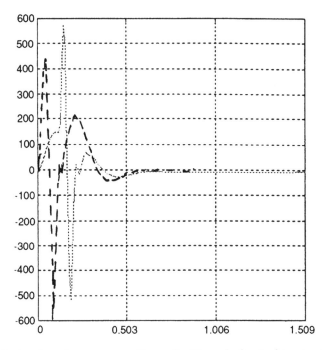

Figure 6.33. System impulse response with woofer 35 mm in front of tweeter (· · · · ·) and with the tweeter 35 mm in front of the woofer (– – – –) [14]

Cored Inductors

In the case of an inductance in series with an LF unit, peak currents greater than 30 A are possible. With the sizeable inductance values required for third-order 8 Ω networks at crossover frequencies in the 250–500 Hz range, it is very costly to produce air-cored components of sufficiently low loss, due to the winding resistance. Alternatively winding resistance may be exploited — to provide additional control of crossover filter Q or to reduce electromagnetic damping in an over-fluxed driver.

The negative peak current drawn by a loudspeaker driver near resonance may be rather higher than anticipated due to the back e.m.f. generated by moving mass. This can appear sufficiently delayed to sum with a following programme transient. The main series inductor to a bass driver may need to handle a short-term peak current much higher than the r.m.s. power input to the driver would suggest [8].

Tests using simulated programme indicate that a number of 8 Ω rated impedance, two- and three-way systems can show a transient or dynamic impedance of 25%–30% of their rated value. (See Chapter 9 on 'Impedance'.)

By incorporating a magnetic core the inductor winding length may be considerably reduced, but the core itself must not be allowed to enter saturation in service. Where ferrite cores are used, the ampere-turns rating is proportional to the core diameter for a given saturation point, the latter usually taken to be a 1% distortion level. If low distortion ratings for the system are important (this is particularly relevant in view of

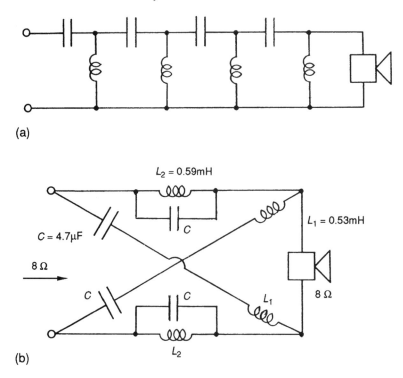

Figure 6.34. (a) High-pass ladder delay network; (b) all-pass symmetric delay network (for 25 mm dome, 200 mm bass, $f_c = 3$ kHz). Network is for $8\,\Omega$ but is mis-terminated in practice (delay $= 152\,\mu s \equiv 5.2$ cm) (courtesy Tannoy)

the odd rather than even harmonic nature of magnetic core distortion), then the core flux must be derated by as much as 50%.*

If an inadequately rated inductor core is driven to saturation, then the incremental inductance decreases. This is due to the failure of the core flux to continue to rise linearly with the increasing current through the component. Such a loss of 'dynamic' inductance may produce a sudden reduction in the system impedance, particularly in low-pass sections. This is likely to induce premature overload in the driving amplifier and consequently increased distortion. The overall result is a characteristic 'cracking' sound often incorrectly attributed to the amplifier.

In 100 W systems, ferrite cores of 19 mm diameter are adequate for inductors up to 5 mH, while at lower power levels, e.g. up to 50 W, 12 mm cores are satisfactory and the low power, two-way 25 W systems commonly utilize cores in the 9.5 mm diameter range.

Well-designed conventional air-gapped transformer type cores are advisable when large inductance values with both low loss and high current ratings are required. Ferrite cores can show additional problems in critical applications. Some grades exhibit notable hysteresis and can mask fine subjective detail, while magneto-striction is another side-effect whereby the inductor core vibrates as the applied current varies, with recovery to quiescent state accompanied by a delayed electrical distortion signal

*See 'Critical Aspects'.

induced in the output current. BBC experience so far suggests that plain, ordinary grade silicon iron laminated cores are subjectively favourable, while good results have been obtained from permalloy and similar specialized magnetic alloys.

Capacitors

Electrolytic capacitors have acquired a poor reputation due to historically wide tolerances, poor stability and a high loss factor at the upper audio frequencies. These criticisms are now less justified, since excellent electrolytic capacitors specifically designed for crossover use are now available. More recently high performance versions have been developed, which offer low loss and greatly improved tolerance and stability.

Life tests on normal grade ($\pm 20\%$) components have shown long-term stability of better than $\pm 2\%$ (two years service), while for a premium they can be obtained to the 5% tolerance advisable for the more critical applications. In the case of high performance systems the capacitors may be selected to a 2.5% tolerance by the speaker manufacturer. The usual rating of 50 Va.c. allows a reasonable safety margin with 8 Ω based systems, up to 100 W programme.

In networks where heavy equalization is involved, peak voltages may rise considerably above the input value and the network should be properly analysed to assess the required ratings; in some cases up to 150 V a.c. may be necessary. Again in the highest quality applications, loss factors such as variable dielectric memory, and the absorption factor of reversible electrolytics, may not be good enough. Tests have shown that under music-related impulse conditions (rather than steady state sine wave excitation) a bipolar electrolytic (reversible) cannot decide how to self-polarize and on the initial transient it may show a degree of asymmetric rectifier-like behaviour. Several designers now bypass electrolytics with smaller 5%–10% of value plastic film types. Many prefer not to use electrolytics at all.

Where lower loss and closer tolerances are required, plastic film capacitors provide the answer. Some varieties have the additional advantage of being self-healing in the event of a transient voltage overload. Values of 20 μF and above are costly, and are usually built up of combinations of smaller components. Nevertheless where a high quality, high power 50 μF or 80 μF capacitor is required, the solution is to employ one of the special types designed for the purpose.

Film capacitors can also exhibit physical vibration and self-resonance; indeed one proposed HF unit employed an ordinary capacitor as the motor element. Their recovery after transient excitation is a potential source of delayed resonances, though these are at present considered rather less serious than those in drive units and enclosures.

While plastic film capacitors are widely used in critical applications such as mid and high-frequency crossover sections, the large values in the bass section are often still of the electrolytic type. At low frequencies the loss factor is usually sufficiently low, and the residual internal resistance may be accounted for in the design.

Circuit Geometry

Due to the presence of potentially high currents in the inductors and the possible mutual coupling of thoughtlessly orientated adjacent inductors, the construction and layout of a crossover network is highly important. If inductors are closely spaced (separation less than 20 mm), then they should always be positioned at right angles to minimize the interaction. If a printed circuit board is employed, then the conductor foils should be of adequate breadth to ensure a low resistance, and in addition the track layout should be designed so that common return paths for the filter sections are avoided. The designer should also note the potential for an inductor to induce current in a conductor track beneath it.

Critical Aspects of Crossover Components

In critical applications screening of the inductors has been proposed (transformer-type and pot cores are self-screened) and physical separation of high- and low-pass sections has proved worthwhile. With high peak currents circulating in a complex crossover, sufficient foil widths and thickness are also necessary. Likewise, for electrical connection to the system, strong recessed binding posts are to be preferred to spring clips and the like.

Here some general observations cover design aspects which can help improve sound quality. While several designers are obsessed with air-core inductors on the basis that magnetic 'core' distortion is absent, they fail to realize their disadvantages of bulk, cost, stray flux radiation and low Q. Low resistance, high inductance value air-core inductances used for low impedance designs can be massive affairs with 50 m or more of winding. Subjective clarity is believed to be impaired by such long windings. The sensible use of magnetic cored or even full transformer-type inductors (with appropriate anti-saturation air gapping) can greatly reduce winding length and also improve Q values. Iron dust cores are more expensive than ferrite and at low signal levels give moderately raised distortion figures of 0.1%–0.15% as compared with 0.02%–0.05%, while at high levels the distortion with ferrite rises more rapidly into saturation. To its advantage the iron dust type saturates more gently and this is considered to be less audible subjectively, given the usual presence of non-linearity from a moving coil bass unit at typical power levels (15–100 W). With bonded end caps added forming a 'dumbell', the larger iron dust cores offer similar permeability (inductance gain) to the simple ferrite cores. Transformer cores (E and I or equivalent patterns) in 50% Mu metal (B-grade Radiometal) perform very well using appropriately thin laminations, and allow very short winding lengths to be used. Such inductor forms also have the advantage of a low stray field but are costly.

Where resistors are required in a crossover, e.g. for damping purposes, where possible the wire gauge in an associated inductor may be chosen to incorporate all or most of the required resistance. The use of resistors or 'attenuators' is often detrimental to subjective clarity. Whenever possible the respective driver's sensitivities and/or impedance should be adjusted by original design to provide the required matching.

In top quality systems any unnecessary devices added to a crossover may reduce performance, for example, non-linear protection elements. The claimed benign properties of such devices cannot be taken for granted.

The choice of crossover capacitors can be important and specially made copper leaded polypropylene capacitors are now available in a wide range of values up to 120 μF and are becoming more widely used. Their use generally results in more detail in the mid-range and a greater treble purity.

Component Non-Linearities in Greater Detail

The standard components used for loudspeaker crossover networks perform well and millions of reliable products have been made using electrolytic capacitors and normal cored inductors. Where very high standards are demanded better performance is possible, at a price. Often the design decisions are based on careful subjective assessment, where measurement does not provide the full picture.

In a given system the unresearched replacement of standard components with low-loss types may not be a good idea since the system alignment may well be adjusted to account for losses, for example the Q values of a given resonant network used for compensation, or the relative levels of two frequency bands, or the intrinsic sound quality of the parts themselves. The casual upgrade of a crossover network can worsen the results.

In a context where the amplifier that is used to drive a loudspeaker has been subject to extensive research with regard to the sound quality of its component parts in addition to the fundamentals of good circuit design, it is logical that the equivalent components used in a matching loudspeaker system be subject to the same scrutiny.

Capacitors

In controlled subjective testing, usually high quality stereo precisely arranged, different makes and grades of bi-polar capacitor may be distinguished. In a familiar reference audio system with well selected speaker drivers, preferences may well emerge. In such tests obvious component losses may need to be accounted for. Some designers have reported audible differences with regard to directionality for crossover components and, where possible, directionality, e.g. the start of windings for both coils and inductors, should be identified and once decided, kept to in production.

There is no mystery to these differences. For example, electrolytic capacitors suffer from small but identifiable errors which, according to type, vary proportionately with frequency and include distortion, dielectric losses and series resistance. Subjectively there may be mild colouration, a loss in sensitivity, frequency emphasis and a mild impairment in dynamics and clarity.

Film capacitors have far lower losses both for their dielectric and equivalent series resistance, but still show differences in subjective sound. Each of the common film materials has its own character which might either aid a given system, or might have a negative effect. For example, polypropylene is generally favoured on the grounds of measured losses which are very low (df at 0.003 or better) and reliably good sound. Yet in crossover networks there are still audible variations according to make and type, i.e. the working voltage. The higher the rated volts the better, but cost is also proportional to voltage. Build quality matters, as in all things — the tightness of the wind, the termination method, and whether the conductive layer is foil or metallization. Recently, more complex geometries have been introduced to reduce unwanted self-inductance, e.g. by distributed windings. Listening tests are essential to check whether such enhancement deliver better sound in the application. Polystyrene is the best sounding film but is impossibly costly in sizes suitable for power networks. And yet in some systems relatively economical polyester film capacitors (df 0.05 typically), regarded by critics as somewhat 'forward' and hard sounding, may provide the right tonal balance with a particular design.

What is known about film capacitors is their microphonic tendency. When transmitting audio power they vibrate due to the piezo property of the film, while conversely vibration from the loudspeaker itself can cause mild colouration to be induced in the electrical output of the component. An anti-resonant wrap or jacket can be helpful. Low vibration mounting methods, full resin potting of the crossover, or removal of the crossovers from the cabinet can all be helpful in minimizing these effects.

Resistors

Resistors also exhibit low order non-linearities, for example due to their temperature coefficient, the variation of value with temperature. Higher power components tend to sound better than low power types, while those with very low 'tempcos', e.g. the bulk foil designs, are significantly better sounding but are hugely expensive. Resistors do have a 'sound', typified as a loss in clarity and immediacy, and where a speaker can be designed by adjusting coil impedances and/or magnet flux to avoid series resistance attenuators, the sound is invariably better. This is one of the advantages of active drive. Note also that speaker distortion increases with series impedance since it is expressed in the input current to the voice coil.

Inductors

Inductors have their own problems. If air core, considered the most neutral sounding owing to the absence of a magnetic core, note that the long winding length required to achieve larger values, say over 1 mH wire has a 'sound' and the longer the path, the greater is the effect. At the amplifier output, enthusiasts strive to employ short speaker cables for greater clarity. Thirty metres of ordinary coil wire inside the speaker, in the inductor to the bass unit, seems to defeat the object. Part of the problem lies in the indifferent dielectric properties of the enamel used to insulate the so-called 'magnet' wire. It does not make good speaker cable and experimental trials with PTFE coated wire coils tend to confirm this.

Other factors concerning air core inductors include the large stray magnetic field. This requires careful orientation with respect to other components. The typically higher winding resistance reduces system sensitivity, and the significant skin effect and self-capacitance, these generally less important with the shorter windings typically present when using magnetically permeable cores. Audiophile grade inductors are available, e.g. flat ribbon windings that minimize capacitance and skin losses, or the use of Litz (individually insulated multistrand wire) conductors.

Inductor Cores: Distortion Data

Cores have had a bad press, mainly because of their rough-sounding distortion which occurs when they are strongly overloaded. With good design this should never happen and with a sensible choice of core, type and size, distortion can be very low, while the winding gain allows typically one-third of the copper length for a given value thus reducing wire losses of all kinds.

Cores can be of ferrite, the audio grades perform very well either as self-screened pot construction, or open bars, or of powdered iron dust or of iron transformer laminations, as an E or I section or as a transformer core, linearized by an air gap of typically 2–4 mm. Powdered iron has proved popular in recent years owing to its softer overload characteristic, but there are disadvantages too. Permeability is lower, requiring more wire, the distortion at normal levels is higher, mainly third harmonic, and the cores are partially conductive, a few ohms per square, indicating special care with the winding to avoid shorts to the core. Iron transformer cores are acceptable in the bass, but often show serious frequency-dependent loss and inductance, such as a 15% change from 100 Hz to 1 kHz. Fine laminated Mu metal stacks are better (frequently used in BBC designs) and are wire efficient. Audio grade toroidal cores can be very good indeed with a low stray field and high power handling.

Figures 6.35–6.39 show some measurements of distortion for various cores. Taking the large iron dust type first, 45 mm dia. by 45 mm long, with end caps; 100 Hz drive at 20 W, value 8 mH, the distortion is 0.36% with a wide harmonic spread. Saturation is progressive with no sudden break in linearity, but the distortion, mainly third, does not improve much at lower levels, and is typically 0.25% over a wide range of frequencies and powers (Figure 6.35).

In Figure 6.36, using crossover grade ferrite, 20 mm dia. by 75 mm long, the distortion is still as expected from a magnetic material, third harmonic, but is much lower in level, 0.04% at the 20 W test, remaining around 0.03% over the range of

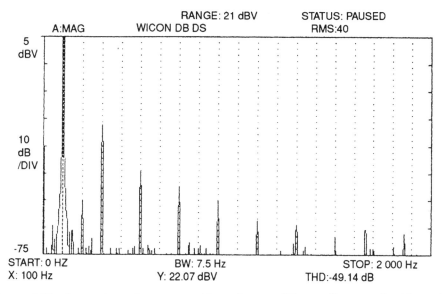

Figure 6.35. Large iron dust core with end caps (45 mm dia. overall, 45 mm long). Distortion = 0.36% with a long harmonic spread. Saturation is soft and progressive, but distortion remains at the 0.2%–0.3% level throughout

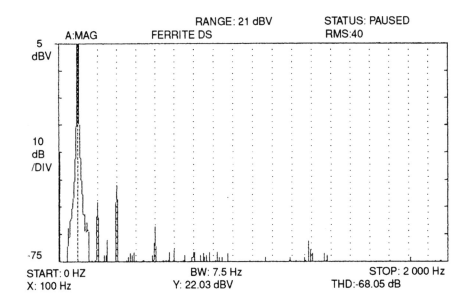

Figure 6.36. Crossover grade ferrite core, 20 mm dia., 75 mm long, thd 0.04%, typically 0.03% over the frequency range. Similar second and third, mild fifth harmonic at −82 dB

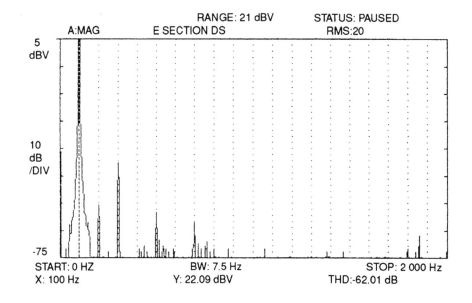

Figure 6.37. 22 mm thick silicon–iron E section laminations 40 × 66 mm. Better than 0.1% thd, mild upper harmonics, typical levels of 0.08%

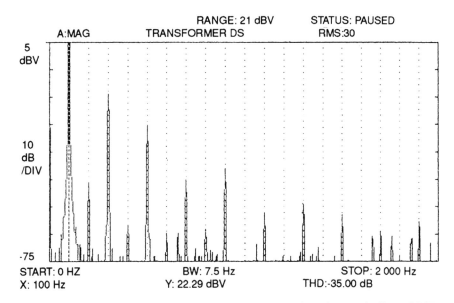

Figure 6.38. 8 mH, 20 W delivered to 8 Ω at 100 Hz. Distortion typically −35 dB overall, steadily increasing with level. Laminations 24 mm deep, 5 cm × 6 cm frame, silicon–iron with 1.5 mm air gap, E and I core. Thd at 20 W, 1.5% predominantly odd order with large harmonic range. (Audio grade permalloy or Mu–metal is far superior, typically 0.1%, and up to 50 W better than 0.4%)

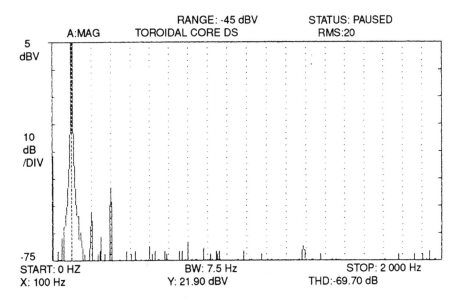

Figure 6.39. Toroidal core is 16 mm square section, 70 mm outer dia. made of good ferrite. Note low distortion of 0.038%, similar second and third. High saturation level > 50 W. Typical distortion is 0.035% wideband

powers and frequencies. Desirably, third and second harmonic levels are not too dissimilar and higher harmonics are negligible, e.g. fifth is $-82\,dB$. No saturation was observed at over 100 W.

A silicon iron cored audio transformer type, 22 mm thick, 40 mm × 66 mm overall, air gapped, gave 0.1%, a good result at low frequencies (Figure 6.37) but the results were poor at high frequencies. Compare this with a low-cost transformer design using power supply grade iron. The distortion was 1.5% which increased steadily with power, e.g. 7% at 50 W and the harmonics were rough, odd order, reaching out to 2 kHz (Figure 6.38).

Finally, a toroidal core (ferrite), 16 mm square section, 70 mm outer dia., was tested (Figure 6.39). At 20 W the result was an excellent 0.038% with only second and third evident and it varied little with power or frequency. It is wind efficient and so copper losses were also low.

Wiring

Internal wiring should not be taken for granted in a speaker. A tight twist is optimum, thus reducing the influence of stray fields, and it is mechanically more stable. Wiring grade is audible in critical applications and should be selected on similar grounds as that used for high quality external speaker cable. It should not be allowed to rattle around inside the box and should be routed away from the driver magnets. Random movement of internal cables may affect the subjective dynamics.

6.3   THE AMPLIFIER–LOUDSPEAKER INTERFACE

It is accepted practice to assume that an amplifier is a voltage source of effectively zero source impedance [13] and that loudspeakers or loudspeaker systems are linked to the amplifier by an appropriate two-wire cable of negligible loop resistance ($<0.2\,\Omega$). For 95% or more situations this is a safe working assumption, while for applications requiring the highest quality and where considerations of clarity and transparency are paramount and where bass dynamics and treble purity are highly valued, a more critical approach to the matching of amplifier and loudspeaker can be rewarding.

Aspects such as current drive are dealt with in Section 6.6, while this section will examine the possibilities available with passive systems where the fundamental aspects of design and performance remain as the designer intended.

Valve/Tube Amplification and Influence on Speaker Performance

In recent years there has been a significant revival in the use of thermionic power amplification. The products of the older more established manufacturers tend to be satisfactorily powerful, push–pull designs with a moderate output resistance of under $0.5\,\Omega$ and which match well to a wide range of loudspeaker systems. However, there is an increasing proportion of designs with lower values — even 0 dB — of negative feedback, determined on grounds of sound quality, combined with output stages of the

SE (single ended) type, generally a single output valve (tube) operating in class A. Power delivered to an 8 Ω load for these designs is typically in the range 6–15 W at a peak distortion of between 2% and 5% (if of low order) and with a relatively high output impedance of between 1.5 Ω and 4 Ω. On the face of it, this is a most unlikely specification for good hi-fi performance and contradicts many of the premises established for accurate speaker design in regard to a low value of source impedance feeding a loudspeaker. Nonetheless, a number of these designs have found significant favour for their sound quality and encompass a wide range of prices, including the luxury end of the market. If these 'revival' SE amplifiers remain popular, then a new class of loudspeaker must be devised to make best use of them. On first inspection a classic horn system is appropriate, e.g. a Klipsch or Voigt three-way, because these designs provide a very high sensitivity of well over 100 dB/W. There will still be some interaction between the amplifier source resistance and the variable impedance of the speaker with frequency. Combinations of amplifier and speaker need to be explored by the customer to check on compatibility. In practice, the ideal speaker would be a direct radiator of high 96 dB/W sensitivity plus a nominal impedance of 10 Ω with the design carefully adjusted for an impedance variation of less than ±2 Ω over a suggested 50 Hz to 15 kHz frequency range. A consequence of the high sensitivity would be a larger than usual enclosure size for the expected bass response, probably 4 ft^3 for a room-matched 30 Hz extension. Such an enclosure would have to be carefully built, cross-braced and loaded, in order to achieve the low colouration now commonly enjoyed with smaller enclosures of lower sensitivity but similar bass extension.

Bi-wiring and Multi-wiring

The term bi-wiring arose when extensive experiments showed that with many two-way systems a small but worthwhile improvement in sound quality, independent of timbre or frequency response, could be obtained by electrically separating the crossovers for the two frequency ranges and providing separate two-wire cables back to the amplifier terminal for each frequency section (Figure 6.40). This technique is proving increasingly popular with high quality designs and can provide the sort of gain achieved by replacing a budget amplifier with one of superior quality.

Polyamplification

As a simple extension to multi-wiring for the separated crossover input ports of a loudspeaker system, where finances permit, similar improvements in sound quality have been observed by using separate power amplifiers to drive each frequency channel. Under the general term 'polyamplification', for a given stereo channel, power amplifiers are provided for each crossover port. All the amplifier inputs for that channel are connected in parallel, without the need for pre-filtering. Improvements in clarity, stereo focus, precision and bass transients are evident. One explanation is that while each amplifier section is fed the whole frequency range, audio power is only drawn in the frequency range appropriate to the passive crossover section to which it is connected.

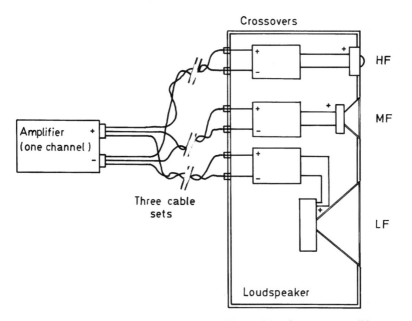

Figure 6.40. Arrangement for passively tri-wiring cables from an amplifier to a three-way loudspeaker system. Generally provides improved clarity and focus. (Japan Pat. 1976, Toshiba)

Amplifier distortions are in general dependent on the frequency range, complexity and power delivered. If the power delivered to the load is filtered, intermodulation is reduced, whether or not the filter is before or after the amplifier (but less so if after the amplifier).

Polyamplification offers a simple route to multi-amplified 'active' systems without the need for custom-designed electronic active crossover filters, although the full benefit of an active system is not attained.

Speaker Cable Practice

For maximum benefit from bi-wiring or other aspects of high quality sound reproduction some simple rules have been established for speaker cabling.

1. Cable runs should be as short as possible.

2. Lengths for the left- and right-hand channels should be equal, any spare cable should not be wound into a coil; it should be folded non-inductively (zig-zag) and secured neatly.

3. Virtually any kind of conductive wire will work but subtle sound quality differences are present between cables. These are significant in critical applications. Some care and expense on cable is justified and, in critical market areas, up to 20% of the loudspeaker system cost may be allocated to cable. For the industry as a whole the cable allocation is unfortunately set at typically 1% of the price of the speakers.

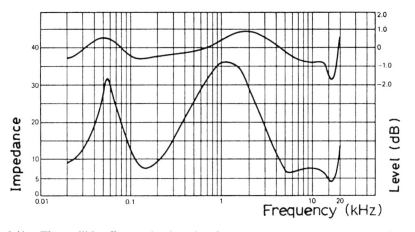

Figure 6.41. The audible effect on loudspeaker frequency response (upper curve) due to cable impedance reflected on the impedance of the loudspeaker (lower curve). The low grade cable used was bell wire, 10 m at 0.1 Ω/m of loop resistance, a total loop inductance of 6.5 μH and negligible capacitance

Cable Design and Sound

It is easy to establish that cables can and do sound different simply on the basis of their conductivity and the effect of their series impedance (mainly resistance and a small series inductance) upon the generally non-uniform load impedance presented by most loudspeakers. With zero amplifier source resistance and zero cable resistance the frequency response of the loudspeaker will be as defined by the designer. Introduce some series resistance, say 1 Ω, and a nominal 8 Ω speaker will lose a just audible 1 dB of sensitivity. In addition, if it has the usual complex impedance varying over the frequency range, then this will be reflected as errors in the axial frequency response.

In the case of the example shown in Figure 6.41, the effect is to increase the relative output at the upper bass resonance, decrease the 200 Hz 'power band' present in the low mid-range and elevate the treble. Some sectors of the hi-fi fraternity have experimented with single strand speaker cable down to 0.4 mm diameter. In lengths above a few metres such thin grades add significant loop resistance, impair bass dynamics and impart response changes which outweigh any benefits. Thus, a sensibly low cable resistance is desirable in high quality applications, a loop resistance of $\leqslant 5\%$ of the nominal impedance is desirable. If the expense is justified in the quality of the loudspeaker, 2% is a worthwhile goal. Low inductance is also important since 10 m of typical parallel 'twin' cable adds 6.5 μH, a further 0.8 Ω at 20 kHz.

Effect of Damping Factor on Speaker Low Frequency Alignment

Low frequency alignments are generally calculated for zero source resistance including amplifier and cable. For drivers with good mechanical Q the damping is dominated by Q_{ES} and in the simple case is given by

$$Q_{ES} = \frac{2\pi f_s M_d}{(Bl)^2}$$

ignoring Q_{MS} and thus Q_{TS}.

With R_E at a typical 6 Ω and Q_T at 0.707, maximally flat,* consider an amplifier and cable combination with a damping factor of 16 for an 8 Ω rating, i.e. a source resistance of 0.5 Ω. Q_T must rise by around 11% to 0.77 and results, by itself, in little change in local frequency response. For example, the tiny peak due to underdamping is only 0.12 dB high, with a slightly larger variation at lower frequencies of perhaps 0.6 dB. On the other hand, the series resistance will attenuate the rest of the frequency range, with an audible difference on an A/B test. Some small variation in frequency response will also be present, reflecting the usual variation of a speaker's load impedance with frequency.

Interestingly, in tests with high quality audio systems using a range of matched cables, where the significant variable was limited to loop resistance only, the amplifier had a low source impedance and the speaker had a powerful, well-damped and extended bass response, cable resistance showed a greater effect than anticipated.

Unexpectedly, it proved possible to characterize sound quality differences in the range 0.2 Ω down to 0.5 Ω of total cable resistance, with the most solid and articulate low frequency sound related to the lowest resistance. The physical weight and rigidity of the cable was also proportional to lowered resistance, so it is possible that some electromechanical factor was also at work here. Accelerometer analysis has shown that at high currents speaker cable conductors do move relative to each other if not sufficiently restrained.

Loudspeaker cable does matter and many types have become available, largely designed on an empirical basis and, in consequence, marketed with a considerable degree of conflicting pseudo-technological claims which do the designers little credit. However, careful subjective analysis indicates that significant sound quality differences between cables do exist which are independent of loop resistance and impedance. From a review of a data base of 100 cable types, it can be shown that conductor type and purity, its state of annealing and any surface plating are relevant. In addition single or multiple stranding, including the use of differential strand diameters or separately insulated strands and/or more complex strand winding methods are influential. Geometry plays a part, for example flat twin, twisted pair, coaxial, planar ribbon and tube.

Finally the dielectric quality of the insulating medium matters together with mechanical properties such as stiffness and damping. For one cable design, a wrap of lead tape was added to increase the mass and damping to a high-power speaker cable, with a clearly audible result.

Cost-effective results can be obtained using 0.8–1 mm high-purity, single-strand conductor with polyethylene insulation arranged as a tight twisted pair. Such a cable offers low dielectric loss, a mechanically rigid construction, low inductance and moderate resistance, together with a well-defined geometry and conductive path.

---

*For the enclosure/system.

The Sound of Metal Conductors

A simple proof of the audibility of sound quality differences between silver and copper conductors was devised. A small high performance two-way speaker, which in manufacture was built to high precision and close tolerance using normal copper conductors, was replicated with the electrically conducting parts made in silver (99.99% purity or better). During its development individual components were built and auditioned, revealing interesting differences, but it proved impossible to anticipate the effect of executing the whole signal path in silver, this including the speaker cable.

Several aspects of sound quality were altered, with general agreement that there were not merely differences but genuine improvements from the substitution of silver for copper. Most interesting was the apparent reduction in mid-range colouration. A degree of previously cone-related colouration and upper mid 'hardness' showed improvement, indicating that a major component of this unwanted sound was due to the conductors. With this came an improvement in mid-range clarity and transparency. Transients were more articulate, while the bass sounded more precise and better focused. In the treble, the tweeter appeared to have a purer, sweeter sound, with reduced 'grain' and related subjective distortion.

AST

The name AST has been coined by Yamaha for an amplifier-speaker matching concept which may well prove popular. Standing for 'Active Servo Technology', it is hard to argue for either term, 'active' or 'servo'. The kernel of the idea is the adjustment of the accompanying amplifier's output impedance to some suitable low negative value in order to remove from the low frequency design equation the d.c. resistance component in the system which is otherwise dominated by the loudspeaker motor coil. With suitable design of bass driver, efficient small-box systems with reflex loading may be produced where the vent or port contribution is unusually effective due to the high electromagnetic coupling achieved between the amplifier and the bass diaphragm. Working a small vent hard requires sophisticated duct design. The several techniques employed to reduce windage noise and secondary resonances include the use of a felt-lined, semi-rigid duct made of rubber which is beneficial with respect to duct standing-wave pipe modes. Airflow shaping is also used to smooth the port exit (Figure 6.42(a), (b) and (c)).

The matching amplifiers, separate or integrated, are fitted with a socket to accept plug-in modules specific to a given speaker model which set the correct negative output impedance and also the necessary heavy bass boost equalization. Preliminary results indicate a superior power and clarity in the bass for the size of enclosure using this AST technique. Ultimately the alignment accuracy will be partly dependent on factors such as the length and resistance of the speaker connecting cable and on motor coil resistance changes with temperature. This is significant due to the heavier than usual power input at low frequencies required to drive the port to the higher than usual output level. Thermal changes in the driver can be modelled in the amplifier synthesis.

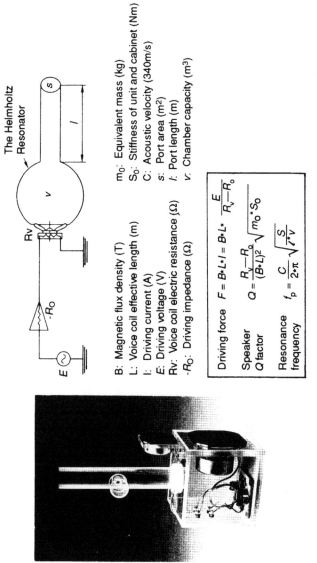

The Helmholtz Resonator

B: Magnetic flux density (T)
L: Voice coil effective length (m)
I: Driving current (A)
E: Driving voltage (V)
Rv: Voice coil electric resistance (Ω)
-Ro: Driving impedance (Ω)

m_0: Equivalent mass (kg)
S_0: Stiffness of unit and cabinet (Nm)
C: Acoustic velocity (340m/s)
s: Port area (m²)
l: Port length (m)
v: Chamber capacity (m³)

Driving force $\quad F = B \cdot L \cdot I = B \cdot L \cdot \dfrac{E}{R_v - R_o}$

Speaker
Q factor $\quad Q = \dfrac{R_v - R_o}{(B \cdot L)^2} \sqrt{m_0 \cdot S_0}$

Resonance
frequency $\quad f_p = \dfrac{C}{2 \cdot \pi} \sqrt{\dfrac{S}{l \cdot v}}$

Figure 6.42. (a)

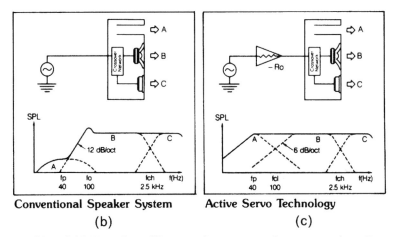

Conventional Speaker System
(b)

Active Servo Technology
(c)

Figure 6.42. (b) and (c) Use of amplifier negative output resistance to reduce Q_r to allow for equalized overdrive of the LF range achieving extended bass response from a small enclosure (after Yamaha AST)

Computer-aided Crossover Design

Programs [9–11] have been devised to synthesize crossover networks according to the target function corresponding to the desired acoustic outputs from the drivers. The drive unit parameters are entered — R_{dc}, L_c, f_0, etc. plus data corresponding to the intrinsic axial frequency responses. Trial networks may be proposed, according to the program, for second, third or fourth order and in effect the computer, via an iterative, successive approximation process generates an equalizer/crossover network. The resulting acoustic output corresponds to the required target function, e.g. fourth-order Butterworth (Figure 6.43(a),(b).

In the example the driver response is far from ideal in the intended crossover region and has been corrected by an equalizer which also includes the required crossover characteristic. Other refinements include the substitution of nearest standard value components with a second computer run to establish how close this constrained solution comes to the ideal. The effect of all tolerances may also be examined (see Figure 6.43(b)). (See Appendix A for software.)

6.4 ACTIVE FILTER CROSSOVERS

The consumer market at present shows a reluctance to accept loudspeakers which employ active crossovers with accompanying multiple power amplifiers, though this is likely to be overcome in the future. However, in theory this technique offers the greatest scope for the advancement of the loudspeaker art and a number of designs are already in production. Important benefits include:

1. A reduction in intermodulation distortion in the accompanying amplifiers due to their operation over a narrower bandwidth.

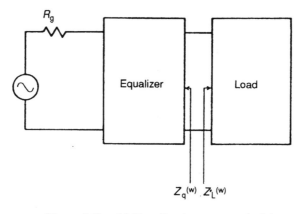

Figure 6.43.   (a) Equalizer/crossover principle

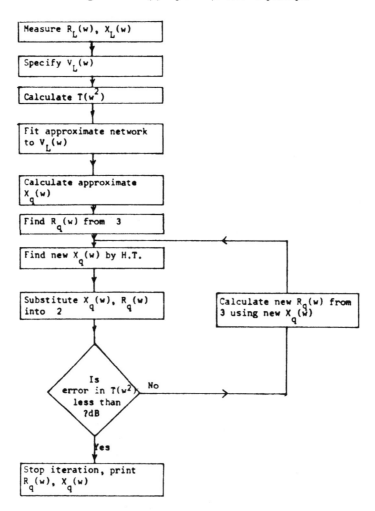

Figure 6.43.   (b) Flow chart for synthesis (after Jones [10])

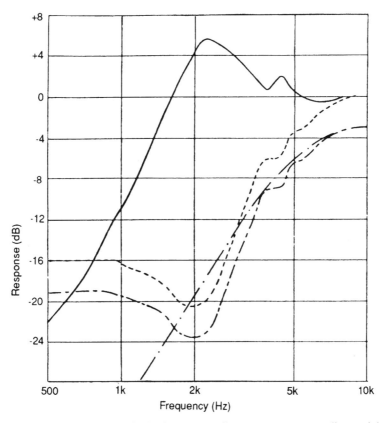

Figure 6.43. (c) Computer synthesized motor-coil response to equalize axial response irregularity at crossover. — · — Required speaker response (second order); ——— actual speaker response; — — — motor-coil response; ——— — sensitivity corrected motor-coil response (after Jones [10])

2. Subjectively, the performance of well-designed active systems exceeds expectation when comparison is made with the single amplifier/passive crossover alternative. Characterizations of 'louder' and 'clearer' are frequently made, and are believed due to the reduction in 'stressful' loading on the individual amplifiers. For example, when a main amplifier clips or enters distortion, as may occur during a momentary powerful bass transient, the distortion harmonics will be clearly reproduced by the treble driver in a passive system. In contrast, the electronic configuration keeps the bass amplifier distortion to the bass driver and the treble range remains clear and undistorted. (This is in fact a special case of the intermodulation improvement noted above.) This quality advantage still holds true for two-way systems with a typical crossover at 3 kHz, provided that the bass/mid-unit has a reasonable intrinsic rolloff above the crossover point. If not, a simple passive low-pass filter could be fitted between it and the respective drive amplifier.

3. Bass equalization may be readily incorporated in the active crossover. This is valuable if the low frequency alignment requires equalization.

4. The association between driver and amplifier may be beneficially extended to include the LF driver in the feedback loop of the matching amplifier (this results in the so-called 'motional feedback' or 'servo bass' designs).

5. Variations in driver sensitivity may be easily controlled via low-level gain control potentiometers.

6. Because each power amplifier feeds a single driver the overload protection thresholds may be more precisely set than is possible with the normal crossover and single 'universal' power amplifier.

7. The power amplifiers are directly connected across the terminals of each driver. The units are thus driven from a constant voltage source which will tend to suppress the fundamental resonance via electromagnetic damping (the degree depends on the driver Q. With treble units the absence of the usual series capacitor component in a matching passive crossover avoids the resonance problem previously discussed in Section 6.2, although the response falloff due to the motor coil inductance will still remain.)

   In addition, the amplifier output impedance can be made negative, if required, providing further control at the fundamental resonance.

8. Electronic filters provide a considerable variety of equalization, response shape and phase characteristics which would be unwieldy if not impossible to realize with passive networks. In addition, electronic filters may be easily adjusted during production.

9. Delay stages may be provided electronically in the signal paths to specific drivers allowing equalization of time delays which may exist between units in the system when optimally mounted. Such compensation is useful for the maintenance of symmetrical directivity in the lateral plane over the several important octaves near the crossover frequencies. Time-delay correction can also facilitate minimum phase design [4].

10. Active filters potentially have lower distortion than passive ones, due to the elimination of cored inductors and the use of high quality film capacitors.

A full treatment of active filters is not within the scope of this book. Details of the subject are well covered [5] but some basic circuits will be examined here together with examples of current practice.

An active crossover system (Figure 6.44) consists of low-level filter sections, gain control and equalization stages and a power amplifier for each filter/driver.

Theoretically an active filter is a more versatile device than its passive counterpart and response functions may be synthesized, which would not only be difficult but might actually be impossible to realize with passive networks. Its great advantage lies in the elimination of the inductor which is the least satisfactory electrical component. With the use of integrated operational amplifier units and sensible value precision capacitors, e.g. 10 nF to 1 μF, rather than the 1–100 μF required for the passive case, almost any filter characteristic may be readily and economically synthesized.

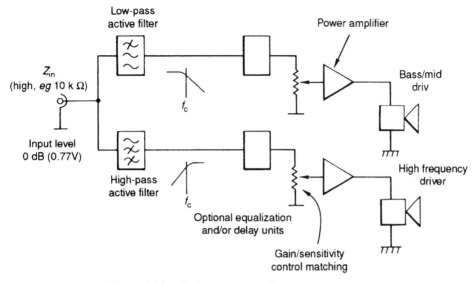

Figure 6.44. Basic two-way active crossover system

Second-order Low-pass Filters

Let us examine an active and a passive form of a low-pass 12 dB/octave filter with a 500 Hz cutoff frequency f_c. In the active form a suitable network comprises an integrated circuit amplifier together with two small capacitors and two resistors (Figure 6.45).

Though the cost of the power supplies cannot be ignored in the active case, it is usually shared with the power amplifiers and other sections of the crossover. The passive

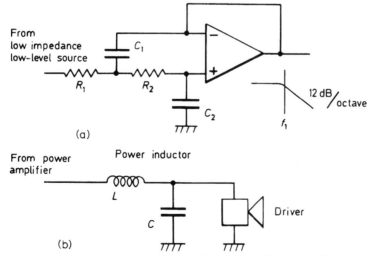

Figure 6.45. (a) Low-pass second-order active filter. (b) Passive power filter, second-order low-pass

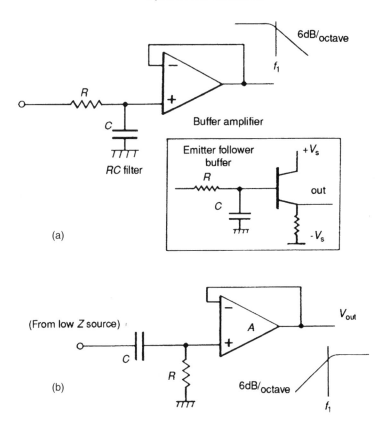

Figure 6.46. First-order RC filter; (a) low-pass form, (b) high-pass form (with buffer amplifier A)

counterpart (Figure 6.46(b)), while avoiding complications of power supplies, conversely requires a large inductor, 2 mH for example, with a correspondingly large matching capacitor. If the latter is built to the same stability and accuracy as the active filter component, then the total cost may be several times greater than that of the active form.

High-pass Filters

Figure 6.46 shows a high-pass second-order form of Figure 6.47(a).

First-order Low-Pass Filters

Simpler first-order filters are conveniently obtained with *RC* networks using an operational amplifier as a high input impedance, low output impedance buffer to preserve the theoretical responses. A simple unity-gain transistor stage, the emitter follower, may be used in less critical applications (Figure 6.46(a)).

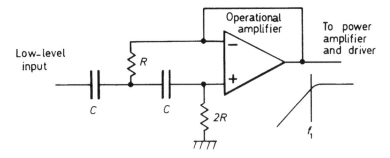

Figure 6.47. High-pass active filter, second-order

$$f_c = \frac{1}{2\pi RC\sqrt{2}}$$

More recently amplifier standards have improved to a level where a simple emitter follower imposes a significant impairment. High quality op-amps for audio use are to be preferred. At the highest quality level the designer needs to devote as much care to filter and power-amp design in an active speaker as he would to a good pre and power amplifier combination.

Higher Orders

Stages may be cascaded to provide almost any rolloff rate, but the cumulative effect of the rolloff frequency error must be noted, each stage response being additive. This also provides a means of controlling the initial slope and shape of the response via the location of the individual $-3\,\text{dB}$ rolloff points (Figure 6.48).

Driving Impedance

All the filters should be driven from a low source impedance. In a multiple-way system, an additional input buffer amplifier (Figure 6.49) will provide this, as well as imparting high input impedance to the entire active filter system.

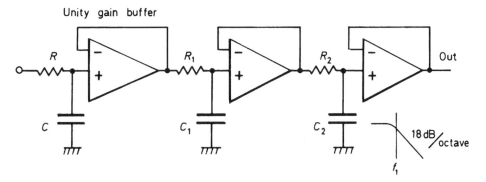

Figure 6.48. Third-order low-pass filter, via cascaded first-order sections

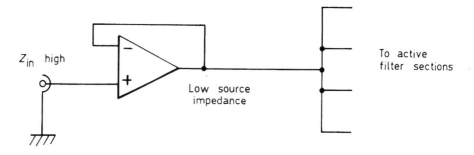

Figure 6.49. Input buffer (may be combined with audio bandpass input filter)

Equalization for Driver Sensitivity and/or Network Losses

Combinations of passive and active *RC* networks may provide almost any required broad acting equalization. Where considerable lift is necessary, the clipping headroom of the following power amplifier driving the loudspeaker must be considered in the context of the spectral content of typical programme.

In an active crossover transmission, loss in the equalizer networks may be readily recovered by adding an amplification stage or by adjusting the amplification of the corresponding op-amps in the filter to some suitable low value (Figure 6.50).

The following are typical examples of equalization.

Correction for Rising Driver Axial Response

Examining the circuit in Figure 6.51, at low frequencies below f_1, C represents a high impedance compared with R_3 and the network is essentially a resistive '*L*' attenuator composed of R_1, R_2 and R_3. At high frequencies, above f_2, C is low in impedance compared with R_3 and R_2 and the attenuation is established by the '*L*' attenuator R_1 and R_2. Between f_1 and f_2 the slope of the response may approach 6 dB/octave when R_2 is low and R_3 is very high, and may be reduced by any degree through adjustment of R_2 and R_3. The approximate 4 dB/octave slope required by the above example is not difficult to achieve over a limited span. If the second rolloff at f_2 is not required, then R_2 is simply reduced to zero.

A perhaps more elegant solution (Figure 6.51(b)) employs shunt feedback around a gain stage; with appropriate values the input impedance can be kept at a higher level. Where it proves necessary to mute the crossover to prevent transient noises at switch-on, a simple FET delay muting circuit can be incorporated.

Correction for Premature Driver Rolloff

Premature driver rolloff can be compensated by using a boost network such as that shown in Figure 6.52.

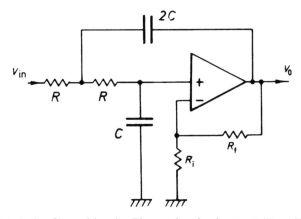

Figure 6.50. (a) Active filter with gain. The passband gain $A = V_0/V_{in} = (R_f + R_i)/R_i$ (if R_i is set, then $R_f = (A-1)R_i$

In this example the network is simply a resistive attenuator composed of R_1 and R_2 at

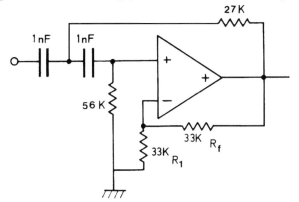

Figure 6.50. (b) A 2.4 kHz high-pass second-order network with 6 dB gain $= (R_f + R_f$

low frequencies where the impedance of C is high. At some frequency where the reactance of $C(Z_C = 1/2\pi f)$ equals R, the response will have risen by 3 dB and will continue to rise by 6 dB/octave thereafter. The slope may be reduced if necessary by adding a further resistor in series with C.

Other Irregularities

With good quality drivers it should rarely be necessary to apply a correction slope steeper than 6 dB/octave, and in practice this equalization usually is provided for final tuning and balancing rather than drastic compensation.

Rather quicker acting equalization may, however, be used, for example in conjunction with a calculated bass rolloff characteristic. A driver may be chosen on the basis of high

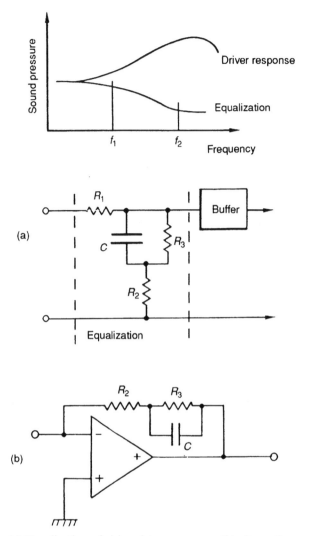

Figure 6.51. (a) Equalization of rising driver response, (b) alternative equalization with shunt feedbacks

mid-band efficiency which often implies a high Bl product with consequent overdamping at the bass resonance. Such a driver used in a sealed box might produce a system resonance at 40 Hz. Optimally damped, the output would be only -3 dB at that frequency. Consider a large magnet, high efficiency model, overdamped, such that a 6 dB loss occurs at 40 Hz, the response beginning to fall gently from as high as 100 Hz. The normal sealed-box rolloff of 12 dB/octave from 40 Hz has become a two-stage slope, 6 dB/octave from 100 Hz to 50 Hz, and approaching 12 dB/octave below 40 Hz. If amplifier headroom and programme considerations permit, bass lift at a 6 dB/octave rate may be applied to restore the low frequency output. A suitable network is shown in Figure 6.53(a). C_s is chosen to reduce the gain at subsonic frequencies (e.g. below 30 Hz) and hence to prevent the bass boost from continuing below that frequency.

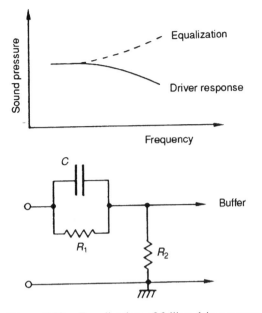

Figure 6.52. Equalization of falling driver response

More complex equalization may be used along these lines, one commercial example forming a tenth-order network in conjunction with the LF loudspeaker system.

Linkwitz [8] described a useful configuration (Figure 6.53(b)) which can be used for all driver/box systems, LF mid and HF, to take control of the system resonance, its Q factor and subsequent below resonance rolloff. An HF unit may thus be used at lower frequencies than normally considered practicable. In the example the circuit provides for a 24 dB/octave fourth-order acoustic output at a 1.5 kHz crossover frequency, despite the 25 mm HF driver possessing a fundamental resonance at 800 Hz at a Q of 0.8. The first stage comprises a damped twin 'T' notch filter providing phase and amplitude control of the unit's resonance. In addition a feedback-generated phase and amplitude equalizes the signal to provide a 12 dB/octave rolloff in the acoustic response. This is controlled with a 12 dB/octave filter, the second stage providing a described 24 dB/octave target function. Similarly the first stage can be used to control a high-pass system resonance and extend the response as desired.

In the Linkwitz HF example the driver was paired with a 110 mm mid-range unit, whose radiating plane appears 42 mm behind the HF. A three-stage HF delay network was used to synchronize the drivers allowing for the correct phase connection to their terminals, each stage delaying the treble output by 22 μs or 14 mm (see Figure 6.54).

Higher-order Filters by Direct Synthesis

Active filter theory is a versatile and powerful tool, and can provide numerous desirable filter characteristics. Ashley and Henne described a two-way third-order active Butterworth filter operating at 318 Hz which employed a single integrated circuit

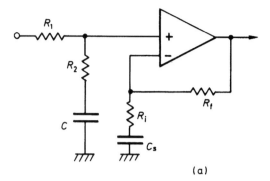

(a)

Figure 6.53. (a) LF equalization. The attenuation produced by step-down response is compensated by the gain network R_f, R_i, in which C_s is incorporated to provide a subsonic rolloff

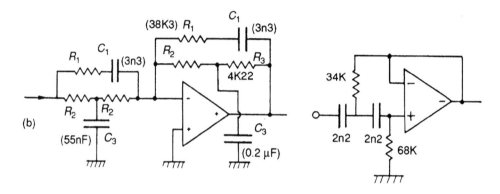

Figure 6.53. (b) Bass and fundamental resonance equalizer/compensator (after Linkwitz [8]). (Values in parentheses correspond to $f_0 = 800\,\text{Hz}$, $Q = 0.9$, crossover provides 24 dB/octave final rolloff at 1.5 kHz)

incorporating two operational amplifier blocks [6] (Figure 6.56(a)). With a total count of 14 small components plus an integrated circuit, and assuming that the power supply is derived from an accompanying power amplifier, this realization seems highly effective from both a cost and performance point of view.

This basic circuit may be scaled to fit almost any application. However, a buffer amplifier stage is a worthwhile addition to drive the two filters, since the total input impedance is well below 10 kHz. This may be compared with the equivalent passive filter from Figure 6.55(b).

Another example of a three-way active crossover circuit is given in Figure 6.56 and a complete system in Figure 6.57.

Gain Limitation in Active Filter Amplifiers

The rolloff rate in an active filter is only maintained as long as there is sufficient gain present in the loop. If the gain is not constant with frequency, when used at the higher

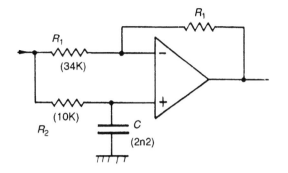

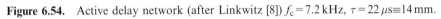

Figure 6.54. Active delay network (after Linkwitz [8]) $f_c = 7.2\,\text{kHz}$, $\tau = 22\,\mu\text{s} \equiv 14\,\text{mm}$.

$$\tau\,(\text{time delay}) = \frac{2R_2 c}{1 + (2\pi f R_2 c)^2}$$

where $RC \leqslant 1/20 f_c$, $f_c = $ crossover frequency

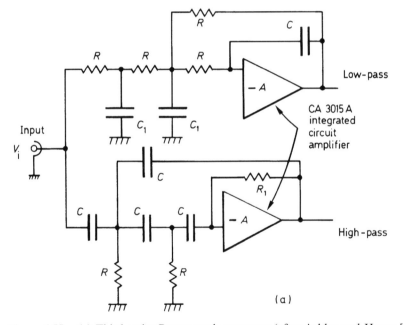

Figure 6.55. (a) Third-order Butterworth crossover (after Ashley and Henne [6])

$$\left.\frac{V_0}{V_i}\right|_{(LF)} = \frac{0.5}{s^3 + 2s^2 + 2s + 1} \qquad \left.\frac{V_0}{V_i}\right|_{(HF)} = \frac{0.55^3}{s^3 + 2s^2 + 2s + 1}$$

with f_0 at 318 Hz,

$R = 8.25\,\text{k}\Omega \qquad R_1 = 90\,\text{k}\Omega$
$C = 0.022\,\mu\text{F} \qquad C_1 = 0.15\,\mu\text{F}$

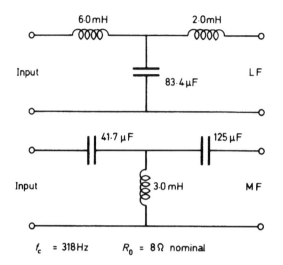

Figure 6.55. (b) Passive constant resistance Butterworth equivalent of Figure 6.56 (a)

audio frequencies, then the loop gain must be sufficient to maintain the performance. Up to now the active filter treatment has assumed perfect op-amp performance, at least over the audio band. This implies that the characteristically high input impedance, low output impedance and high gain of such a block are uniform over the working frequency range. The 741 integrated circuit is adequate for low frequency work, but a wider bandwidth type is recommended for mid and high frequency use, e.g. an OPA606 or equivalent.

When working near gain limits, operational amplifiers will understandably produce higher distortion. Some also suffer from crossover effects, since class A/B output stages are often used. The existence of these distortions should be borne in mind, and for highly critical applications a designer may prefer to use special amplifier units or even discrete circuitry in his active filters.

Several commercial high performance active crossovers employ special class A output, high-gain operational amplifier units of wide bandwidth and low distortion.

A Low-cost Example

An easy entry into active system design is afforded by modern high power monolithic operational amplifiers with output ratings up to 50 W into 4 Ω, delivered by a single TO220, five-lead package. The best examples have decently low distortion, class A/B complementary output stages and include protection against short circuit and excess temperature. It is possible to include an integral active crossover, connected directly to the respective loudspeaker driver terminals, which includes some measure of driver equalization. On the assumption that the high frequency unit is the more sensitive, alignment of the treble to mid-frequency balance may easily be attained via a small resistive attenuator placed at the input to the treble section. A low source of driving

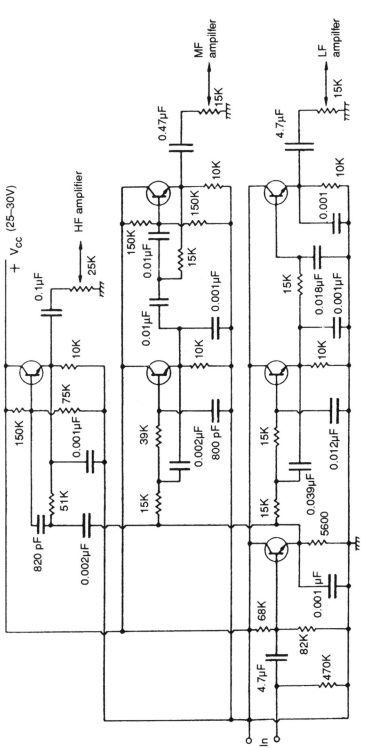

Figure 6.56. A simple three-way active crossover ($f_1 = 500$ Hz, $f_2 = 5.3$ kHz) (after C. G. McProud). (Unity gain op-amps could be substituted for the emitter followers, see Figure 6.43)

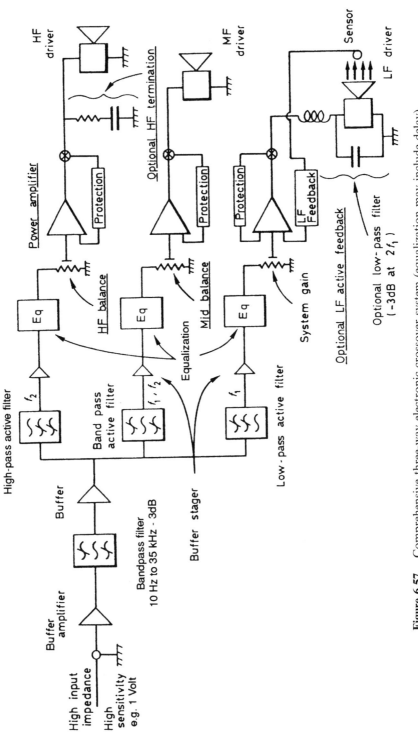

Figure 6.57. Comprehensive three-way electronic crossover system (equalization may include delay)

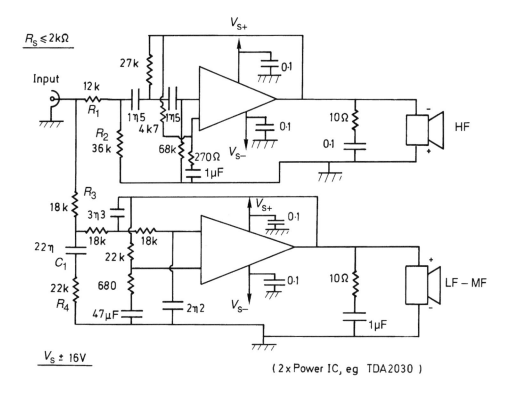

Figure 6.58. A simple arrangement for one channel of an active loudspeaker system using a low-cost power amplifier IC (TDA 2030) where the equalizer and second-order filters are integrated with the power amplifier circuit

point impedance is assumed, e.g. $2\,k\Omega$ or less, and many modern sources such as CD players may be connected directly to such a system, either via their variable output, so often present, or alternatively via a simple $10\,k$ logarithmic volume control. This offers a maximum output impedance of $2.5\,k\Omega$ (this maximum occurs at $-6\,dB$, half-position, consisting of two $5\,k\Omega$ resistors in parallel).

An input sensitivity of 0.3 V is appropriate for full output and a suggested circuit for a two-way system consisting of a 170 mm flared cone bass–mid and a 25 mm dome is given in Figure 6.58.

R_1, R_2 is the input attenuator for the treble section, R_3, R_4, C_1 forms a simple 'step' equalizer for the mid-range; the remaining components set the amplifier gains and establish a second-order, 2-pole crossover, high pass and low pass.

For anything more complex, separate, buffered active filter crossover stages are recommended. As regards power supplies, a 40 VA transformer will suffice for each stereo channel, providing unregulated d.c. supplies on load of ± 16 V (for a TDA 2030).

While the effective power per filtered channel is quite low, it goes a long way when connected directly to good sensitivity drivers and such a simple design can offer good focus, fine clarity and superior bass definition.

Loudspeaker System Alignment

The ability to readily adjust the gain of the various sections of an active filter crossover to account for sensitivity variations in the drive units has already been mentioned. A further facility concerns individual tailoring of the response characteristics. A given equalization may be correct for a typical driver but the performance spread in a batch of units may necessitate individual alignment. By judicious use of pre-set variable resistors in place of fixed equalization components, provision for such adjustment is readily incorporated.

Maintenance and Repair

Speaker systems designed for the most critical applications, such as studio monitoring, must be consistent. If a fault occurs, either due to a blown driver or electrical failure in the crossover, then it must be possible to restore the system to the same standard of performance after repair.

The obvious penalty of an electronic crossover system is the degraded reliability when compared with its passive counterpart. The very presence of pre-set gain and equalization controls adds to the problems, and also allows an inexperienced operator to misalign the system.

A recent active crossover design from the UK company of Boothroyd-Stuart (Meridian) illustrates some of the precautions it is necessary to take to ensure a consistent performance. In this three-way system the drivers are pre-tested for sensitivity and response. Divided into groups for each system, each group is coded and the amplifier crossover matched to it. In the event of a driver failure, a matching coded unit may be used for replacement and if a fault develops in the crossover, a complete new panel is supplied, also pre-programmed to match the serial number code of the defective system.

An alternative approach would involve a careful alignment and calibration procedure for the electronics section, with the relevant code and level settings provided. The system could then be re-calibrated via electrical rather than acoustic measurement in the event of a repair.

6.5 PASSIVE CROSSOVER SCALING

Consider a simple way of deriving numerical values for the third-order Butterworth crossover network. The third-order Butterworth crossover in Figure 6.59 may be scaled to any required frequency by simple multiplication since the values are inversely proportional to frequency. The set values refer to 1 kHz with nominally 8 Ω drivers. To readjust for 4 Ω systems, the capacitance values are doubled and the inductors halved. For 16 Ω systems the inductance values are doubled and the capacitors halved. For 2 kHz the values are halved, for 200 Hz all values are multiplied by 5, etc.

Final system response conformity with the theory relies on the driver input impedance being nominally resistive and constant, and also that the amplitude/frequency response be uniform.

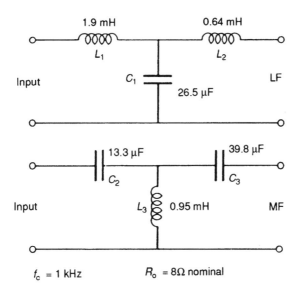

Figure 6.59. Third-order Butterworth, constant resistance

6.6   CURRENT DRIVE

It is normal practice to drive loudspeakers from an amplifier representing a voltage source, this by definition providing a near zero source impedance. Standard low frequency system theory, crossover design, sensitivity matching practice, etc. all relate directly to this approach. From the equivalent circuit for a speaker system (Figure 2.14) it is evident that various series elements of resistance and impedance limit the flow of

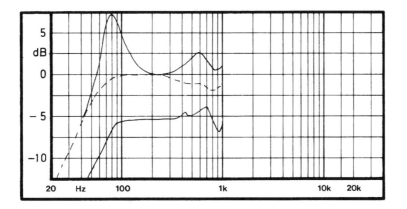

Figure 6.60. Upper trace – simple current drive; dashed – with velocity feedback. Lower trace – same driver under voltage source conditon

Table 6.1. Comparison of distortion for current and voltage drive (after Mills and Hawksford [11])

(a) Harmonic distortion

Conditions	Harmonic distortion (dB)			
	2nd	3rd	4th	5th
100 Hz Voltage drive	− 34.5	− 42	− 52	− 65
100 Hz Current drive	− 44	− 46.5	− 58	− 58
3 kHz Voltage drive	− 29	− 60	> − 80	> − 80
3 kHz Current drive	− 56	− 73	> − 80	> − 80

(b) Intermodulation distortion

Conditions 50 Hz : 1 kHz, 1 : 1	Intermodulation distortion (dB)			
	f2, f1	f, 2f1	f2, 3f1	f2, 4f1
Voltage drive	− 40	− 43	− 63	− 55
Current drive	− 46	− 45	> − 70	− 65

Note: Test conditions, 1A peak drive

current from the amplifier and in the case of a moving coil system, the motor-coil resistance R_{ac} is predominant. The accuracy of movement of the drive unit diaphragm, and consequently the acoustic output generated, depends on the linearity of the real and implied elements in the equivalent circuit. A fundamental non-linearity is present, due to the thermal coefficient of resistance for any conductor (except compensated special resistance alloys such as constantan). For copper, the coefficient is $+0.35\%$ per °C. Thus a $6\,\Omega$ coil operating at 200°C will suffer thermal compression of up to 4.2 dB over a long time constant, the latter dependent upon the thermal mass of the coil and, to a lesser degree, that of the magnet structure.

While this effect occurs towards maximum power levels, other non-linearities are also present, for example motor-coil inductance and its dynamic variation with coil position under heavy excursion. Coil inductance also varies with frequency, complicated by non-linear eddy currents in the magnet structure.

Motor-coil heating has several significant implications in respect of driver sensitivity matching in a multi-way system, for crossover network termination and alignment, and for low frequency system design. Here the vital controlling element Q_T is strongly dependent on R_c. Such series element non-linearity in theory may be greatly reduced by a power source combined with a linear high resistance, providing current rather than voltage and whose pure resistance is then dominant. In its simplest form the realization consists of a high voltage amplifier with an output resistance much higher than the speaker system, 100 times or more. A standard drive unit was measured under voltage and current drive and gave the results shown (Figure 6.60). In the upper graph, current drive has allowed full expression of the motional impedance. Note that the loss in output at higher frequencies due to motor coil inductance is much reduced. Using an accurate simulation for this driver [11], harmonic and intermodulation data for the two conditions of voltage and current drive have been compared (Table 6.1).

The distortion improvements look promising. Further practical use of current drive either requires the use of drivers with very low mechanical Q and/or combined with more advanced electronic techniques to take control of the motional impedance. By the use of a second sensing coil, or some other error-generating transducer, velocity feedback may be applied to achieve any desired low frequency alignment (dotted graph). Velocity feedback also provides some additional linearization which is effective at low frequencies. For the example driver, it reduces the current driven distortion components, e.g. third and fourth, by an average of 10 dB; however, second and fifth harmonics were unaffected.

Velocity feedback can also linearize conventional, voltage-driven low frequency systems and in any case the use of an active crossover, in association with power amplifiers which are closely coupled to the loudspeaker drivers, also helps reduce distortion. Practical active speaker systems using current drive require the use of a feedback power amplifier where the loudspeaker return lead passes via a sensing resistor, e.g. 0.5 Ω, in order to provide a source of negative current feedback. Current drive system design is not trivial and requires a very high open loop gain if good bandwidth and linearity are to be achieved. Stability problems are common. A sophisticated if complex solution employs an open loop buffer to isolate the transconductance or current amplifier from the load [11].

If distortion reduction is the main objective of current drive, careful design of magnet systems can provide worthwhile improvements with voltage drive while a combination of higher sensitivity and larger, higher power capacity, motor coils will help mitigate the effects of coil heating. Likewise, the choice of low frequency alignments and crossover formats with a reduced sensitivity to R_{dc} changes are helpful.

Hybrid Design

Time and again the attractions of open type transducers such as ribbons, electrostatics and open baffle configurations make themselves felt. Often the sound is as open and un-boxy sounding as the physical build implies. Understanding that these speakers have limitations at low frequencies, and taking account of price and size constraints, there is a temptation to try to build a conventional direct radiator box system for the bass with the alternative technology for the mid and treble. Great care is required in the voicing and blending of the outputs, mainly because of the dissimilarity between the omnidirectional radiation of the low frequencies compared with the typically dipolar radiation of the upper range. The changeover is difficult to disguise, and in addition the box system may have greater audible colouration than the technically superior (faster impulse response) upper section. Within their limits, ribbon and electrostatics are extremely linear with almost zero compression. Again this makes matching difficult with the mildly compressive nature of a moving coil bass system.

The most successful examples employ active bass systems of high or even feedback compensated linearity helping to match the remaining performance profile.

A well-designed ribbon transducer can possess a beguiling transparency and purity, but making it loud enough over a wide enough frequency range is the problem, and usually results in elongated structures which set new problems in terms of directivity.

6.7 DIGITAL LOUDSPEAKERS

The idea of a truly digital loudspeaker has been proposed where digitally coded audio is used at a sufficient power level to directly actuate an electromechanical array capable of conversion directly into sound pressure, an acoustic digital to analogue convertor. Given that huge problems exist in creating such a device with sufficiently low quantization and spurious noises and with good directivity, much can still be done in the field of digital loudspeaker engineering. In fact DSP holds a tremendous potential for the future. With the wide dissemination of digital audio systems throughout professional and consumer applications, for music recording, compact disc, satellite broadcast and the like, sophisticated signal processing systems are now available in the digital domain. A digital audio communications interface has been adopted for professional applications, called AES-EBU, which incorporates the SPDIF (Sony–Philips digital interface) consumer version, the latter widely used in CD player systems. Digitally coded audio can be conveyed by wire or by an optical fibre. Thus, an integrated, active loudspeaker design fitted with a digital decoder can use such an optical link as the only connection to the audio control equipment and signal sources, working exclusively in the digital format. Such use avoids ground loop problems and can reduce interference.

A digital signal processor is essentially a fast dedicated computer realized as a single integrated circuit or as a number of circuits which may be programmed to compute a variety of audio control functions in real time according to designed software code. Almost any filter imaginable can be achieved including the powerful anti-alias filters required for A to D and D to A conversion. Digital signal processor (DSP) units are now of manageable size and may be fitted directly to an active, electronic speaker design.

DSP can offer a variety of filter and tone control functions. In addition it can process in the time domain, adding the programmed acoustic signatures of a variety of real or synthetic environments, the complex reverberation appearing as if it were present on the original recording. A large number of such processors have sold well and are used to take control of room acoustics and enhance the sense of being present at an original performance. In their optimum exposition such systems feed the processed reverberation to an array of six matched loudspeakers, four arranged in an elevated position near to the room corners while two are placed conventionally and are responsible for the main stereo sound stage.

DSP Potential

The potential of DSP enhanced loudspeakers is very considerable and deserves some space. Let us assume that the electro-acoustic conversion is conventional, carried out by established and perfectly satisfactory techniques whether moving coil or otherwise. In its simplest form a DSP unit may be added to a digital audio chain where a digital to analogue unit appears ahead of a normal amplifier and speaker.

Control of the Room–Speaker Interface

The second stage concerns the acoustic interface between the speaker and the listening room. With loudspeaker placement and room acoustics constituting a major variable in the final performance it is possible to use a sensing microphone at the listening position and by employing a specified calibrating impulse signal have the DSP sections learn the room errors, the major standing wave modes, local boundary interferences and the overall reverberant characteristic. Given the defined speaker placement the DSP can then synthesize the inverse correcting signals to be added to the loudspeaker input to generate a sound field at the listener position which is much closer to the ideal.

Control of Loudspeaker Characteristics

The third stage exploits recent fast DSP systems to take control of the full bandwidth frequency response of the loudspeaker itself. Assuming that the given loudspeaker system is basically well behaved with satisfactory consistency through the crossover ranges, the whole axial frequency/impulse characteristic may be acquired and used in the DSP calculating section to generate a near perfect output in both frequency and time, and consequently also in terms of phase. If desired, a linear phase characteristic may be obtained while the DSP correction allows for any desired low frequency alignment, fixed or user adjustable, and can also provide excellent pair matching and consistency over a long production run. Such a speaker could be adaptable to a wide range of usage, for example, when programmed for an extended low frequency response it is applicable for medium loudness classical music working, or alternatively when adjusted for a narrower bandwidth it can offer a superior power handling for the reproduction of rock or popular material at higher sound levels. Thus, speakers could offer several personalities.

DSP Error Correction

DSP can include first-order correction for delayed resonances and the resulting colouration. Much as processing for echo and reverberation is possible, the speaker's own mechanical resonance may be partially corrected by predictive compensation at the signal drive to the loudspeaker.

The DSP section may also be programmed to correct for distortion and compression in loudspeakers. For example, the thermal characteristics of the motor coils can be modelled. By using the music programme history and the present volume level, the temperature cycling can be predicted and a correction applied to the thermal resistance 'compression' in the drivers. Likewise the natural variations in distortion with frequency and level may be modelled as a set of equations applied to the DSP and then inverted. Thus complex dynamic equalization of basic loudspeaker non-linearities may be added to the speaker input, further perfecting its acoustic output (see the subsection on Vented-box distortion in Section 4.4, Chapter 4).

Digital Active Loudspeakers

The increasing availability of digital audio signals and digital signal processors (DSPs) is opening up the field of active loudspeaker systems where each driver has its own DAC and power amplifier. With these elements in place, constituting a relatively straightforward piece of electronics, a preceding digital signal processor may be given a wide range of responsibilities. The potential advantages were apparent as early as 1982 when Hawksford built a system using what would now be regarded as a very primitive 8 bit processor operated without the benefit of dither linearization for the audio computations.

Digital Speaker System Design

While these are exciting and powerful opportunities, digital technology can be still more intimately involved in speaker design. The DSP may be used to generate the specific crossover and equalization characteristics for individual drivers in a multi-way system. A speaker designer than has a very powerful design tool since a speaker prototype can be made using the ideal choice of drivers mounted in an enclosure chosen for its optimum structural and directivity properties, one where the acoustic output of the drivers has the opportunity to blend properly both on- and off-axis. By using a multi-channel digital decoder and amplifier arrangement, the crossover function may be put entirely in the digital domain. Here the individual drivers may be corrected in amplitude and phase and also may be easily time aligned with respect to each other. Such a 'perfected' pair of drivers may then be crossed over at ideal frequency points using any desired slope. While high-order, high-slope crossovers are difficult to execute in passive power filters due to losses, tolerances and the like, virtually any filter order is possible using a digital signal processor. Some research done to date suggests that high slopes with narrow crossover regions can work very well, helping to achieve the best polar and off-axis response consistency.

Thus DSP technology has the ability to refine multi-unit speaker system design to a high degree together with the potential for extremely accurate performance, orders of magnitude better than conventional analogue-based designs.

Time shared techniques may be used so that one DSP can handle the crossover requirements of several drivers, these signals finally separated digitally before reaching the individual digital to analogue convertors in each frequency band. Filtered bandwidth operation also helps to improve convertor performance by reducing broad band intermodulation products in the D/A convertors.

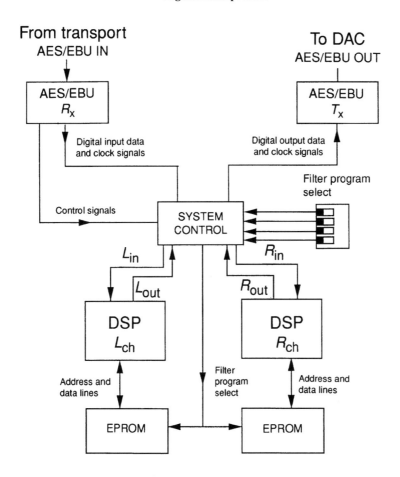

Figure 6.61. Digital correction filter for loudspeaker (courtesy of Hawksford)

The Impulse Response Objective

Viewed simply, this design objective is an ideal impulse response delivered to the acoustic space occupied by the listener(s). This can be described as 'focusing the response' via the equalization. While the detail is more complex, simple digital equalization/correction (Figure 6.61) may be achieved by measuring the existing impulse response, applying some sensible truncation and windowing of the data, inverting the function, and apply the required correction in the digital signal path leading to the main power amplifier — this simple form excluding the active speaker case.

Phase and Frequency Response

If we look at the fundamentals, the acoustic responses of the drivers may be measured, inverted and used in the crossover synthesis to provide the target frequency response for

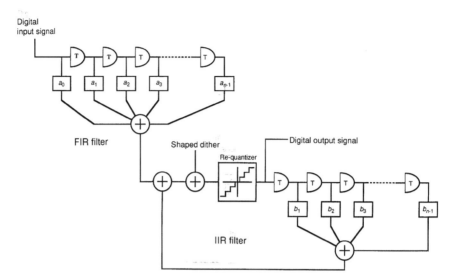

Figure 6.62. Infinite impulse response digital filter (courtesy of Hawksford)

the entire system. Delay and phase errors may be readily compensated for, with the result that a near perfect transfer function is possible, linear phase, excepting the inevitable effects due to the finite low frequency bandwidth.

Filter Order for the DSP

With regard to response correction, some designers have attempted to compensate for every last fraction of a decibel, employing high-order filters to achieve it. Two factors indicate that this is of little value. First, the definition of a driver's characteristic response must be an approximation, and indeed the axial result may not be sufficiently general to describe its tonal balance under room and cabinet diffraction conditions. A response average covering the forward solid angle of say 15°–20° may be more effective. By implication, the required digital correction is also a generality and attaining perfection for a single measurement axis is irrelevant.

Secondly, there is accumulating empirical evidence that subjective factors are present in the design of DSPs which are not fully quantified in theory. For example, band stop filters used prior to a DAC may have similar stop bands and dynamic range but still show sound quality differences.

With digitally processed loudspeakers, DSP filters of moderate order are preferred to high order, which precludes fine detail compensation. Auditioning suggests in any case that very fine correction is not superior to a more general compensation except in anechoic, single-axis presentations. IIR filters tend to be preferred to FIR types in speaker applications and they match the relatively long impulses of loudspeakers well.

Low frequencies present a special problem in terms of complexity. Equalization is generally avoided. For example, with a standard 44.1 kHz sample rate the filter poles are very critical, the terms are very long, and decimation would be required. Fortunately, the low frequency performance of most speakers is relatively well ordered and predictable; the required electronic correction is best and most economically achieved in the analogue domain, e.g. with a corrector in the power amp feed to the low frequency driver. For the same reason digital equalization for room boundaries is quite costly in processing power and in such schemes it is good practice to separate the more critical upper frequency range from such processing to maintain overall transparency.

Resolution

For a domestic reproducer of 105 dB maximum level it is possible to operate at 16 bit resolution, particularly if pre-emphasis is employed on the more critical treble section. It can reduce noise via the real increase in dynamic range possible, i.e. up to 2 bits (provided that the de-emphasis is in the analogue domain, i.e. in the power amp section). Otherwise, the DACs should run at 18 bit plus to ensure that hiss is not audible from the drivers. In quiet domestic conditions better than 100 dB of signal-to-noise ratio is a worthwhile target. The digital processing must be executed at a respectable number of bits sufficient to preserve the original signal detail, provide the dynamic range for the filter coefficients, and to allow proper rounding and noise shaping to the word length required by the chosen DACs (Figure 6.62).

System Details

Volume control

With effective digital and analogue muting there is no reason why a DAC of wide dynamic range should not be operated directly coupled to the power amplifiers. Only where low-level linearity or noise is a problem need analogue gain ranging be required; this is ganged or synchronized through each power amplifier and held under central microprocessor control. A good digital gain control may be incorporated in the DSP, of say 56 bits or better computation, with dither. Gain control is not a trivial operation if transparency and overall signal quality is to be maintained over a wide range of operating levels (see Figure 6.63).

Filter order for crossover filters

In theory, any filter order may be chosen, while in practice, after extended trials, fourth-order Linkwitz–Riley is considered effective, once details of delay compensation have been addressed. A minimum phase transfer function is achievable with this type. Other details include the potential for tone controls, stereo balance, augmented by gain and inter-channel delay steering, boundary correction and overall tonal balance.

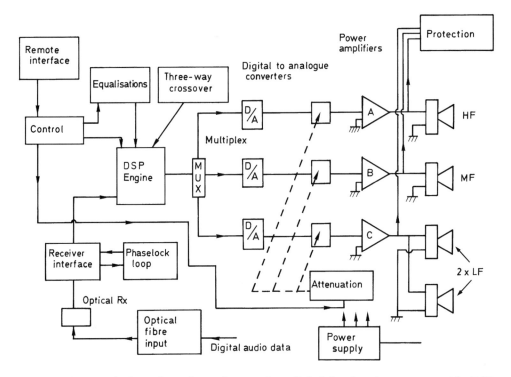

Figure 6.63. Practical configuration of an active, digital loudspeaker system with DSP equalization and crossover characteristics determined by a digital signal processor. Coarse attenuation of typically 6 dB steps is performed in the analogue domain, fine control of volume in the digital domain. Both are programmed via the system control. Employing optical data interfaces, no ground loops are generated in the overall stereo chain

Performance

DSP techniques are now sufficiently mature for a view to be taken concerning sound quality. Provided that a given loudspeaker is properly constructed, has well chosen and consistent drivers, is fitted to a low resonance, low diffraction enclosure, a digital active design can provide a significant and absolute gain in performance over both the equivalent passive implementation and the other undeniably superior, active analogue type.

The global correction of driver characteristics, including relative delays, improves the subjective accuracy and coherence through the crossover range. The transfer function is more perfect, which can be heard as greater stereo image depth, greater resolution of perspective, a more natural timbre, and reduced colouration. Transient sounds are more focused, more dynamic, and more natural. Image focus is tighter, stage width enhanced; the image floats more freely from the loudspeaker locations. The listener is less aware of individual drivers, and instead hears a more plausible blending of acoustic energy.

A short digital signal path, aided by the prefiltered drive to each DAC stage, can have exceptional clarity; unexpected detail may be retrieved in good systems used with high quality digital source programme.

DSP Speaker Dynamic Range

The question of system sensitivity and dynamic range requires consideration. Much as one would like to connect the D/A convertors in a 'digital' loudspeaker design directly to the final power amplifiers, there may be dynamic range difficulties. Typically digital audio replay is 16 bit with a 96 dB dynamic range or equivalent signal-to-noise ratio. However, loudspeakers may have to work over a wider dynamic range than this despite a 60 dB limit for typical programme dynamics. For example, most medium level loudspeaker systems can attain short-term programme-related sound levels of 110 dB at 1 m. On the basis that typical recorded programme has a dynamic range of 50–60 dB and a background noise level of −80 dB or so, relative to full modulation or MSB (the most significant bit in the digital audio code), the user could set the programme dynamic window anywhere in the speaker's working range by using the volume control on the matching amplifier. Set to full volume the speaker may be called upon to work over a 50–110 dB range. Alternatively, if set at the lowest, the speaker must operate from near inaudibility at 0 dB to a hushed peak level of 60 dB s.p.l., or its equivalent from a stereo pair of loudspeakers in the listening room. What part of the dynamic range is used for a digital chain deserves consideration.

REFERENCES

[1]   Harwood, H. D., 'Some aspects of loudspeaker quality', *Wireless World*, May (1976)
[2]   Hughes, F. M. (a pseudonym for M. Colloms), 'A group test of thirty pairs of commercial loudspeakers', *Hi Fi For Pleasure*, **4**, July, September and October (1976)
[3]   Benson, J. E., 'An introduction to the design of filtered loudspeaker systems', *J. Audio Engng Soc.*, **23**, No. 7 (1975)
[4]   Linkwitz, S. H., 'Active crossover networks for non-coincident drivers', *J. Audio Engng Soc.*, **24**, No. 1 (1976)
[5]   Allen, P. E., 'Practical considerations of active filter design', *J. Audio Engng Soc.*, **22**, No. 10 (1974)
[6]   Ashley, J. R. and Henne, L. M., 'Operational amplifier implementation of ideal electronic crossover networks', *J. Audio Engng Soc.*, **19**, No. 1 (1971)
[7]   Linkwitz, S. H., 'Loudspeaker system design', *Wireless World*, **84**, No. 1516 (1978)
[8]   Martikainen, I., Varla, A. and Otala, M., 'Input requirements of high quality loudspeaker systems', *Audio Engng Soc. 73rd Convention* (preprint 1987) (1983)
[9]   Adams, G. and Roe, S., 'Computer aided design of loudspeaker crossover network', *J. Audio Engng Soc.*, **30**, Nos. 7/8 (1982)
[10] Jones, A. and Hawksford, M. J., 'Computer-aided design of lossless crossover networks', *Audio Engng Soc. 65th Convention*, pp. 1589 (1980)
[11] Mills and Hawksford, 'Amplifiers for current driven loudspeakers', *J. Audio Engng Soc.*, **37**, No. 10 (1989)

[12] Colloms, M., 'The Amplifier/Loudspeaker Interface', Ch. 5, *Loudspeaker and Headphone Handbook* (Ed. Borwick, J.), Butterworth (1988)

[13] Carlsson, B., 'Driver offset-related phase shifts in crossover design, Part 2', *Speaker Builder*, **2**, 26 (1995)

BIBLIOGRAPHY

Aarts, R. M. and Kaizer, A. J. M., 'Simulation of loudspeaker crossover filters with a digital signal processor', *J. Audio Engng Soc.*, **36**, No. 3 (1988)

Ashley, J. R. and Kaminsky, A. L., 'Active and passive filters as loudspeaker crossover networks', *J. Audio Engng Soc.*, **19**, No. 6 (1971)

Bank, G. and Hathaway, G. T., 'Three-dimensional energy plots in the frequency domain', *J. Audio Engng Soc.*, **30**, Nos. 1/2 (1982)

Borwick, J., *Loudspeaker Handbook*, Butterworth (1985)

Bullock, R. M., 'Loudspeaker crossover systems', *J. Audio Engng Soc.*, **3**, Nos. 7/8 (1982)

Bullock, R. M., 'Satisfying loudspeaker crossover constraints with conventional networks — old and new designs', *J. Audio Engng Soc.*, **31**, No. 7 (1983)

Catrysse, J., 'Feedback circuits for loudspeakers', *J. Audio Engng Soc.*, **33**, No. 6 (1985)

Colloms, M., 'Some practical aspects of loudspeaker design relating to perceived stereo image quality', *Hi-Fi News*, **24**, No. 6 (1979)

Cooke, R., 'Monitoring loudspeakers', *Hi-Fi News*, London

Dickason, V., *The Loudspeaker Design Cookbook*, Marshall Jones Co., 3rd edn (1987), 5th edn. (1995)

Garde, P., 'All pass crossover systems', *J. Audio Engng Soc.*, **28**, No. 9 (1980)

Greiner, R. A., 'Amplifier–loudspeaker interfacing', *J. Audio Engng Soc.*, May (1980), also *Loudspeakers Anthology*, Vol II

Hillerich, B., 'Acoustic alignment of loudspeaker drivers by nonsymmetrical crossover of different orders', *J. Audio Engng Soc.*, **37**, No. 9 (1989)

Hilliard, J. K. and Kimball, H. R., 'Dividing networks for loudspeaker systems', *J. Audio Engng Soc.*, **26**, No. 11 (1978)

KEF Electronics Ltd., 'Model 105', *Keftopics*, **3**, No. 1 (1978)

Kelly, S., 'Network niceties and notional loudspeakers', *Hifi News*, pp. 34, 35, 39, August (1984)

King, M. W., 'Activating your loudspeaker crossover', *Audio*, USA, April (1972)

Kurigama, J. and Furukawa, Y., 'Adaptive loudspeaker systems', *J. Audio Engng Soc.*, **37**, No. 11 (1989)

Linkwitz, S. H., 'Loudspeaker system design', *Wireless World*, **84**, No. 1516 (1978)

Linkwitz, S. H., 'Passive crossover networks for non-coincident drivers', *J. Audio Engng Soc.*, **26**, No. 3 (1978)

Marshall-Leach, Jr., W., 'Electroacoustic-analogous circuit models for filled enclosures', *J. Audio Engng Soc.*, **37**, No. 7/8 (1989)

Read, D. C., 'Active crossover networks', *Wireless World*, 574–6, December (1973) and 443–8, November (1974)

Rettinger, M., *Practical Electroacoustics*, Thames & Hudson, London (c. 1955)

Schuck, P. L., 'Design of optimised loudspeaker crossover networks using a personal computer', *Audio Engng Soc. 72nd Convention* (preprint 1950) (1982)

Schuck, P. L., 'Design of optimised loudspeaker crossover networks using a P.C.', *J. Audio Engng Soc.*, **34**, No. 3 (1986). (Small programme correction, p. 563, *J. Audio Engng Soc.*, **34**, No. 7/8.)

Small, R. H., 'Constant voltage crossover network design', *J. Audio Engng Soc.*, **19**, No. 1 (1971)

Small, R. H., 'Speaker crossover networks', *Electronics Today International*, p. 64 et seq., October (1972)

Thiele, A. N., 'Air cored inductors for audio', *J. Audio Engng Soc.*, **24**, No. 5 (1976)

Thiele, N., 'Another look at crossover networks', *Audio*, USA, August (1978)

Waldman, W., 'Simulation and optimisation of multiway loudspeaker systems using a personal computer', *J. Audio Engng Soc.*, **36**, No. 9 (1988)

Wall, P. K., 'Active and passive crossover networks with no transient distortion', *Proc. Audio Engng Soc. 50th Convention*, London, March (1975)

Wall, P. K., 'Active loudspeaker crossover filter', *Elector*, **3**, No. 6 (1977)

7

The Enclosure

The enclosure is a structure that supports the loudspeaker drive units and may provide designed acoustic properties of radiation pattern and driver acoustic loading. The term is usually taken as referring to an enclosed box where some form of loading is applied to the bass driver. However, an enclosure can be open backed or conceal a horn or other system folded into a compact structure.

While the vast majority of systems are enclosed box types, the open baffle has a steadfast following. Subjectively it is appealing, demonstrating a literally 'open' quality. While large area baffles are necessary for satisfactory bass, some designers have combined 'open' mid-treble systems allied to an enclosed box, low frequency section. It should be possible to produce a large area, moving-coil-based open baffle by employing an array of modular drivers. Due consideration will need to be made of the flexural modes of such a panel and choice of both structural material and form could prove crucial in such a design. Webbed cast alloy sections suggest themselves which could be partially disassembled for transport and bolted up to form the complete radiating panel on delivery. With power sharing over the array, individual drivers need only be low power, sensitive units of a few watts capacity.

In practice the enclosure exerts a considerable influence over the sound of a complete system. A moving coil driver radiates energy by vibrating a diaphragm assembly whose reaction will simultaneously excite the driver chassis causing it to vibrate in sympathy. The chassis must therefore be rigidly clamped to a strong panel to prevent it from moving, and hence reducing the cone output. Nevertheless, some energy will inevitably be imparted to the panel. To this may be added the sound energy radiated from the diaphragm incident upon the panel. This energy is then dissipated by the enclosed structure as unwanted acoustic output, the severity and decay time of the enclosure panel resonances constituting a major factor in loudspeaker pulse decay properties and in the consequent subjective colouration.

7.1 ENCLOSURE MATERIALS

A simple box structure will possess a series of resonant modes due to torsion and panel flexure, their frequency and magnitude dependent on the Q of the material and also its thickness and density. As resonant modes are less likely to be excited with increased

Table 7.1.   Some enclosure materials (see also Table 5.2)

Material	Density (ρ) (kg/m $\times 10^3$)	Young's modulus (E) (kN/m$^3 \times 10^{10}$)	E/ρ (m/s $\times 10^7$)
Lead	11.3	Negligible	—
Steel	7.7	20	2.6
Mazak (die casting alloy)	6.0	10	1.7
Aluminium (alloy)	2.6	7.5	2.9
Concrete	2.6	N/A	—
Brick	1.8	N/A	—
Sand	1.5	N/A	—
Bituminous damping	1–3	$\simeq 0.1$	—
Polythene	$\simeq 1$	0.1	0.1
Chipboard (high density special)	0.9	0.33	0.37
Chipboard HD	0.81	0.27	0.33
Birch ply (best quality)	0.78	0.86	1.1
MDF (best tough fibreboard)	0.75	0.33	0.44
Oak	0.72	Grain dependent	—
Plywood (average)	0.67	Grain dependent	—
Mahogany	0.67	Grain dependent	—
Chipboard (three layer)	0.650	0.22	0.34
Pine	0.45	Grain dependent	—
Fibreboard (Celotex light)	0.32	0.02	0.63
Polyurethane structural foam	0.3–0.5	0.15	1.3 (typ.)

Note: The E values given are static. Under dynamic flexure the E value for many materials will alter.

panel thickness and density they will become correspondingly less obvious and thus it is well worth noting the relative properties of the various materials used in cabinet construction (Table 7.1).

Wood and Wood Based Composition Materials

Traditionally most loudspeaker enclosures have been constructed from wood. It has many advantages, not the least of which is that for its mass it is relatively non-resonant. Most countries have an established cabinet industry, often linked with furniture manufacturers, and thus enclosures can usually be supplied at a realistic price with a wide range of wood grades and thicknesses to choose from. It is an easy material to work with and historically labour costs were relatively low.

Chipboard is the densest of all the wood materials and as it comes in reasonably priced uniform sheets which resist warping and can be easily veneered, it is not surprising that the great majority of wood cabinets today are constructed from this material. The thickness used is scaled to the volume of cabinet required. Below 30 litres, 12 mm is suitable, while a 30–60 litre enclosure generally employs 18 mm, and larger cabinets, 25 mm. Chipboard is also available in various densities, the grade known as '600' being the one generally employed; preferably in 'three layer' variety, with fine texture outer skins covering a coarser interior. This has the lowest flexural Q of almost any undamped wood material.

Other woods may also be employed for enclosure construction, the results depending on the grade selected and the size of cabinet involved. Solid woods such as well seasoned afromosia have been used in the past, and high quality birch plywood is also frequently encountered as an alternative to chipboard, although both its Q and more particularly its cost are greater.

Now coming into favour is a dense high quality hardboard or compressed bonded fibreboard. One commercial example is called MDF (medium density fibreboard), its fine texture machines well with no chips and its mechanical properties are similar to plywood with the advantage of better damping at high frequencies. Chipboard can be frequency selective and can 'ring' significantly well into the treble range.

Concrete

With enclosures of very high mass such as those built of thick concrete or brick, it is difficult to excite the resonant modes. However, such enclosures are generally assembled on-site and as such are not as practicable for normal manufacture and supply. However, in the past thin-wall cast concrete enclosures have been suggested, but unfortunately the effect of reducing the wall thickness means that the advantage of high mass is lost, the enclosure will thus prove quite resonant and will have poor resistance to impact. It is, however, possible to produce a successful concrete enclosure through choice of suitable filler and additives to reduce the Q and by employing suitably designed internal strengthening beams cast into the structure to control the resonant modes.

Other Materials

Provided that the structure is well damped there is no objection to using metal for cabinet construction, and a few high quality designs have been produced that employ steel or aluminium alloy front panels. Welded steel enclosures have been used for systems that are subject to arduous duty, such as in the field of public address.

Moulded plastic cabinets will make increasing economic sense for quantity production, in view of the accelerating price costs of both labour and natural wood. In fact, further research in cooperation with major producers is urgently required to develop suitable synthetics that possess the necessary acoustic properties and can be easily moulded into cabinet shells. For example, certain grades of expanded polyurethane have been tried by several manufacturers, notably B & O, Wharfedale and KEF.

On the composite front several resin materials are finding some success for the more costly loudspeaker enclosures. In one example the casing is a matrix of chopped rubber strands for improved impact resistance and damping, stone chips for mass loading and reinforcement bonded in a catalysed polyester resin. Gravity moulding is appropriate while the mould interior may be pre finished in a gel coat to improve the appearance. For critical markets it is usual to gloss finish the exterior in a multi-coat polyester lacquer buffed between coats for a mirror finish. Recently aluminium has been utilized both in a thin-wall bituminous damped form and box section extrusion for the smaller enclosures.

Systems have been produced using aluminium skinned honeycomb-cored panels, fabricated and bonded with epoxy adhesives. With a panel thickness of typically 12 mm, the low mass results in panel resonances one or two octaves higher than conventional woods. In consequence, the decay time is rapid and it is possible to bring the decay time of the enclosure to a range comparable with the drivers themselves. Transient response is then very good and the usual 'wooden' colouration is avoided in such enclosures. Resin fibre honeycombs are also suitable, but these and aluminium types are expensive; as raw boards they cost 5 to 10 times as much as chipboard. Injection moulded, expanded polyurethane structural plastic is a potential alternative, but the tooling cost is high, ruling out its use for the lower volume high quality systems where it could be most beneficial. Research on inexpensive phenolic laminate skinned panels with plastic foam cores is also showing good results.

Mixed Materials

While it is general practice to build enclosures in a common material such as chip/ particle board or MDF, mixing different materials is another line of attack which can help reduce the audible resonance colouration signal of the enclosure panels. For example, an enclosure shell (sides, top and bottom) can be folded up in one grade of chipboard while the back and front panels can use the alternative of MDF, also of a different thickness, to change the sound of these surfaces relative to the main shell. MDF is often favoured for the front baffle because it machines well for recessed driver mountings and also allows for any contouring intended to reduce diffraction.

In advanced enclosures, where exceptional local rigidity is required, e.g. in the region where the drive units are fixed or where floor or stand fixings are located, more costly materials may be used, for example high density laminated phenolic panels for driver baffle and base plate, or even of machined plate or cast metal alloy. Designers generally underestimate the influence of vibration coupled from one driver frame to another and the change from a wood composite to phenolic or metal for the driver baffle can provide a surprising improvement in audible definition if the additional cost can be tolerated.

A bonus provided by a driver panel in metal is the improved window area behind the driver thanks to the thinner section made possible by the very high stiffness of the metal compared with the usual wood.

In a given cabinet some designers achieve mass loading, stiffening and resonance control by lamination, generally with a selected visco-elastic bonding agent with interior slabs of slate, marble, or steel, or alternatively a plasticized plaster or similar concrete.

7.2   ENCLOSURE RESONANCES

The volume resonances of a box enclosure system, including the Helmholtz resonances, are dealt with in Chapter 4 on acoustic loading. The following discussion concerns those internal and undesired resonances which may contribute to colouration.

Unless an enclosure is spherical or ellipsoidal, all or some of its sides will consist of plane surfaces clamped at their edges. Such a clamped panel will have its own acoustic output when forming part of an energized loudspeaker, derived from sound energy within the enclosure. The output consists of standing-wave modes at higher frequencies and pressure modes at those frequencies where the wavelengths exceed the internal enclosure dimensions. Adjacent panels may be similarly excited by vibrational energy from the drive unit chassis.

Theoretically a clamped panel has a well-defined vibration series in both longitudinal (volume stiffness) and bending modes. A further mode is due to the panel mass resonating with its own and the enclosure's air volume stiffness. Stevens [1] found that in a typical reflex cabinet this latter resonance appeared at almost twice the fundamental enclosure resonance, a condition verified over a range of tuned system frequencies. Fully clamped, the panel resonance frequencies are given by

$$F_r = \frac{12}{2\pi}\left(\frac{B_s}{\rho}\right)^{\frac{1}{2}}\left(\frac{7}{2}\left(\frac{1}{A^4}+\frac{4}{7}-\frac{1}{A^2B^2}+\frac{1}{B^4}\right)\right)^{\frac{1}{2}}$$

where $B_s = [Et^2/12(1-c^2)]$, the bending stiffness; $\rho =$ density in kg/m^3; A, B are the respective mode order; $c =$ velocity of longitudinal sound waves in the material; $t =$ panel thickness.

For an edge supported or clamped panel

$$F_r = \frac{\pi}{2}\left(\frac{B_s}{\rho}\right)^{\frac{1}{2}}\left(\frac{A^2}{L^2}+\frac{B^2}{W^2}\right)$$

or

$$F_r = 0.48 \times ct\left(\frac{A^2}{L^2}+\frac{B^2}{W^2}\right)$$

the remaining resonance series is given by

$$F_r = \frac{nc}{2}$$

where $l =$ panel dimension related to the mode calculated, n an integer from 1 to ∞ (see Figure 7.1)

The main difficulty in the theoretical analysis of enclosure panels is in defining the boundary conditions. The results are so complex that so far enclosure colouration cannot be predicted very well and designers have to rely on a combination of experience and trial and error in enclosure design.

Major resonances are amenable to computer-aided analysis. Using a combination of modal and finite element analysis, work has been done on enclosure panel vibration and it is possible to illustrate the results using animated graphics (Figure 7.2). Such is the variability of the resonance mix that major changes in the subjective colouration of high quality systems can result from a change in one enclosure dimension of as little as 5%, or even by a small change in adhesive.

7.3   MAGNITUDE OF UNDAMPED PANEL OUTPUT

Many speaker designers have laboured by trial and error to control cabinet resonance by such methods as the use of high density constructional materials and internal

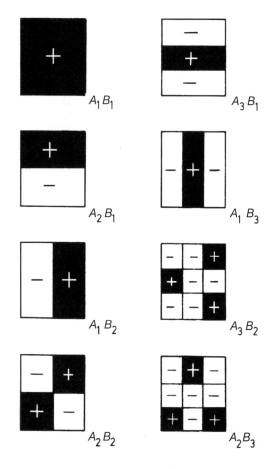

Figure 7.1. Panel vibration modes to third order; motion of shaded areas opposite to motion of unshaded areas

bracing, or loading the panels with ceramic tiles and sand. However, their efforts have met with only moderate success.

Sand is an awkward medium to work with as it requires a retaining panel to hold it in place and has the further undesirable effect of causing the weight of an enclosure to rise dramatically. However, this treatment can be quite effective due to the added mass and also to the high loss imparted by the vibration of the individual particles.

Rank's Leak/Wharfedale division hold a patent for an intriguing variation on the theme of filled cabinets, namely a water 'sandwich'. The enclosure is presumably a double skinned synthetic moulding, the intention being that the purchaser should fill the cabinet on delivery. An obvious advantage for the manufacturer is the greatly reduced transit weight of the partially completed system. With suitable design a similar enclosure could also be sand filled on installation, as is already done with some models of loudspeaker stand.

Resonant modes can be modified by increasing the thickness of the panels or by attaching bracing to them, but although these measures may displace the resonance to

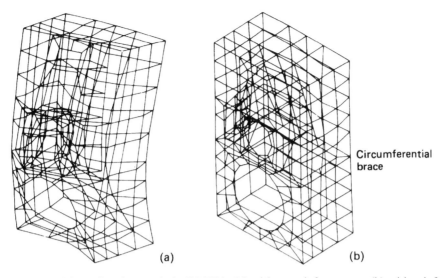

(a)   (b)

Circumferential
brace

Figure 7.2. Cabinet vibration analysis (284 Hz); (a) without reinforcement, (b) with reinforcement (courtesy Technics)

more subjectively acceptable frequencies, they usually have little or no effect on their magnitude. Beam coupling of two opposite panels will only effect the fundamental bending resonance, where a major stiffening will result, suppressing this mode. Barlow [3] measured the sound output of square birch-ply panels excited by a driver mounted on the inside surface. A remarkable discovery was made, namely that the output at certain of the resonances approached the level achieved by the driver with the panel absent, thus indicating that the panel was almost wholly acoustically transparent at these frequencies (Figure 7.3). However, this is an exaggerated case because enclosure panels tend to be rectangular rather than square, and the listener is rarely on the panel axis. Nevertheless Stevens [1] has shown that for a typical 50 litre enclosure built of 18 mm chipboard, radiation from an undamped rear panel may have peaks which are only 10 dB below the front axial output (Figure 7.4).

In a normal sound field the output of the six cabinet walls will contribute to the primary forward radiation. Harwood has noted a working Q of up to 100 for cabinet panels made from several varieties of wood. Subjectively derived evidence has shown that these resonances are clearly audible and may have durations of half a second or more. Clearly the choice of panel material alone is not likely to reduce either the Q or the reverberation time to a level where it becomes unobtrusive.

Variation of Young's Modulus with Frequency

With high Q, materials with a 'crisp' handling quality in thin sheet form, the stiffness may vary little with frequency, although almost all materials show some falloff in the upper range of the audio band. Metals show the least variation, for example aluminium and alloy cored honeycomb. Glass and carbon fibre reinforcement is good while many plastics with a more crystalline molecular structure maintain low loss to high

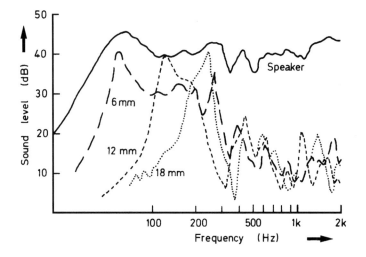

Figure 7.3.   Sound output of Birch-ply panels (after Barlow [3])

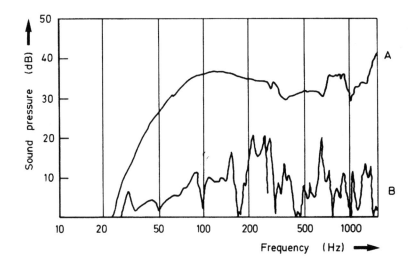

Figure 7.4.   A, frontal output of cabinet; B, sound output from rear panel of cabinet (after Stevens [1])

frequencies. Most natural materials, including wood or organic fibre composites, show substantial variation in stiffness and Q with frequency.

Static measurement may relate poorly to the conditions of use. For example, grade 600 chipboard will remain close to its static value up to 300 Hz or 400 Hz, but falls 25% by 1 kHz and is down to one-half by 3 kHz and one-quarter by 8 kHz. Plywood constructions are similar and this behaviour means that computation of the higher order resonant modes is complex. Detailed testing of material properties over a wide frequency range may be necessary, allied to FEA methods for problem-solving.

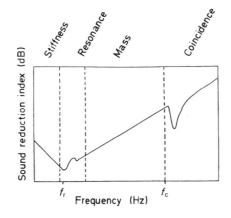

Figure 7.5. General characteristics of attenuation versus frequency for a panel including coincidence f_c

Coincidence

Coincidence may also play a part in panel acoustic output. For a given panel there will be a first critical frequency where the propagated wavelength in air equals the bending mode wavelength in the panel. Further coincidences are possible as the higher panel modes align at submultiples (see Figure 7.5).

For a panel, the angle of incidence may be critical in aligning the crest-to-crest air wavefront spacing relative to the bending mode. At coincidence the panel is easily coupled or driven and appears nearly transparent, the incident acoustic wave reappears on the other side. The critical frequency is given by

$$f_c = \frac{c_{air}^2}{2\pi}\sqrt{\left(\frac{P_s}{D}\right)}$$

where P_s is the mass per unit area and the flexural stiffness

$$D = \sqrt{\frac{Et^3}{\rho(1 - U^2)}}$$

The critical frequency reduces with increasing stiffness D indicating that with stiff light panels more internal absorption may be required to prevent the egress of sound from within the box. For high quality 18 mm plywood f_c is typically 1.2 kHz, rather higher than the dominant bending modes. With lighter panels the air volume/panel resonance will also be at higher frequency.

7.4 AUDIBILITY OF RESONANCE

It is worth noting that research conducted into the audibility of resonances relative to a given response curve indicates that low Q resonances can be the most obvious, $Q = 1$

being the critically prominent value [4]. Although this might seem to contradict the findings above, the two can in fact be viewed as separate and distinct cases. The audibility threshold work was done using a single resonance, whereas the enclosure effects represent a series of resonances. The word 'series' is highly relevant in this context, where it refers to a regular mathematically related sequence. It would appear that the human ear is highly sensitive to response irregularities based on a series [5], the latter being an accurate term to use when defining enclosure resonances. Fryer [4] has shown that under optimum conditions, a resonance whose amplitude is 20 dB below the steady state curve is detectable in classical music, and up to 30 dB below in steady noise excitation. This is confirmed in Steven's work [1] as well as by other practical work conducted on test enclosures.

In a specific example at fairly low frequencies a resonant Q of 20 placed at 145 Hz with decay R_t of 300 ms, was clearly audible even when the peak was buried 10 dB below the steady state response (see Reference 48, Chapter 9 and p. 373).

7.5 RESONANCE CONTROL, DAMPING MATERIALS AND BRACING

If a method of resonance control or 'damping' is applied, the Q of the intrinsic panel resonance can be greatly reduced and hence the choice of panel material is of little consequence. This means that previously little used substances, such as injection moulded plastics, can be employed for cabinet construction.

Whereas in the case of a single enclosure high mass was a positive advantage, with the damped cabinet it is actually detrimental. Resonance control employs the principle of dissipation through friction. A given panel should be laminated with a special layer of comparable mass per unit area, this mass equivalence ensuring that a good mechanical match is achieved between the two layers to permit the effective transfer of vibrational energy from panel to the damping material. The layer is often a bituminous impregnated felt or card used for vibration control (largely in connection with automobile body panels). These materials have a high frictional loss in bending. A thin enclosure of low mass is clearly more effectively controlled by a given density of damping pad than is a thick one (Figures 7.6 and 7.7).

Fairly pure lead sheeting is effective material for use in loading and damping enclosure walls, as it has exceptionally high density and plasticity. Its main drawback is its high cost, although it could well prove worthwhile for small cabinets of moderate internal volume. Certain grades of heat-cured polyurethane (monothane) are potentially effective vibration absorbers. Other damping materials include Celotex or fibreboard which is generally attached to a panel using a flexible adhesive such as PVA, and will produce some damping, particularly of the higher frequencies. Automobile underseal loaded with a sand filler is also fairly effective if applied thickly. However, a word of caution is necessary with these latter substances. They rarely dry out completely and can contain a volatile solvent which may soften the cones and bonding of plastic drive units. This may result in a premature failure of the adhesive holding together the cone and surround, leading to the disintegration of the unit. Another pitfall is pad ageing, where flexibility is lost over a period of time. Pads which are not

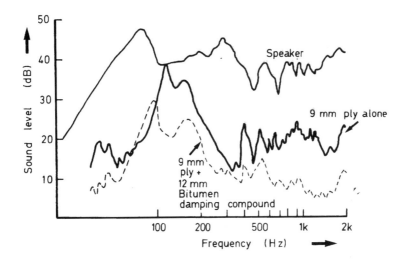

Figure 7.6. Two results are apparent from the application of damping. First, the amplitude of the resonances has been lowered by about 10 dB, and second, the added mass has somewhat reduced the fundamental panel resonant frequency. This in itself can prove advantageous, as the resonances which are likely to occur in the critical mid band may as a result be shifted to a less aurally sensitive region (after Barlow [3])

properly bonded to the panels (staples are not enough) are ineffective and often become detached in service.

One of the more useful advances in the area of panel damping materials was made in 1972, when a patent was filed in Germany for a synthetic medium consisting of a heavy mineral-loaded thermoplastic. Dunlop (UK) have also investigated damping systems, and have found a soft foam sheet with a high mass counterlayer very effective [6,7] (Figure 7.8 and Tables 7.2–7.6).

Placement of Damping Pads

The dense damping pads appear to work by a combination of bending loss due to panel flexure and by the absorption of surface waves in the panel. A well-built enclosure will have strengthened corners and seams, and hence the panel edges will possess a high mechanical impedance where the absorption will be poor. Harwood [5] suggests that a 50% control layer coverage of a panel in the central area will produce the most effective damping for a given quantity of material.

Bracing

With the difficulties involved in damping many designers are exploring more carefully the control of major enclosure resonances by bracing techniques. Softwood is almost useless, but hardwood and metal reinforcements can be effective. One design employs an array of steel rods in tension, another a complex array of internal hardwood bracing, not unlike a cathedral roof support.

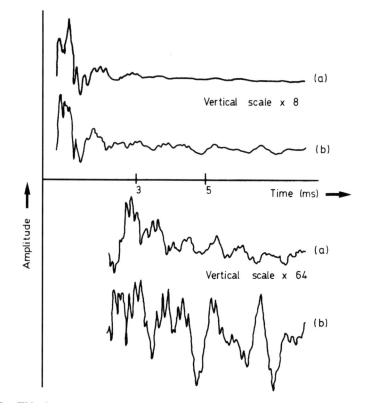

Figure 7.7. This shows two contrasting delayed impulse responses for a 100 mm plastic bass mid range drive unit in a 7 litre closed box. The output is that which is mainly heard through the cone, although the panels also make some contribution. Curve (a) employs a 12 mm chipboard construction with 12 mm of bituminous panel damping applied; curve (b) uses hardboard of 6 mm thickness with no damping (courtesy KEF Electronics)

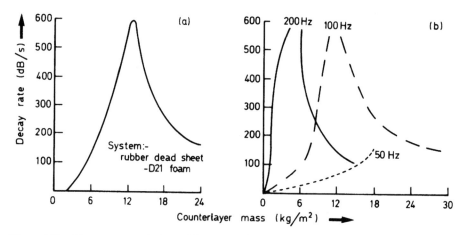

Figure 7.8. Decay rate variation with counter layer mass and frequency in laminated damping layers (rubber counter layers on foam base) (after Payne *et al.* [7])

Table 7.2. Results of decay rate versus surface density using samples 1.6 mm thick

Material	Surface density (kg/m²)	Decay rate (dB/s) Free	Bonded
Aluminium	0.66	10	4.5
Polyurethane 'Chipfoam'	1.27	44	1.5
Felt	1.58	30	2.5
Latex foam	1.63	22	6
PVC/plasticizer/china clay (100/75/200)	2.50	40	3
PVC/plasticizer/whiting (100/75/200)	2.51	13	2
'Dead sheet'	4.8	50	2
PVC/platicizer/molybdenum powder (100/75/500)	5.0	59	2
Lead sheet	14.7	164	15

Table 7.3. Commercially available damping products

Material	Surface density (kg/m²)	Temperature (°C)	Decay rate (dB/s)
Bitumen	1.48	Ambient	4
		+5	5.5
Coumarone indene resin	1.45	Ambient	5.5
		+50	10.5
Polyethylacrylate/c. black	1.45	Ambient	2
		−10	23.5

Table 7.4. Performance of foams

Foam code	Type	Bulk density (kg/m³)	IYM (MN/m²)	Bulk stiffness (kN/m²)	Decay rate (dB/s)	Performance
	Test bar	—	—	—	2	Poor
D7	Polyether	21.5	20.6	12.9	3.5	Poor
DE300	Polyether	23.4	11.7	10.8	4.5	Poor
S12	Polyether	29.3	6.0	5.1	4.0	Poor
D1	Polyether	16.5	18.9	4.9	5.5	Poor

In choosing a reinforcement, its location needs consideration with respect to major modes. Clearly a half-height circumferential brace is of little value as it will fail to control the first panel mode A_2B_1 (see Figure 7.1). Diagonal placement to couple as much unsupported panel area as possible will maximize the benefit. Symmetry emphasizes dominant mode output, hence dissimilar panel dimensions, width, length, height, help to disperse modes; a tapered profile in one or more planes is even better. [2]

Recent subjective data also suggest that damped thin-wall enclosures may suffer from some low frequency loss and that rigid boxes give firmer, better defined bass.

Table 7.5.   Attenuating properties

| Counterlayer | Foam layer | Decay rate (dB/s) at 100 Hz | |
		Bonded	Non-bonded
Rubber (4.9 kg/m^2)	D7 polyurethane foam	3.5	183
Rubber (4.9 kg/m^2)	D21 polyurethane foam	214	186
Rubber (4.9 kg/m^2)	SBR latex foam	236	150

Table 7.6.   Effect of varying foam thickness (D21 foam)

Foam thickness (mm)	Decay rate (dB/s) at 100 Hz
2	7.5
4	34
6	51
8	190
10	407
12	194

Modification of boundary stiffness may also provide a significant contribution. One engineer hit upon the elegantly simple expedient of recessing the back panel of an enclosure by 3–4 cm. Internal volume is affected little, but the rear seams are dramatically stiffened by the effective 'T' section girders formed. In a particular Wharfedale example, a modestly priced loudspeaker demonstrated notably improved enclosure colouration by this small change in construction.

Barlow [3] investigated curved surface structures which, in theory, are much stronger than plane panelled equivalents. Experiments using a multi-ply cardboard cylinder showed that this form was extremely rigid in terms of both expansion and compression waves, and potentially offered very low colouration. It should be possible to produce moulded or cast panels with plane mating edges and curved surfaces. These could be designed on modular principles to produce different sizes of enclosure.

Curved Walls

Acknowledging the pioneering work by Barlow on the dramatically improved stiffness of cylindrical cabinets, some manufacturers have incorporated curved sections for the main panel areas allied to strong solid or hardwood framing sections. In one example, the curved section was made of laminated hoops of solid walnut with steel strip reinforcement (Sonus Faber Guarneri), while in another, the curved side walls are formed of a stressed carbon fibre composite, this enclosure also including an alloy panel for precision, low resonance mounting of the drivers (Wilson Benesch ACT).

A disadvantage of pure cylindrical or spherical enclosures is the dominant radial mode standing wave in the interior volume which may require excessive absorbent.

7.6 STANDING-WAVE MODES

A speaker enclosure is nothing less than a small room. When the internal space is driven by a loudspeaker a whole range of standing waves will be produced over that driver's operating frequency range. If spherical, the enclosure has the highest degree of symmetry. The frequency f of the fundamental mode present depends on the internal diameter defined by the half wavelength expressed as

$$f = c/l$$

where c = speed of sound, i.e. 314 m/s; e.g. for a 0.6 m diameter sphere

$$f = \frac{314}{2.06} = 261 \text{ Hz}$$

and this mode will be strong due to the high degree of symmetry.

If the enclosure is rectangular, then the primary mode is based on the longest dimension, e.g. if 0.6 m in height, then $f = 261$ Hz.

In addition there will be a series of higher modes based on integer multiples. Other modes will be present at frequencies dependent on the internal width and depth of the enclosure.

Given that a driver cone is partially transparent acoustically and that it is neither possible nor desirable to try to totally absorb the acoustic energy inside the box, two factors affect the loss in sound quality resulting from these standing-wave resonances being heard via the cone. One is the internal proportions and the resulting distribution of resonances, including their relative magnitude, and the other is the location of the drive unit (or bass port) on the relevant enclosure panel.

Dominant standing waves are not greatly affected by the usual non-parallel cabinet geometries. Large in wavelength in comparison with the non-parallel elements, the standing-wave energy at lower frequencies generally sums over the surface, working with the dominant raw dimensions. Severely non-parallel surfaces are necessary for a significant reduction.

For the fundamental half-wave mode it will be most strongly excited (and read by) a driver placed at either height extreme. The second-order mode will be most severely coupled by a driver position half-way up, this suppressing the fundamental. Theoreticians often argue for placement between the one-half and one-third positions for minimum excitation and transmission of modal energy. Higher frequency modes are then successfully damped by moderate foam linings or fibre stuffings.

Note that excessive damping can "dry up the sound" and increase system losses at low frequencies.

As with a room, if possible the ratio of internal dimensions should follow some form of the 'golden ratio' thus avoiding harmonic coincidence for the three dimensions. If styling considerations dictate an acoustically unsuitable slim enclosure, internal partitions may be used to achieve the desired ratios, these disguised within the enclosure.

The peaky nature of the sound output from a panel is exacerbated by the presence of a series of discrete standing-wave modes inside a plain rectangular enclosure.* These

*Also a listening room. See 'Room Characteristics', Chapter 9.

standing waves are well defined and may be responsible for distinct 'sub-colouration'. In the case of a small enclosure, this is effectively characterized by the use of the term 'boxiness'.

The standing-wave frequencies are given by Rayleigh:

$$f_r = (c/2)[(A/L)^2 + (B/W)^2 + (D/H)^2]^{1/2}$$

where c = velocity of sound in air; A, B and D are the mode orders; and L, W and H are the internal cabinet dimensions.

Several steps may be taken to reduce or control standing waves. The first is to avoid symmetry in cabinet construction; for example, the worst case is an enclosure where all panel dimensions are equal. In the field of room acoustics, ratios have been calculated to provide the minimum excitation of standing-wave modes, the room dimensions being 2.3 : 1.6 : 1.0, these equally valid in the context of loudspeaker enclosures.

While parallel-piped or conventional rectangular enclosures are the general rule, a cabinet with strongly anti-parallel sides will stagger the modes and hence reduce the internal resonances. For example, it is often beneficial to design the sub-enclosure for a mid-range unit so that the plane of the rear panel is angled with respect to the front, as this will help to suppress the main front/back reflection. Long tubes have also been used with success in this application; well packed with absorbent they form a graded termination to the rear cone energy. Even if a cabinet has dimensions that are inharmonically related, as recommended above, the magnitude of the standing-wave resonances can still be appreciable.

A test enclosure of internal dimensions $570 \times 320 \times 266$ mm, 48 litres in volume, and tuned to 40 Hz by a tube 203 mm long and 51 mm internal diameter, possessed prominent standing-wave modes at 300, 510 and 620 Hz with a typical Q of 40. The duct had its own tube resonance at 750 Hz ($Q = 40$), in addition to the fundamental bass resonance at 40 Hz ($Q = 1.5$). To control such resonances, internal treatment by acoustic absorption is essential in a high performance system.

Where an absolute minimum of audible internal resonances is required, the quantity of absorbent required can be so high as to impair bass efficiency. At this stage it makes sense to close the box or employ a non-acoustically transparent radiator instead of an open port.

Suitable materials include lossy fibre blankets such as fibreglass and mineral or rockwool pads, plus polyester, cellulose and bonded acetate fibre wadding, the latter commonly known as BAF. Speaker enclosures of the sealed-box variety may employ self-supporting volume fillings such as wool-felt, resin-bonded fibreglass or long-haired wool, these also providing a degree of damping at the fundamental resonance (Figure 7.9).

However, if a speaker is optimally designed, little or no damping of the fundamental resonance should be required as this factor will have been accounted for in the driver system analysis. In such a case, the absorbent wadding need only be fixed to the enclosure walls, a well bonded material such as BAF or polyester* wadding being the most suitable. Thick carpet underfelt is also useful.

*Needs to be used in a densely packed roll to be effective.

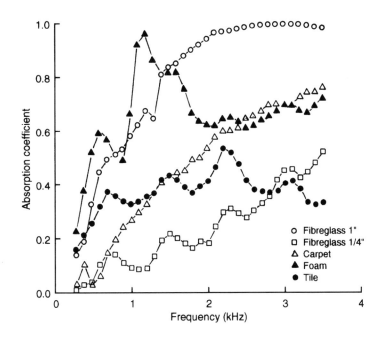

Figure 7.9. Sound absorption capacity vs frequency (after *Speaker Builder*)

High level non-linearity is common with linings which can move at low frequencies. This can be seen as a discrepancy in the amplitude/frequency responses taken under steady state and transient conditions [10] (see Figure 7.10).

In recent years, open-cell synthetic foams have found favour, in particular certain acoustically absorbent grades of polyurethane. This material is also used for absorption wedges in many anechoic chambers. In small enclosures 12–18 mm thicknesses are effective and in the larger designs of 60–100 litre volume, blocks 50 mm thick are usually satisfactory. Some manufacturers have also used foam sheets with a wedged surface moulded into the structure. (Synthetic foams may represent a fire hazard,* and the location of the crossover and adjacent linings deserves consideration.)

Optimum Placement of Absorbent

The region of maximum standing-wave energy is in the enclosure 'volume' rather than at the sides, and thus the absorption material is ideally placed in the cabinet 'space' rather than attached as sheets to the inside surface of the panels. Since the filling should not be allowed to move, some system of internal support rods should be installed. Some grades of absorptive foam may, with appropriate system design, be used as self-supporting blocks completely filling the enclosure.

---

*Fire retardant grades are now available.

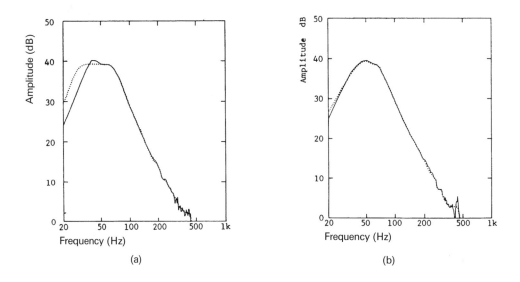

Figure 7.10. The effect of moving fillings or linings within an enclosure (hysteresis in driver suspensions has a related behaviour). In (a) the response of a third-order bandpass enclosure with a loose filling was measured by 'steady state' sine wave excitation (dotted line) and by a transient or dynamic method, the FFT of an impulse response. The difference is considerable between 20 Hz and 50 Hz. Under impulse testing the linings do not have time to move. With a slow sine wave sweep they act as additional resonators of indeterminate transient response. (b) The effect of locking the linings into place whereby a good agreement between the two test methods is now shown with similar consistency anticipated for the bass sound quality

7.7 DRIVER-CONE TRANSMISSION OF INTERNAL RESONANCES

Standing-wave modes and volume resonances may not only be heard via the cabinet panels but also through a driver cone. A thick, heavy cone will be less acoustically transparent than a thin one. Furthermore the output from a cone will be proportional to its area, a small cone allowing less of the internal cabinet energy to escape. With large cones, in shallow enclosures, where the front to rear mode is strong and relatively high in frequency, a layer of absorption immediately behind the driver has proved beneficial, and additional acoustic obstructions such as polystyrene sections have been attached to larger cones to increase their opacity.

A speaker cone will obey the mass law for sound transmission above a certain break frequency around 5 kHz, and the transmission is fairly constant at lower frequencies except where resonances occur. In the example given in Figure 7.11, longitudinal modes in the cone area are attributed to the poor attenuation in the 1–5 kHz region, while the average loss is 16–18 dB. At the driver's fundamental resonance the attenuation will of course approach zero. (Figure 7.12 shows a transmitted box reflection.)

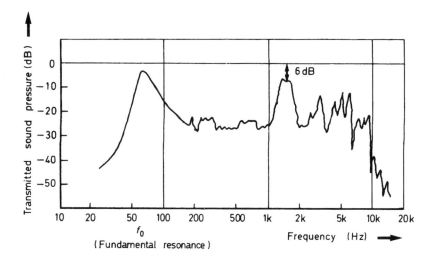

Figure 7.11. Transmission curve of a 170 mm chassis Bextrene cone driver (after Fryer [4])

Effect of Cabinet Filling on Frequency Response and Sound Quality

Enclosures generally have defined quantities of absorbent stuffing or foam linings to absorb the internal acoustic energy from the back of the driver; at low frequencies they serve to moderate the system Q and help to increase the effective volume of the enclosure [see low frequency, p 132]. In the mid range, standing-wave modes inside the enclosure are controlled. While this makes engineering sense, some audio critics have suggested that such Q moderating additions and absorption generally detract from sound quality and have advised their removal from enclosures, with the claim that loudness, dynamics and rhythm all benefit. In Figure 7.13 the upper curve shows the errors introduced when the usual absorbent acoustic filling of the enclosure, otherwise shown in the lower trace, is removed. The resulting standing waves inside the enclosure are transmitted through the cone and result in peaks and dips which, as expected, follow a harmonic series. As such these are easily recognized aurally as a colouration with an identifiable pitch. The measured frequency response is seen to be marred by additional $+3$, -5 dB variations, and is certainly louder since more sound energy is radiated. However, this will generally be of impaired fidelity. Removing the designed linings or fillings is not recommended, and in any case may alter the low frequency alignment and thus the whole balance of the loudspeaker design. In certain cases an enclosure with an irregular internal geometry may suffer minimal standing-wave problems and require little or no absorption. Its low frequency design will take account of this omission. This question raises an often neglected aspect of enclosure fillings, in that their behaviour may not be well defined at low frequencies, particularly under high power operation. Unless specifically constrained, fibrous stuffings may move about in a random and erratic manner in response to pressure variations. Mathematically this constitutes an error in timing relative to the driving system, and depending on the degree, the perceived sense of rhythm may well be impaired (see Figure 7.10).

Cumulative decay spectra

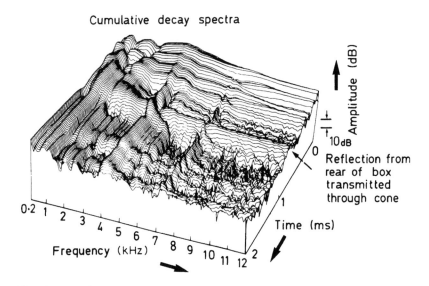

Figure 7.12. Output of 110 mm moving coil bass/mid range unit in 7 litre closed box showing ridge at $t = 0.9$ ms due to box reflection (courtesy KEF Electronics)

7.8   CABINET CONSTRUCTION

Cabinet Size

The unwanted sound output of a cabinet structure is proportional to its surface area. Small cabinets are less of a problem in this respect and hence panel damping is not so critical. Conversely, such an enclosure will, by comparison with larger cabinets, suffer from internal reflection modes of both higher Q and frequency. These observations suggest that for rigid enclosures of up to 15 litres, internal volume absorption is important, while for larger sizes, the increasing panel surface area means that panel damping assumes greater significance.

Large enclosures of 60–100 litres volume utilizing the thin-wall, highly damped technique, may suffer a lack of rigidity at low frequencies, due to the flexure of the large panel area absorbing bass energy at the main enclosure volume resonance. With medium sized 40–50 litre enclosures there may need to be a compromise between mid-frequency colouration and cabinet rigidity at low frequencies. Cross-bracing is useful for increasing the wall stiffness of such a cabinet without greatly modifying the colouration characteristics. Front to back bracing is also worth employing where the front panel is weakened by a significant area of drive-unit aperture.

Subjective results indicate that the designer must find a compromise between the articulate, precise sounding low frequency register potentially possible with highly rigid enclosures, often infinite baffle, and the need for uncoloured transparent mid-frequency reproduction. The rigid enclosure may have poorer mid-region panel resonances, 500 Hz–1 kHz, while the thin-wall, damped alternative with 200–400 Hz modes may soften the impact of bass transients. Only careful system and enclosure design will provide a workable solution to both these requirements.

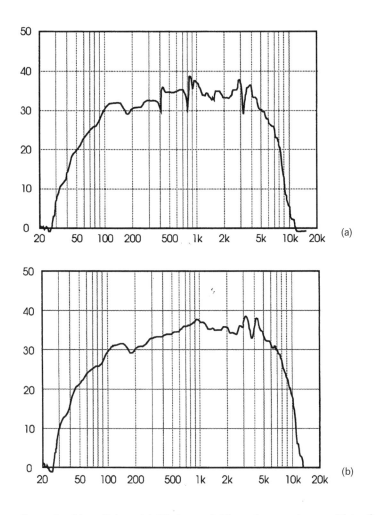

Figure 7.13. Effect of cabinet lining. (a) Undamped 25 cm deep enclosure. Note the strong reflections (dips) in the undamped curve at 400 Hz and multiples of this frequency. Sound reflects from the back of the cabinet and is radiated through the cone. (b) Damped 25 cm deep enclosure

The performance of commercial systems suggests that cabinet colouration is generally worse with the larger systems unless they are very carefully designed. Interestingly, designers are often unaware that enclosure colouration may be viewed as a signal-to-noise ratio. High driver sensitivity provides a greater level of wanted direct output to frame reaction vibration and consequently to induced cabinet colouration, and can sound 'cleaner' in consequence. The lighter driver diaphragm reduces magnet reaction vibration. The result is a better direct-to-cabinet noise ratio.

Corner Joints

Modern cabinet-making techniques can produce accurate mitred joints with the use of 'V' groove cutting machines, and a speaker designer may be tempted to accept these as satisfactory. While this may hold true in the case of small, thick walled enclosures, a more substantial edge clamping technique is required for larger cabinets. The success of panel damping to some extent depends upon the rigid coupling of adjacent panels, so that one helps to dissipate energy in the other. The front panel is often weakened by both drive-unit apertures and ports, and because it is usually undamped, it must be rigidly coupled to the enclosure shell in order to adequately transfer and dissipate the unwanted energy.

In an enclosure, all internal seams benefit from well glued battens or corner pieces made from a tough grade of wood about 18 mm in cross-section. However, chipboard or plywood offcuts are also suitable, the aim being to increase the adhesive contact area between adjacent panels at the seam. If the front or rear panels are removable (many modern designs have permanently fixed panels with access to the crossover and wiring obtained through the bass unit aperture), then a generous quantity of screws should be employed to securely clamp down the panels.

Front Grilles

Most loudspeakers have some sort of acoustically transparent covering to protect the drive unit diaphragms and to screen them from view, although with a few domestic designs the visual effect of the drivers has been exploited and a grille omitted. Likewise many professional systems, particularly the horn types, do not require protection, nor are visual considerations very important in this context.

The grille material can have a considerable effect on sound quality and must be carefully selected. A less than acoustically transparent covering will not only absorb sound energy, particularly in the treble range, but will also provide a partially reflecting surface adjacent to the drivers, thus producing further colouration. It is not easy to produce a fabric which is sufficiently transparent in acoustic terms and visually opaque. Light machine-knitted 'stretch' polyester fabrics are suitable and special plastic monofilament woven materials are also produced.

This problem is alleviated if the contrast of the driver panel assembly in reduced; for example, by painting both the drivers and the panel black. If a fabric is employed it must be firmly attached under tension, spaced from the accompanying panel, or it may flap and produce spurious noises. A thin layer of open-cell foam, approximately 4–6 cm thick, may be fitted under the grille cloth at its perimeter in order to space the fabric away from the driver panel and impart some resilient tensioning.

An additional complication arises with a reflex enclosure which has a high air velocity at the port opening, as the close proximity of a decorative grille fabric may well disturb its operation. The fabric is usually stretched over a light wood or metal frame whose side members should be substantial since the fabric tension can easily warp a frail structure. The 'step' which results from positioning the produced grille baffle adjacent to the drive units must be given due consideration so that the reflections do not cause colouration and disturb the polar response.

Large pore open-cell foam slabs have been used in place of fabric grilles, and while these have a remarkably low loss on axis, off-axis some reduction of treble energy can occur, due to the sound traversing a thicker section of material. This could conceivably prove an advantage with some drivers, where the off-axis polar response needs to be curtailed. The appeal of foam grilles is largely aesthetic, as a wide range of surface

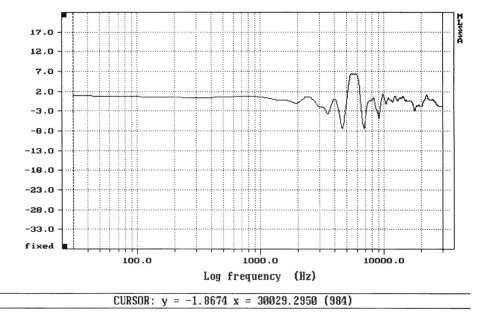

CURSOR: y = -1.8674 x = 30029.2950 (984)

Regent errors introduced by grille

Figure 7.14. Effect of an unsuitable grille geometry (12 mm plain step). Peak response deviations are ±7 dB

contours and colours are available. Their self-supporting nature means that they can be directly fixed to the cabinet by means of a suitable grade of multi-hook plastic strip such as 'Velcro'. A further advantage concerns the fact that the 'step' of a wooden grille frame is avoided.*

Fabricated grilles of wood slats are generally unsatisfactory, since they tend to act as diffraction gratings, but open-weave metal mesh has been successfully employed, and if backed with a thin black fabric can be visually effective. Tests on a large number of commercial systems continue to show very poor grille design, and many of these systems are quite superior with their grilles discarded. Not only is the axial response in the 2–20 kHz range disturbed by a poor grille, but also a loss of stereo focus and detail can be heard (Figure 7.14).

The Matrix Enclosure

Practically, the idea of the porous, solid foam enclosure has been realized mainly using conventional cabinetry in the form of the Matrix series by B & W. This type of enclosure may be regarded in two ways, either as a fabricated cellular version of the solid foam design, or as a cabinet with high order bracing where the bracing is conceived as a set of pierced interlocking planes which form a three-dimensional honeycomb matrix within

---

*These grilles can suffer premature ageing in tropical environments.

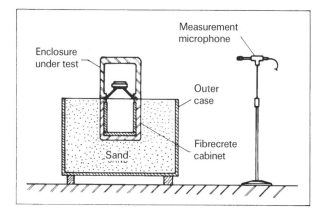

Figure 7.15.   (a) A test fixture for measuring the acoustic output of enclosure panels (B & W)

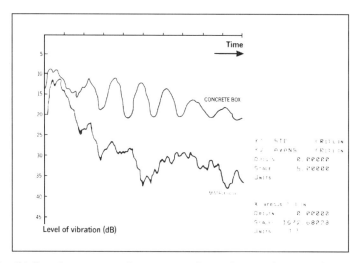

Figure 17.15.   (b) Impulse response decay comparison of a matrix type of enclosure with a
commercial Scandinavian concrete enclosure using the test fixture

the enclosure. All points and all panel surfaces are effectively coupled together. The
bracing planes are of medium density fibreboard strongly glued at each interlocking join.
The volume matrix locks tightly into a normal MDF or particle board shell and the
result is a singularly rigid, inert structure. Both the honeycomb and matrix techniques
are only suitable for more costly designs given that the enclosures cost three to five times
that of a conventional construction (see Figure 7.15).

Low Mass Enclosures

There is an alternative approach to those traditional high mass and high rigidity designs
which generally employ damping and or bracing to control panel resonance. One

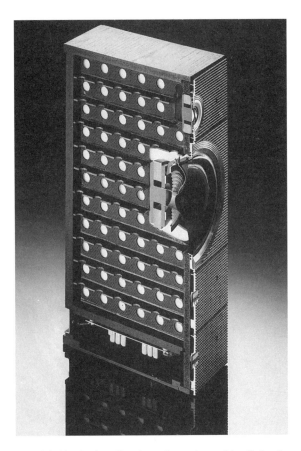

Figure 7.15. (c) Physical realization of matrix multi-cellular bracing (B & W)

alternative is the low mass enclosure constructed with the intention of minimizing stored energy. If the time or transient signature of such an enclosure, designed using ultra-rigid low mass engineering honeycomb, can be made short enough, then it is possible to bring it into the time frame of the main drive unit where it becomes part of the overall system response and equalization. Low stored energy is equivalent both to low colouration and good transient response. Speakers designed using such costly cabinet techniques have indeed shown low colouration, good transparency and well-defined transients.

The most successful form so far has been a 12 litre rectangular box made of 20 mm aluminium honeycomb laminate, 'V' grooved and assembled with liberal quantities of catalytic cured epoxy adhesive. The front panel is reinforced by the addition of a 10 mm alloy plate. Some resistive loss is deliberately introduced at one seam since the high Q and high Young's modulus of these aircraft-type structural panels can result in a mild, high order resonance in the upper mid range.

In a later variation an internal bracing panel is fitted, also fabricated in honeycomb alloy with suitable apertures, which provides further structural improvement thus

Figure 7.16. A high performance system using a lightweight aluminium honeycomb laminate enclosure with internal bracing designed for a minimum of stored energy and maximum rigidity (see also Figure 7.22 for dedicated stand)

removing the need for the resistive seam. Extremely stiff at low frequencies, these enclosures have a characteristically tuneful 'open' bass (Figure 7.16).

However, the low panel mass results in some acoustic panel transparency in the mid range. To prevent energy from the back of the driver diaphragm from exiting via the panels, the interior is filled with layers of graded absorptive foam providing a sort of transmission line termination for acoustic energy within the enclosure.

Variations on the honeycomb theme include a paper board honeycomb called Torlyte and also a Melamine skinned structural polyurethane foam. Another suggestion involved a larger semi-solid enclosure of porous foam, lightly skinned. Effectively this has no panels and thus no resonances. It simultaneously contains the low frequency pressure waves and also terminates the mid-range energy.

A State of the Art Enclosure Design for Minimum Audible Resonance

Where price is no object, costly materials and techniques can be employed to generate the finest results. In one system example, the WATT by Dave Wilson, the enclosure benefits from many techniques to achieve a remarkably inert result. The following details are all considered influential, including the small size (approximately 9 litres) which naturally improves strength and also results in a small enclosure surface area with reduced acoustic radiation. The enclosure itself is a truncated pyramid; as a result the panels are non-rectangular and the internal surfaces anti-parallel. The latter minimizes internal standing-wave modes while the former helps to disperse and moderate the usual plate resonances present in conventional enclosure panels. In addition the interior is lined with anechoic grade foam supplemented by a volume filling of polyester fibre.

The enclosure panels are cut from a dense, naturally inert composite—an acrylic, heavily loaded with ceramic and a mineral powder—which may be machined like marble. High frequency panel modes are controlled by a highly resistive bituminous laminate on the inner surface while the remaining fundamental resonances are handled by heavy, 20 mm thick lead slabs bolted into position on elastic mountings to provide tuned, seismic damping. Furthermore, the side panels are extended at the rear to form small triangular 'wings'. A massive alloy bar is bolted up between these wings, horizontally disposed and providing a stressed reinforcement for these largest radiating surfaces. Finally, the finished mass of approximately 25 kg provides a heavy inert foundation for the two-way driver lineup to perform at its best. The performance attained in this enclosure design is an object lesson in the continuing importance of enclosure colouration in box speaker design.

Both mechanical impulse tests and listening have shown that this quality of enclosure has a dramatic effect in improving sound quality, particularly with transients, subjective dynamics, stereo focus and depth; as such it shows that despite considerable improvements, we still have a long way to go in the field of commercial enclosure design. However, this performance is achieved at high cost, approximately 15 times that of a normal enclosure of this size.

Coupled Vibration Between Drivers

Over the past few years several designers have become aware of subtle effects resulting from the interaction between driver outputs due to coupled vibration. An early technique pioneered by KEF involved the resilient mounting of LF and MF driver frames, decoupling them from the enclosure to reduce the transfer of higher frequency frame vibrations into the enclosure panels. Cabinet colouration can be shown to be improved in many areas by this technique but at some subjective cost to bass dynamics and to perceived clarity of the driver output. This is probably due to a correspondingly increased level of driver chassis vibration.

In addition to the enclosure panel resonance effect mentioned, vibrational energy coupled between drivers can affect their sound quality. For example, the perception of treble sound quality can be described in terms of 'purity' and 'clarity', in terms of stereo focus and 'grain', or as perceived distortion. While these are mainly a function of treble unit and speaker system design and performance, the subjective sound quality is also partly dependent upon the mechanical integrity of the HF unit, its mounting and the overall vibration level experienced by it. Listening tests have shown that steps taken to reduce transmission of vibration or to isolate the treble unit from bass and mid-range energy do improve HF sound quality.* At the fundamental level, such steps include good enclosure and driver mounting practice. Rigid, low resonance enclosure construction is the first step, followed by vibration absorbing acoustic or visco-elastic seals for the driver frames mating to the enclosure. Mounting plate strength for the treble unit is important, as is the tightness of all the fixing bolts. A loss in dynamic power and transient definition can occur with low tension fixing methods or weak driver frames or plates. Metal is much preferred to plastic for HF unit plates.

Push–pull Driver Mounting

A fairly well-known technique to reduce enclosure vibration involves the use of two equal bass, or bass mid-drivers, preferably rigidly mounted together and driven such that their acoustic output is appropriate for the system design but such that their frame vibration is in opposition. With a cancellation of magnet reaction the composite, coupled driver suffers little or no frame vibration and offers a much cleaner sound when mounted in an enclosure (see Figure 7.17).

Separated Box Enclosures

Further, more interesting techniques for vibration isolation include the divided enclosure or the SBL, the 'separated box loudspeaker'. Here specific drivers in a given system are provided with separate boxes which assemble closely together to form the normal enclosure outline, but designed to avoid effective mechanical coupling between them. In one design the HF unit, mounted on a concealed cantilever stand, has its own sub-enclosure which defines its own radiating properties and its inertial reference plane. This technique frees it from vibration emanating from the main driver and enclosure section which operates over the rest of the frequency range. An alternative solution has the HF driver mounted on a heavy MDF panel suspended on four diagonally disposed, damped coil springs. This resulted in a free resonance of just a few Hz for the suspended assembly. Above 10 Hz the treble section is well isolated from any vibration present in the main enclosure by this mechanical filter. Any technique which is effective in

*See also SBL design p 136 Figure 4.17, also p 345.

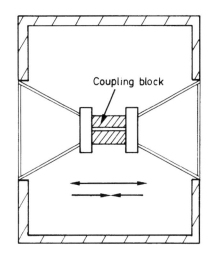

Figure 7.17. A zero reaction composite LF driver. The magnets and/or frames are locked; the drivers operate in electrical parallel so that frame reaction forces are nulled

reducing enclosure vibration and stored energy may be hopeful to clarity, definition and stereo focus (see Figure 4.17).

7.9   DIFFRACTION AND CABINET SHAPE

Because of diffraction effects, the properties, size and surface irregularities of a baffle or enclosure will have a considerable influence on both the measured and subjective performance of a system. At low frequencies, where wavelengths are long, even a large enclosure presents a relatively small obstacle to the radiated energy, and hence, through diffraction, the output is uniformly propagated around the cabinet, i.e. the radiation pattern is virtually omnidirectional.

Common Rectangular Enclosures

With increasing frequency a region is attained where the frontal dimensions of an enclosure become comparable with the wavelength of the sound being reproduced. The enclosure then begins to increasingly direct the sound energy into the frontal or forward plane.

At still higher frequencies the radiation pattern is theoretically hemispherical, bounded by the front panel plane, but in practice it is usually narrower, due to the inherent dispersion pattern of the drive units themselves (Figure 7.18). The characteristic change in radiation pattern with frequency will vary with the size and shape of the enclosure involved. This behaviour is now examined in greater detail.

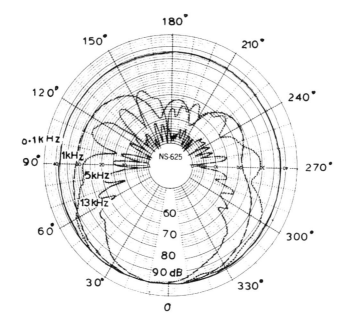

Figure 7.18. Polar diagram from a 30 litre two-unit system containing a 0.2 m cone driver working up to 2 kHz, and a 45 mm dome treble unit. The cabinet measures $50 \times 28 \times 25$ cm. The drivers are vertically mounted with the polar plot taken into the horizontal, i.e. lateral, plane. At 100 Hz the output at both the front and the rear is comparable as the cabinet dimensions are much less than a wavelength. At 1 kHz, the cabinet width approximates to a wavelength, and the rear energy output is well down at -25 dB, the cabinet showing a hemispherical pattern with increasing frequency. However, by 15 kHz, the polar pattern is in fact controlled by the high frequency unit whose diaphragm diameter is the dominant factor (courtesy Yamaha)

The 6 dB Pressure Response Step

Consider an ideal piston drive unit which possesses a theoretical uniform response when mounted in an infinite baffle. If positioned on the front face of a tall tubular enclosure, the resulting axial response would exhibit a distinct step of 6 dB at the transition between omnidirectional radiation at the lower frequencies and forward directed hemispherical radiation at higher frequencies. This irregularity is difficult to equalize and in consequence this cabinet shape is avoided by designers. Olsen's [8] classic set of responses for an identical driver in a series of cabinets shows that the 6 dB step is not present at a single frequency but appears as a series of ripples in the response curve at multiples of the basic frequency, with a peak-to-peak amplitude as high as 10 dB (Figure 7.19).

These curves are valuable to all designers who seek to experiment with unusual cabinet shapes. Olsen appears to have tried most of them, and the resulting responses are worthy of closer inspection. 'A', the sphere, undoubtedly gives the smoothest characteristic. This is not surprising, as its shape is free from sharp

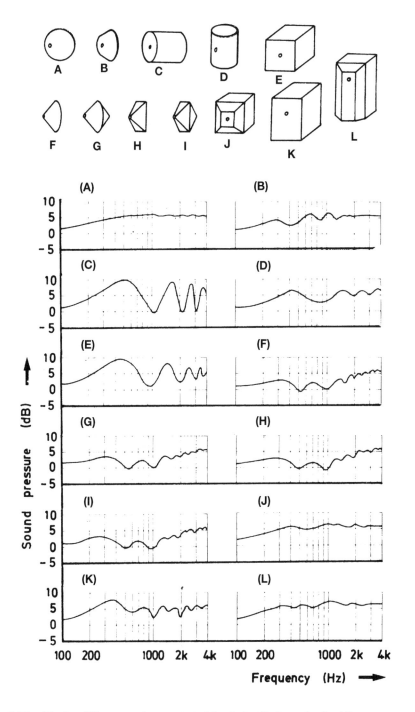

Figure 7.19. Twelve different enclosures tested for their effects on the final frequency response of a loudspeaker (after Olson [8])

discontinuities in the path of the expanding sound field. However, it is possible to achieve a reasonably smooth response with enclosures built of plane panels provided that they are based on an unequal length, width and height ratios, and preferably have chamfered or rounded front edges to eliminate the usual sharp profile (shape 'L', Figure 7.19).

The frequencies at which these major diffraction irregularities occur depends on the cabinet size. Small bookshelf systems might primarily exhibit the effect from 1 kHz to 2 kHz, whereas it would be more prominent from 200 Hz to 800 Hz in the case of large free-standing enclosures. The pattern is also dramatically affected by position; for example, by placing a cabinet against a solid boundary such as a wall or floor.

It might even be possible to make some nominal correction to the axial response of an imperfect driver by suitable choice of cabinet size and proportion, so that the resulting diffraction effects help to equalize the driver.

System design is complicated by the tendency for driver manufacturers to specify the performance of their units in a plane infinite baffle, or a near approximation to it in the primary frequency range. The advantage lies in the elimination of diffraction thus allowing the natural response characteristic to be shown together with the reference sensitivity at the defined distance, usually 1 m.

This is all very well until the driver is placed in a real enclosure, on a relatively small baffle area and spaced at some distance from the nearest plane boundary. For development, free space or the simulated gated impulse equivalent for acoustic radiation may be invoked. Now the response looks nothing like the manufacturer's data. At lower frequencies, including the reference region, say 150–300 Hz, the measured sensitivity is almost 6 dB too low since this baffle reinforcement is no longer present and this power can diffract around the enclosure with a near spherical wave front. Over some region of the upper mid range this axial response rises towards the original level but carries additional irregularities due to cabinet shape and local geometry, and is now out of balance with the reference range.

Thus a driver with a falling response on an infinite plane baffle measurement might have a largely corrected response when normally mounted. A flat specified curve* will require strong compensation in the crossover network to equalize the result in-cabinet and this largely explains why the first inductor to the bass-mid unit is generally several times larger than filter theory would suggest.

At low frequencies the local room boundaries come into range as the acoustic wavelength increases. Some of that lost power returns in the form of reflections and standing waves, hopefully well enough integrated to usefully augment the overall response. Ideally this boost comes into play as the speaker system is naturally rolling off in power, generally due to size and headroom limitations. The degree of augmentation depends on the structural properties of the building; European houses and flats are often brick or concrete with only moderate losses at low frequencies. American open-plan timber frame houses absorb and diffuse bass energy rather more and this results in a distinct difference in the preferred low frequency alignments. In the US, designers favour a higher system Q, perhaps 0.8; the UK preference is for a Q nearer 0.5.

---

*On the plane baffle.

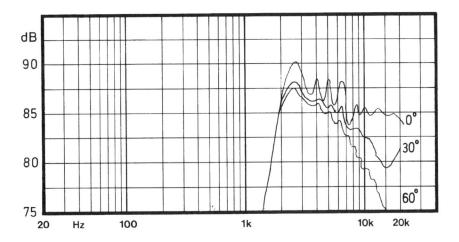

Figure 7.20. Surface mounting effects. Note the lack of off-axis uniformity and the amplitude ripples of ±2 dB (expanded vertical scale)

Minimization of Diffraction

When designing a suitable cabinet shape, asymmetrical placement of the drive units is beneficial as it results in unequal path lengths from the drivers to the edge of the enclosure.

Some drive units will have a relatively narrow directivity in the mid band. This directs the energy clear of the cabinet edge and hence avoids diffraction effects at these frequencies. A typical plastic-coned bass mid range unit working up to 2 or 3 kHz in a medium sized enclosure (40 litres) will have an off-axis response which falls rapidly above 700 Hz and hence avoids cabinet edge diffraction effects which might otherwise be expected at the higher frequencies.

The most pronounced mid-band diffraction effects occur with small 50 mm dome radiators whose directivity is almost hemispherical up to 2 kHz. With these units and also in the case of small dome tweeters, a plane surface free of obstruction, cavities or steps is essential if irregularities in the axial frequency response caused by diffraction are to be avoided. This applies also to the design of the front plates and mounting chassis of such radiators, which should present a smooth contour from the edge of the dome to the front panel surface. Even the projecting screw-heads on the chassis can cause measurable changes at the higher frequencies.

Investigation of the geometry for a surface mounted dome has shown that the transducer will be mismatched in the lower part of its frequency range. This mismatch distorts the wave front which is launched near the dome perimeter and produces ripples in the frequency response of typically ±1.5 dB when other sources of irregularity are removed. For a 25 mm HF unit this can extend from 2 kHz to 8 kHz (see Figure 7.20).

Recessing the dome in a simple cavity generally has catastrophic effects on the response and is often warned against. However, if the recess is designed as a shallow contoured horn, properly curved to provide a termination at right angles to the dome

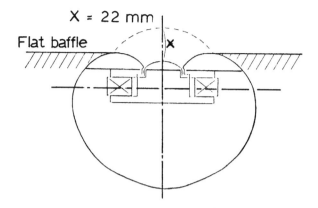

Figure 7.21. The ideal minimum diffraction case is a spherical wave-front horn mounting, shown here in outline behind the baffle. The front section has now been integrated with a plane baffle and results in an improved response compared with the normal mounting

circumference, then the edge of the dome then sees a matched acoustic mirror and the sound wave is launched without distortion (Figure 7.21).

Implicit in this mounting technique is an overall shift in the shape of axial response since the flared recess acts as a vestigial horn thus narrowing the directivity in the lower frequency range where it is effective (see Figure 7.22).

Hayakawa [11] goes into some detail while several transducer designers have evolved similar front plate constructions on a wholly empirical basis.

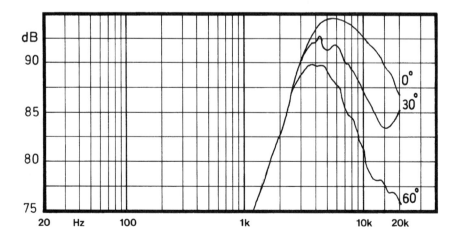

Figure 7.22. Vestigial horn low diffraction mounting and resulting frequency responses. Note much reduced ripple and superior off-axis uniformity (expanded vertical scale) (after Hayakawa [11])

7.10　DRIVE-UNIT MOUNTING

The above considerations also concern the method of drive-unit mounting. For example, a 120 mm chassis mid-range unit should not be mounted behind a 12–18 mm panel if colouration due to diffraction and cavity resonance effects is to be avoided.* Flush mounting in the front of the baffle is recommended for such drivers. This suggestion may be ignored in the case of larger mid and bass-mid units as the front panel thickness is small relative to both the drive-unit apertures and the sound wavelength produced.

A wooden block may be fitted to the cabinet to brace the magnet assembly of a driver possessing a weak pressed-steel chassis. This can reduce both driver and cabinet colouration. Work by KEF and B & W [9] has shown that an appropriately compliant driver mounting gasket can usefully reduce cabinet resonances generated by energy transmission from the driver frame (See p. 344).

The suspended frame resonance is normally tuned by hand adjustment of mounting bolt tension to around 100 Hz. Transmissions of frame vibration to the cabinet above this frequency is much reduced. However, subjective results suggest that you cannot get something for nothing. As with damped thin-wall enclosures, compliant driver mounting seems to detract from the bass transient quality and some critics have noted a loss in upper range detail, attributed to the undesirable movement of the driver frame. In pursuit of greater sound definition designers are returning to rigid driver fixing, often with superior frame forms and employing thicker, reinforced front mounting panels.

Some designs have employed stepped baffles to improve inter-unit time delay compensation. The resulting ledges usually impair the performance far more than any benefit in time-delay terms. In one design the (HF) driver was recessed 6 cm, in a large front aperture lined with absorbent material to reduce the inevitable cavity colouration and reflections. In so doing, a large proportion of the off-axis energy output of the driver was also absorbed. While a uniform axial response was attained the system sounded 'dead' due to the loss of energy in the reverberant field.

Sources of Unwanted Vibration and Resonance

Driver fixing

One of the problems facing the designer is fixing a driver to an enclosure where the latter is generally made of wood or a wood composite and may suffer from inelastic compression and dimensional changes with time, temperature and humidity. Thus a driver fixing which was tight when manufactured may be fairly slack just a few weeks later, particularly after air freighting or a change in climatic region. A loss of designed tension in a screw or a bolt fixing will impair the sound quality, both as regards dynamics and sound quality. Fixings must be chosen which tend to lock in their set position and maintain a high tension for long periods. Both the choice of sealing gasket for the driver and the structural quality of the driver frame are contributing factors. Even today many driver frames distort under their optimal screw tension due to design weakness in the rim.

*The 1.3 kHz 'glitch' in the LS 3/5a.

Other sources of resonance

It is quite common for the crossover network to constitute a resonant mechanical element, adding colouration to the speaker. Heavy inductors may vibrate on a relatively thin printed circuit or terminal board. In addition, the complete network may represent a significant mass that is attached to a section of the enclosure which commonly has a cutout for the terminals and may lack sufficient bracing to prevent local resonance. The location of the components on the crossover board may itself reduce sound quality due to unwanted electro-magnetic coupling between the various sections.

7.11   OPEN BAFFLES

Due to their simplicity, plane baffles were once the most common enclosure encountered. For adequate bass the path length from the front to the rear of a given driver when baffled should be substantial, since it must be comparable with the wavelength of the lowest bass frequency required. A large baffle is a weak structure and thus prone to resonances unless it is well braced and loaded with sand or heavy bituminous pads.

As with enclosures, asymmetry for baffle shape and drive-unit mounting helps to minimize cancellation and diffraction irregularities, and result in some low frequency improvement.

In theory a typical baffle area for adequate bass is $5\,m^2$, but with a suitable driver plus floor augmentation a 30 Hz response is possible at a reasonable pressure level from a $2\,m \times 1\,m$ panel.

Another problem with the simple baffle concerns the rear radiation, virtually equal to that from the front, the presence of which will affect both frequency balance and subjective character. This effect is also present with the other 'open-backed' systems such as the larger electrostatic and magnetically driven film-diaphragm speakers, as well as in units such as open ribbon-tweeters. However, the open baffle does have its advantages. While it is unwieldy in bass form, for mid and high frequency unit mounting it is free of the particular colouration produced by the alternative mid-range sub-enclosure.* The sound of such a baffle is certainly 'open' and free of boxy effects, and several successful high performance systems have used reflex or sealed-box bass sections in conjunction with smaller open baffle mid and upper mid components. Among these are the Bowers and Wilkins DM70, the Strathearn Audio speaker and the Dahlquist DQ10.

Augmentation of the low frequency response is possible by optimum placement of the driver resonance.

Against the advantage of subjectively perceived openness, must be weighed the unpredictable performance of such baffles in different locations, due to the rear radiation component. Another factor is the polar response difficulties when different ranges are integrated at a crossover point. (See Chapter 2, Section 2.1 on 'Directivity'.)

*See also the distributed mode radiator section 2.5.

7.12   LOUDSPEAKER SUPPORTS

A floor or wall mounted system will operate acoustically as if the adjacent surfaces are a continuation of the cabinet and, depending on the frequency radiated, acoustic coupling will occur between the two. The directivity theory indicates that at these lower frequencies (below 500 Hz) each adjacent surface will lift the sound power by 3 dB.*

If a system is balanced to give a natural sound in free space or semi-anechoic conditions, then it will be unbalanced by wall or floor mounting. A rigid open stand about 0.5 m high will transpose the range of lift due to local surfaces to a lower frequency, allowing a reasonably accurate result from such a speaker, particularly in the critical mid-range region.

It can thus be seen that the enclosure is by no means a simple box or baffle and that quite apart from considerations of low frequency driver loading (see Chapter 4) a number of other factors are also crucial in determining the subjective quality of the speaker system of which it forms an integral part.

A stand can in fact be regarded as an integral part of the system, and if this concept is widely adopted it could result in better systems with a more consistent sound quality. Placing the enclosure on a stand couples the two, with the assembly itself resting on the floor. Cabinet vibration can cause resonances in the stand while the floor interface is also a factor. Under heavy LF transients, enclosures can rock and bounce on their support, enough to produce audibly defective reproduction. Rigid coupling of stand to floor, for example with small adjustable spikes, can make a considerable difference, much reducing bounce and rocking. The minimization of rocking also benefits the mid-band stereo focus and detail. Doppler effects are not in question here; rather the accuracy of positional information transients generated by a stereo pair. If allowed to rock even slightly, the natural asynchrony of the two speakers will tend to blur differential transient wavefronts.

Wall Placement

About 20% of systems are designed to be used close to the wall, and given that most designers intend that the listening axis be located roughly between the mid and treble sections of a speaker, some sort of elevated support or stand is mandatory. Strong wall brackets may also be used for the wall aligned designs and units are specially made for speaker support. Several points are worth noting in connection with wall placement. The wall should be as solid as possible to reduce the incidence of coupled resonance and should be free of adjacent obstructions, furniture, etc. which could add spurious reflections and disturb the directivity pattern. Speaker placement on a long shelf or sideboard is inadvisable on grounds of serious coupled resonances. Regarding placement, the speakers should not be located too close to the room corners (greater than 0.6 m) otherwise the undesired room resonances will be driven strongly. Ideally, a 'wall mount' speaker (designed for use against a wall boundary) should also be located on a good floor stand.

*In moderate sized rooms this is nearfield at low frequencies with correspondence of power and pressure levels.

Floor Stands

Even a direct contact, floor mounted speaker may benefit from the addition of a vestigial 'stand'. In its simplest form this may consist of a welded tubular steel frame corresponding to the outline of the speaker base, fitted with suitable coupling hardware such as spikes. To interface the speaker to the frame low profile 'cone' spikes may be used, machined in steel or aluminium. For carpeted and softwood floors, adjustable, locking, fine point spikes made in hardened steel are recommended. These penetrate the carpet, generally causing negligible damage (except to antique Persians!), allowing an intimate vibration-free bond to be made with the floor.

Considering the floor interface first (for a polished hardwood or marble floor, hard rubber plastic domes may be used) the achievement of good mechanical coupling is vital. It greatly increases the stability of the supported system and tends to reduce spurious vibration and audible resonance in the stand. Resulting improvements in bass definition and dynamics, colouration, stereo focus and depth are far from trivial.

These benefits are greatest when the design and construction of the stand is acoustically as non-resonant as possible and is mechanically stable and rigid. Many forms of construction are possible; plastic designs generally proving unsatisfactory. The least expensive and most popular are manufactured in hollow-section welded steel, the most successful of these using three or four heavy vertical pillars and a large base area to improve stability. A small surface area for a stand structure helps to reduce its audible acoustic signature. Successful stands have been built in marble or slate, in mineral loaded, heavy acrylic, in thick section MDF composites and in hardwood. To date, good performance has been achieved using aluminium in the form of properly engineered castings for the top and bottom plates, linked by a massive extruded central pillar(s), generally mineral filled to control self resonance. The pillar is loaded with a dry mixture of sand and lead shot (see Figure 7.23).

While the ideal floor interface has three points, such tripod bases tend to be easily overturned, unless of large base area; four points are generally recommended here. At the top plate the three-point alternative can be used and has the advantage of requiring no adjustment. Small shallow cones on the top plate are effective while a light-weight enclosure may be safely locked to the stand via a clearance-fit captive bolt or a screw system to avoid accidental damage. In a domestic situation one must consider the consequences of a stand mounted enclosure toppling onto an infant.

In some cases an optimum result may be achieved without the use of top cones or spikes. Here small lumps of compressed mastic, Blu-Tack or similar can provide a good interface coupling and also provide a measure of adhesion thus helping to lock the stand and speaker together. Such is the subjective quality advantage of top grade stands that in some rare cases the stand price may equal or exceed that of the loudspeaker used with it.

Given that stands are often neglected at the point of sale it is important for speaker designers to pay attention to the subject. Ideally stands should be tailored for each design. Some loudspeakers are sold with the correct stands included in the package and this practice is to be encouraged despite possible sales floor resistance to the notionally higher price for the resulting system.

The best speaker system performance will only be achieved in conjunction with a compatible, optimally designed and secured stand. It is not often realized what a

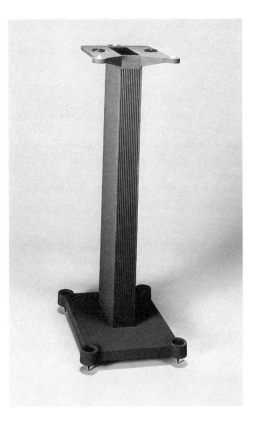

Figure 7.23. An example of dedicated stand and speaker design where the stand complements the free space system mechanically, acoustically and visually (see also Figure 7.16). The top casting has the three point anti-resonance mounting cones in aluminium alloy which engage milled brass cups bonded to the underside of the aerospace honeycomb laminate of the enclosure base panel. Two lateral rubber washers are dimensioned for a clearance fit for captive bolts to the enclosure to prevent it from being accidentally knocked off the stand. The column is of extruded aluminium alloy with a heavy filling of lead shot and sand. Top and bottom plates are alloy castings. Floor locking and levelling is achieved by adjustable spikes with a thumbwheel lock

positive contribution the stand can make in terms of system colouration and stereo focus. Comparing the performance of a smaller stand mounted system with a larger floor mounted example, the former benefits from the air space beneath and a reduced cabinet area, consequently reducing colouration. A stand allows the three local boundary distances to be dissimilar resulting in the smoothest room response (for example, a 40 cm stand places a bass-mid driver at 55 cm above the floor; the back wall may then be set at 75 cm distance and the side wall at 95 cm) (see Figure 4.1). However, for the floor mounted alternative the acoustic loading to the floor boundary is enhanced, often resulting in a deeper notch in frequency response due to the floor reflection. The taller floor cabinets may also suffer from increased internal standing-wave modes and, if not carefully designed, may have a poorer off-axis response, mildly impairing stereo image focus. Nevertheless, floor standers are a popular and growing type.

REFERENCES

[1]   Stevens, W. R., 'Sound radiated from loudspeaker cabinets', *Proc. Audio Engng Soc. 50th Convention*, London, March (1975)

[2]   Beranek, L., *Noise and Vibration Control*, McGraw-Hill

[3]   Barlow, D. A., 'Sound output of loudspeaker cabinet walls', *Proc. Audio Engng Soc. 50th Convention*, London, March (1975)

[4]   Fryer, P. A., 'Intermodulation distortion listening tests', *Proc. Audio Engng Soc. 50th Convention*, London, March (1975)

[5]   Harwood, H. D., 'Some factors in loudspeaker quality', *Wireless World*, **82**, No. 1485, p. 45 et seq. (1976)

[6]   Terosow Werks GmbH, Heidelberg, Damping Material, British Patent 1 310 241

[7]   Payne, E. W., Wilson, P. I. and Wragg, W. T., 'Latest damping materials design out panel noise', *Design Engineering*, pp. 54–57, May (1976)

[8]   Olsen, H. F., 'Direct radiator loudspeaker enclosures', *J. Audio Engng Soc.*, **17**, No. 1, 22–29 (1969)

[9]   Adams, G. J., 'New developments in L.S. Design — B&W 801', B&W Publication (1979)

[10] Fincham, L. R., 'A bandpass loudspeaker enclosure', KEF Electronics Ltd. (1979)

[11] Hayakawa, J. et al., 'Improvement in dome loudspeaker characteristics using spherical-wave-front horn baffle', *J. Audio Engng Soc.*, **36**, No. 7/8 (1988)

BIBLIOGRAPHY

Ballou, G., *Handbook for Sound Engineers*, SAMS, 2nd edn (1991)

Briggs, G. A., *Cabinet Handbook*, Wharfedale Wireless Works, Idle, Yorkshire (1962)

Iverson, S. K., 'The theory of loudspeaker cabinet resonances', in 'Loudspeaker Anthology', *J. Audio Engng Soc.*, 1–25 (1979)

KEF Electronics Ltd., *You and Your Loudspeaker*, KEF Electronics Ltd. (c. 1970)

KEF Electronics Ltd. *Loudspeaker Testing Using Digital Techniques*, KEF Electronics Ltd., March (1975)

Tappan, P. W., 'Loudspeaker enclosure walls', in 'Loudspeaker Anthology', *J. Audio Engng Soc.*, 1–25 (1979)

Tremaine, H. M., *Audio Cyclopedia*, 2nd edn, Howard Sams, New York (1974)

8

Home Theatre and Surround Sound

Growing in popularity, many domestic audio systems are becoming increasingly focused on a mixed media installation, combining hi-fi loudspeakers allied to direct view and projection video. Home theatre, or home cinema, covers these systems which include surround sound capability from video and non-video programmes. Stereo, two-channel sources, range from traditional vinyl disc and analogue tape, predominately cassette, to digital tape (RDAT), DCC (cassette-based data reduced digital tape), CD, MD (mini disc, data reduced) and radio tuner. There is also videotape, laser disc, DVD, broadcast television and video disc, plus cable and satellite transmissions. An increasing proportion of software, particularly video based is multi-channel encoded, and, where the installation gives surround sound, a full sound field is available all around the audience. In addition, audio processors are available which can extract and/or synthesize surround sound information from suitable sources.

The most popular home theatre format at present is five channel, and typically includes a centre channel speaker; this often called the dialogue channel because it generally carries the vital speech components of a movie and must also provide a basic level of intelligibility.

The development of a sound stage, with characteristics of width and depth, is the responsibility of the front left and right audio channels. These are generally conventional high quality speaker systems, often full-range types flanking, or placed relatively close to, the TV monitor or the projection screen. In some cases the latter is acoustically transparent and the centre channel speaker may be hidden behind it. A superior blend of sound and visual image results, with better entertainment value (Figure 8.1).*

In basic installations a pair of small ambience speaker systems is located near the back of the room, preferably elevated to reduce localization effects. It is desirable that the ambience channels do not draw undue attention to themselves; excessive level or proximity to the listeners can easily disturb the sound field balance. The rear channel speakers may reproduce a common, mono ambience channel, which may or may not be

---

*See Section 2.5, combination speaker and screen.

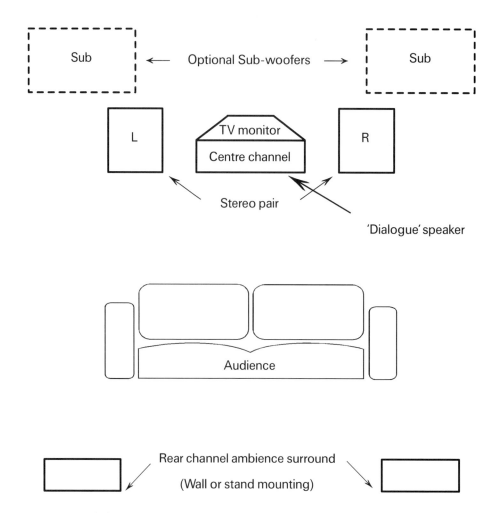

Figure 8.1 Basic home theatre/surround sound system. A five-speaker layout using speakers with good low frequency performance for front left/right. Thus the dotted sub-woofer(s) are optimal, only required if the 'stereo' pair have limited bass*

served by individual amplifier channels, as with Dolby PRO-LOGIC. The more recent systems for digital discrete, multi-channel encoding including DTS and the industry standard Dolby AC-3 can deliver discrete signals on five channels or more with the right processing (up to eight in some instances).

With smaller, neater speaker arrangements, though with more limited bass power and extension, a sub-woofer may be added to reinforce and extend the low frequency range. This is particularly effective in view of the powerful low frequency effects commonly found on film sound tracks. In fact, sub-woofers are generally recommended for all good home theatre installations.

*Ambience speakers may also be mounted on the side walls.

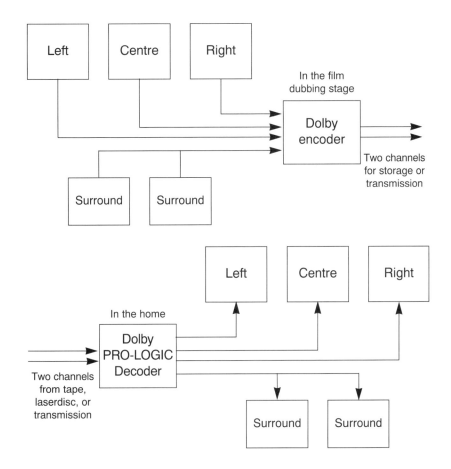

Figure 8.2. Dolby Surround/PRO·LOGIC. Four channels are matrixed/encoded to two channels which will reproduce as stereo. When decoded, especially with PRO·LOGIC steering circuitry, improved separation is achieved for four output channels. The optional fifth channel for the sub-woofer is not shown (after Harley)

The popular surround system is founded on the Dolby PRO·LOGIC coding of sound tracks, and matched decoders are available over a very wide price range and are often conveniently incorporated into multi-channel amplifiers. A specially filtered sub-woofer feed may be available (Figure 8.2).

Many considerations arise with multi channel systems which are covered below.

8.1   STEREO COMPATIBILITY

Very good multi-channel systems may be configured, but there is always some degree of conflict with the requirements for the best quality stereo reproduction. The best sound for a pair of high performance speakers will be obtained when they are the only speakers of significant size in the listening room. Both precision and power in the bass is diluted in the presence of unwanted additional speakers.

Furthermore, in the case of a direct view television placed in the centre stage position, it frequently contains its own speakers and it also constitutes a large acoustic obstacle, reflecting and redistributing the sound field formed between left- and right-hand speakers. Also the speaker placement for optimal two-channel stereo is generally wider than that defined for good video-based multi-channel working. However when a TV monitor is replaced by a projection screen, this has rather less impact on the local acoustics and may be set farther back away from the stereo pair (Figure 8.3).

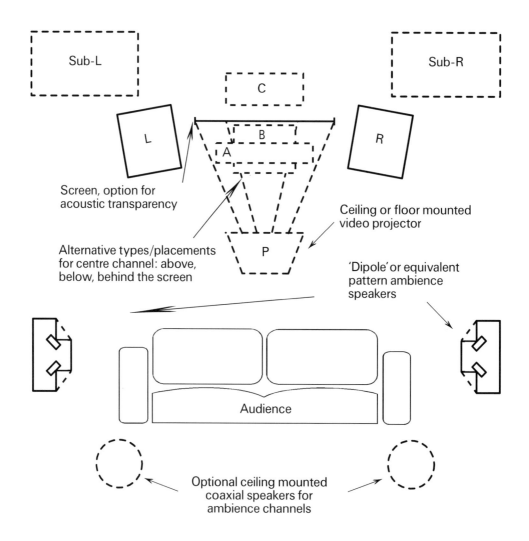

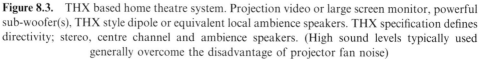

Figure 8.3. THX based home theatre system. Projection video or large screen monitor, powerful sub-woofer(s), THX style dipole or equivalent local ambience speakers. THX specification defines directivity; stereo, centre channel and ambience speakers. (High sound levels typically used generally overcome the disadvantage of projector fan noise)

In addition, given the present state of the art, the fidelity of multi-channel electronics, processors and amplifiers is not as high as that of discrete component stereo systems. Some compromise in respect of two-channel high-fidelity performance is inevitable.

8.2 MULTI-CHANNEL ADVANTAGE

It may be argued that multi-channel working conveys additional information and that the greater sensory experience resulting from surround sound working balances the loss in absolute fidelity when compared with pure stereo.

Greater versatility will be required from such systems as the new digital coding systems are introduced, such as the discrete channel Dolby AC-3. By comparison PRO-LOGIC is a two-channel system using matrix techniques for coding the additional directional information (Figure 8.4).

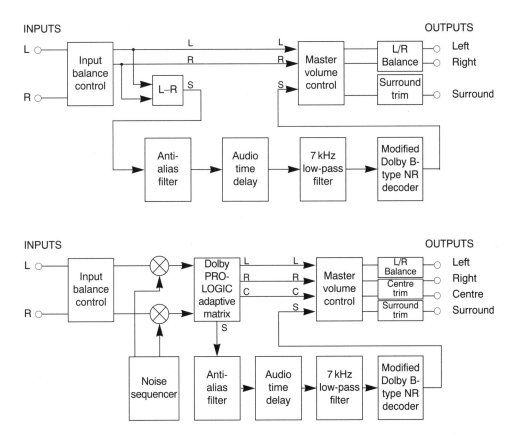

Figure 8.4. Comparing the passive decoding of Dolby Surround with the active matrix system used for PRO·LOGIC for home theatre. Note that both systems include bandwidth limiting for the surround channel, specified at 100 Hz to 7 kHz −3 dB. In practice, this reduces the demands made on rear channel speakers; a single full-range driver may be sufficient (courtesy Dolby Laboratories)

8.3 THX

The THX sound laboratory has set specifications and provides design recommendations to aid in the generation of a tonal balance and a sound field in the home which
more closely mirrors professional cinema practice, and which in theory makes for a
closer sonic match with the film production intentions. Higher sound system cost and
performance is the result, together with some divergence from normal Dolby surround
practice. THX compatible decoders provide shaped audio signals while speaker design
also has some special features which will emerge in the following sections.

8.4 SPEAKER DESIGN

Due regard must be taken of the specific requirements and acoustic environments for
the different types of speaker system in a multi-channel installation. Varied they may
be; nonetheless the objective is a well-distributed sound field of uniform quality and
tonality, such that no speaker or room region draws undue attention to itself and thus
impairs the stability of the surround sound effect.

Equipment critics have found many instances of poor matching of the sound
characters of the various speakers in a multi-channel system.

Design rules follow good speaker engineering practice, but in addition, incorporate
some specific factors which are outlined below. In addition, top quality systems
conforming to the THX™ specification have their own set of standards and systems
must qualify via direct evaluation at the THX laboratory.

Stereo, Left and Right

Almost any good stereo speaker, floor or stand mounted, will serve for the front stereo
pair. If the application omits a sub-woofer, then full-range speakers with more powerful
bass are advised, probably of floor standing design. Given their likely proximity to each
other and to the monitor, also potentially not very distant from the rear wall, the bass
alignment may be set somewhat overdamped, e.g. a Q of 0.5, thus avoiding any excess
of energy in the 50–100 Hz range. If not, the reproduction of normal television speech,
whether mono or two channel, may sound unnaturally heavy and boomy. This may
induce early aural fatigue even if the results are impressive on the movies. For THX
working, a vertical driver array is favoured, so reducing the reverberant contribution
from the floor and ceiling reflections.

The mid-range tonal balance may also benefit from some subtle adjustment to take
account of the proximity of the TV monitor casing, and its likely supporting cabinet.

Rear Channel

For the simple reproduction of ambience effects an extended frequency response is
unnecessary and compact slim-line enclosures are favoured. These are typically sealed-
box types with a bass driver and treble unit (in some cases just a 'full-range' single cone
driver) with an alignment balanced for wall boundary location, possibly a downwards

directed polar response. A maximum level of 100 dB, an 85–90 dB/W sensitivity and a 90 Hz to 12 kHz bandwidth is often perfectly satisfactory if the general quality, colouration standard, etc. match the primary system. In this instance the speaker bandwidth need only be a little wider than that imparted by the processor (Figure 8.5).

In some installations ceiling mounted speakers have been used successfully, these usually coaxial with the attendant advantage of a symmetrical radiation pattern. Complications include the acoustics of the ceiling and the lack of a defined enclosure, which results in some variations in performance. The advantage is almost perfect concealment, while the additional height above the listeners aids dispersion of the ambient sound field (Figure 8.6).

With digital multi-channel, the rear channel feed may be discrete, well localized, of higher quality and full bandwidth. For top quality systems superior 'ambience' speakers will be worthwhile (Figure 8.7).

Centre Channel or Dialogue Speaker

For projection systems, acoustically transparent screens are available and any suitable high quality speaker may be concealed behind it.

Where a TV monitor is involved, the dialogue speaker should be placed centrally to optimize the acoustic/optical alignment, and since the image cannot be obstructed it

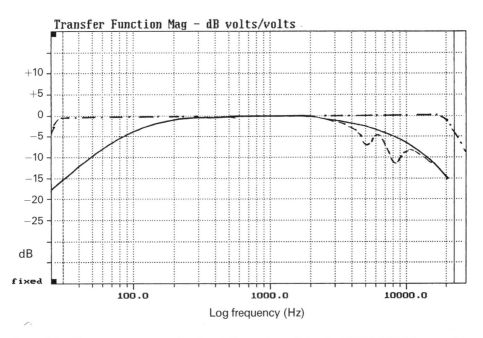

Figure 8.5. Frequency responses for decoded rear channel signals. PRO-LOGIC (————) has a 100 Hz to 7 kHz bandwidth, THX augments with extra filtering (- - - -) while AC-3 digital decoding offers discrete rear channels with an additional user option, full bandwidth 20–20 kHz (-·--·-)

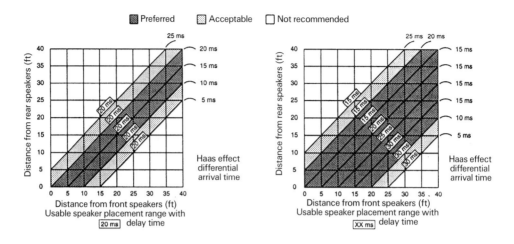

Figure 8.6. Recommendations for speaker placement (rear channel relative to front) to allow good frontal localization but avoid 'echo' effect from ambience channel feed. (20 ms is a standard delay; some systems provide programmable delay for greater versatility). (Courtesy Dolby Laboratories)

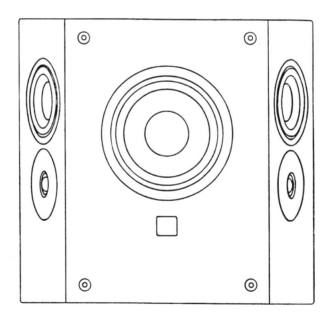

Figure 8.7. An example of a dipole ambience field speaker system. The angled driver sets operate in anti-phase above 300 Hz. The centre driver gives omni-directional radiation below 300 Hz. Drivers are 170 mm, 100 mm and 25 mm. THX application (after Aerial Acoustics Corporation), for side wall placement

must be located directly above or below the faceplate. While a vertical orientation of drivers offers the best angle of horizontal directivity for the audience (many THX speakers conform to this type), aesthetic considerations generally dictate a low profile, slim-line, horizontally disposed enclosure, matching the monitor as closely as possible.

However, a horizontal driver disposition is the least favourable arrangement for audience coverage. Conventional two-driver designs (bass and treble) have been tried, but inevitably the responses in the left direction are unbalanced relative to the right due to asymmetry.

Symmetrical arrangements are favoured, and use a centre, preferably narrow chassis dome tweeter, slimmed by the use of a closed field and thus magnetically screened miniature magnet using neodymium alloy. On each side is a bass–mid unit typically of 80–120 mm chassis size and the overall enclosure height is held in the region of 160 mm.

Factors that help to widen the radiation angle in the crossover range are high slope crossovers, and a lower than usual crossover frequency, e.g. 2 kHz instead of the usual 3 kHz and 4 kHz. The use of a driver baffle structured to angle the bass drivers (also well screened magnetically) away from the central treble unit, e.g. by 10° or 20° helps to widen the acoustic lobe in the crossover range.

Note that these horizontal designs have another notch in their response, this time due to the relative delay, for the two mid sections appearing off-axis. This notch may be 12 dB deep in the 1 kHz region and adds a further complication.

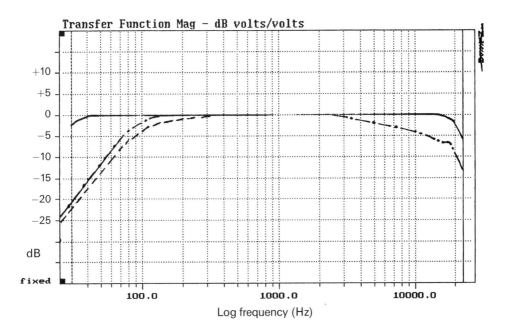

Figure 8.8. Frequency responses for decoded outputs of surround processors, centre channel; PRO-LOGIC offer a full bandwidth, or a high-pass option, −3 dB at 100 Hz (——— , - - - -). THX processing, directed to a tonal quality close to cinema sound, also filters the bass and in addition softens the treble (-·--·-). For the discrete channel digital systems AC-3 and DTS also provide many shaping options according to the decoding design and user settings

A frequency range of 80 Hz to 15 kHz is worthwhile and reflex loading, despite the small enclosure volume, is common to help improve the power handling. The centre channel may need to play loudly, 103 dB or more, and needs an 88–92 dB/W sensitivity with up to 100 W power handling.

Fine centre channel speakers have also been made with concentric drivers, such as the UNI-Q or the ICT. Such drivers have good directivity on all axes.

Specific voicing may be applied to a centre channel speaker. The proximity of the TV screen also requires consideration. In addition, high clarity, intelligibility and articulation are paramount, even at high sound levels. The actor's words must be heard clearly, no matter how complex the mix of sound effects.

On balance there is a trend towards a taut, fast upper bass, occasionally some mid-range and presence-range prominence of a few decibels, and an upper treble which does not draw undue attention to itself. For THX, the narrow bandwidth setting, a range of only 60 Hz to 10 kHz is required (Figure 8.8).

Each designer must make his own decisions on the final balance according to the speaker technology employed, such judgement made using real world, multi-channel set-ups.

Screening

Magnetic screening is now recommended for the front left and right speakers in addition to the centre channel units, in view of the magnetic sensitivity of the larger direct view TV monitors. Steel plates may be required to line the cabinet sides, and these may be bonded with an appropriate visco-elastic adhesive to improve resonance damping. Even screened speakers may result in some colour shifts when placed on a monitor and this matter needs careful checking at the design stage.

Low Frequency Power

Considerable output is required in the bass for a suitably impressive film playback. It is rumoured that many of the 'foley' originated low frequency sound effects are simply shaped and/or gated bursts of third octave pink noise at 30 Hz! These can reach peak level, which accounts for the THX requirement for 105 dB, 30 Hz at 1 m for moderate listening rooms. Two sub-woofers sound rather better than one — more than their arithmetic sum would suggest. Placement in the room corners generally gives the smoothest and most powerful bass. Good boundary matching is important because it reduces the demands made on the woofer, resulting in improved performance and a higher dynamic range.

Multi-channel Discrete Processors

Speaker systems for home theatre have evolved around matrix processed signals, often band limited to reduce the audibility of spurious processing artefacts. Such practice also makes good economic sense; there is no point in over-specifying the auxiliary speaker systems, centre and rear, if the result is to price the system out of the market.

However, with the introduction of the discrete multi-channel processors, the European MPEG variants, the US designed DTS and Dolby AC-3 systems plus multi-channel (rather than synthesized) surround sound, all the audio channels may have a discrete, wide band, high quality identity.

Potentially, with the right replay system, the sound producer will be able to place a virtual acoustic image anywhere in the listening room, anywhere in the defined soundfield. Even at this early stage, systems such as DTS can reproduce a full circle 'walkaround' with stable geometry, using a sound as complex and as transient rich as a close miked orchestral chimes.

With such a potential for high quality, it is obvious that the more costly home theatre and surround sound systems are going to need better speaker systems for all quadrants of the sound field. Revisions may be anticipated with regard to existing practice, performance specifications, response shaping and required bandwidth to meet these new demands.

BIBLIOGRAPHY

Dressler, R., *The [Near] Future of multi-channel Sound*, Dolby Labs S94/1009 (1994)

Haxley, R., *The Complete Guide to High-End Audio*, Acapella Publishing (1994)

Holman, T., 'New factors in sound for cinema and television', *J. Audio Engng Soc.*, July/August (1991)

Holman, T., 'Home THX: Lucasfilm's approach to bringing the theatre experience home', *Stereo Review*, April (1994)

Smyth, M. and S. 'DTS coherent acoustics: The future of audio', *Widescreen Review*, **14**, **16**, **18**, Parts 1, 2, 3

9

Loudspeaker Assessment

Evaluation is a fundamental part of the creation of any new product. In the case of a loudspeaker, although the subjective quality must be the ultimate arbiter of performance, subjective testing is arduous and fraught with difficulties. Figure 9.1 gives a suggested hierarchy of the stages which may be involved in the complete evaluation of a loudspeaker system, this also including objective measurements.

Preceding chapters have outlined the technical aspects of loudspeaker engineering, and to a reader versed in the exact world of mechanical science, speaker technology might appear equally well defined. However, if this were the case, laboratory measurement alone would suffice to describe the entire performance of a speaker, including its sound quality. In fact, the present state of our knowledge is fairly limited as regards the significance of various errors and their audibility thresholds, and despite much research into this aspect, subjective listening tests still represent a vital part of loudspeaker assessment. In themselves, such tests are a difficult proposition, for they involve physiological aspects of human perception, and due precautions need to be taken to ensure sufficiently reliable results.

This chapter covers most aspects of loudspeaker evaluation, both objective and subjective. An outline of the major parameters and distortions encountered with loudspeakers has been compiled, and where information is available, suggested standards and audibility thresholds are discussed.

9.1 LOUDSPEAKER SPECIFICATIONS, STANDARDS AND DISTORTIONS

Amplitude/Frequency Response

As with other items of equipment in the audio field, an adequately uniform response over the audible range is an obvious objective for the loudspeaker designer to aim at. Due to the strong interaction of the speaker's acoustic output with the listening environment, it is necessary to ask which response is the most important — total energy, the forward 2π hemisphere energy or the axial pressure, or perhaps something in-between [1,2]. Each philosophy has its proponents — for example, it has been proposed that for wall mounted systems, a flat energy response in the forward hemisphere

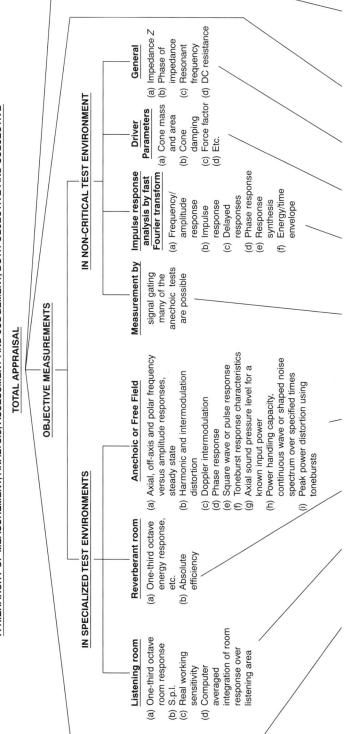

LOUDSPEAKER APPRAISAL

A HIERARCHY OF MEASUREMENT, ANALYSIS, ASSESSMENT AND JUDGEMENT, BOTH OBJECTIVE AND SUBJECTIVE

TOTAL APPRAISAL

OBJECTIVE MEASUREMENTS

IN SPECIALIZED TEST ENVIRONMENTS

Listening room

(a) One-third octave room response
(b) S.p.l.
(c) Real working sensitivity
(d) Computer averaged integration of room response over listening area

Reverberant room

(a) One-third octave energy response, etc.
(b) Absolute efficiency

Anechoic or Free Field

(a) Axial, off-axis and polar frequency versus amplitude responses, steady state
(b) Harmonic and intermodulation distortion
(c) Doppler intermodulation
(d) Phase response
(e) Square wave or pulse response
(f) Toneburst response characteristics
(g) Axial sound pressure level for a known input power
(h) Power handling capacity, continuous wave or shaped noise spectrum over specified times
(i) Peak power distortion using tonebursts

IN NON-CRITICAL TEST ENVIRONMENT

Measurement by signal gating many of the anechoic tests are possible

Impulse response analysis by fast Fourier transform

(a) Frequency/ amplitude response
(b) Impulse response
(c) Delayed responses
(d) Phase response
(e) Response synthesis
(f) Energy/time envelope

Driver Parameters

(a) Cone mass and area
(b) Cone damping
(c) Force factor
(d) Etc.

General

(a) Impedance Z
(b) Phase of impedance
(c) Resonant frequency
(d) DC resistance

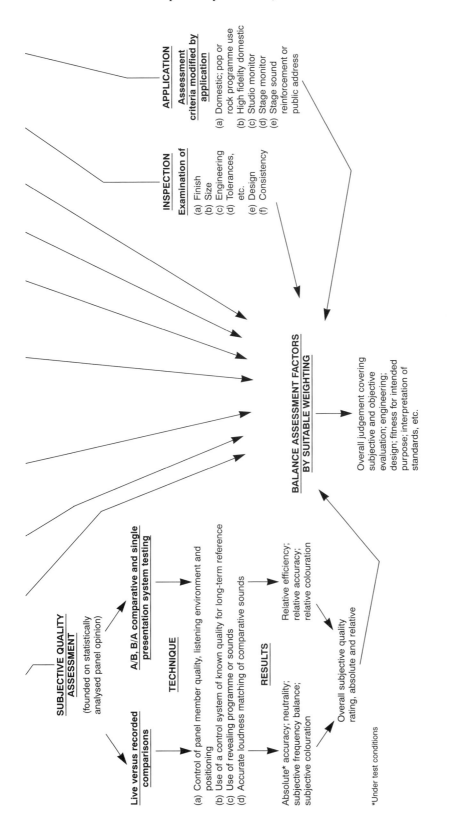

Figure 9.1. Loudspeaker appraisal

produces the most natural sounding result. They are supported in this opinion by some of the Scandinavian manufacturers as well as by several European companies [3,4]. More specifically, their aim is to produce an equal energy response over the frequency range in the listening room, this is obtained with systems designed for shelf mounting. However, work conducted on the subjective quality and accuracy of commercial loudspeakers suggests that the most successful models have been designed essentially for free-field use and are best auditioned mounted on open stands as far away as is practicable from room walls or corners [5]. Such systems appear most natural if their anechoically measured axial response is essentially uniform with frequency, provided that certain other requirements are met, namely that the system should employ direct radiator drivers possessing a good directivity over the forward ±30° angle, and that the levels of distortion and delayed resonance colouration should be low [6,7]. This confirms BBC research findings and validates the basic historic standards for high quality loudspeakers, although it does assume that this type of speaker is fed with an accurately balanced programme. Admittedly this is a relatively uncommon occurrence. As a result, modifications in the response uniformity may be required to suit different microphone techniques, for example.

While a uniform axial output above 150 Hz is an accepted standard, below that frequency free-field or anechoic measurement will not give results which correlate with the response in domestic rooms. It is at this point that the requirements for a high performance domestic system, which is generally used fairly near room boundaries, diverge from those applicable to a large studio or reinforcement monitor which may be used in much larger spaces. A speaker seeking to serve both applications could ideally be adjustable at low frequencies to suit different boundary conditions.

Directivity* (Dispersion)

No worthwhile standard has yet been established for directivity — in other words, the uniformity of response over a defined forward radiating angle. For a direct radiator system used for stereo, a predictably uniform output over the axial 60° solid angle is desirable, particularly in the lateral plane. This confers a usefully wide stereo coverage.

A suggested standard is 'deviation to be held within ±2 dB with respect to the axial curve over ±30° laterally, and ±10° vertically'. This would apply over the 100 Hz to 10 kHz range, one-third octave analysed, and should not be difficult to achieve using modern drive units. Outside of this angle, a rapid reduction in output may be considered a positive advantage, since the reflections from the adjacent walls would be diminished, with a consequent reduction in ambient energy and an improvement in stereo image stability. However, in practice it is difficult to make the off-axis output independent of frequency. An uncoloured speaker with a flat axial response but possessing an uneven off-axis characteristic, may well sound 'coloured' in a listening room since the reverberant energy is derived from the total energy response, including that off-axis.

*See Section 2.1.

Audibility of Response Irregularities

An investigation into the audibility of response irregularities has been conducted in England by Fryer [8].* His tests have confirmed the long held suspicion that low Q deviations of small amplitude and broad extent can be considerably more significant than high Q narrow-band irregularities, mainly due to the reduced probability of exciting the latter. The results of this test may be relevant to steady-state irregularities, since they concern the addition of resonant components whose peak amplitude could be less than the reference level, but which nevertheless produced mild humps in the total response. The ear's sensitivity to such resonances was higher for pink noise than for popular music, with a classical orchestral programme falling somewhere between. Figure 9.2 shows that by using the latter type of material, a resonant of Q of 25 or more could be of similar amplitude to the steady state before being detected, whereas a Q of 1, which might be expected to be less audible is in fact more noticeable. At 2.5 kHz, the region of peak sensitivity, the panel noticed the effect of a resonance $Q = 1$, at a level some 20 dB below the steady-state response. The ear's high sensitivity to broad acting, low amplitude response irregularities (approximately 1 dB), will be examined at a later stage in this chapter.

Harwood has stated that certain other features of frequency response are also responsible for obvious subjective effects, e.g. the presence of a 'series' in a sequence of irregularities. If these are random, then they may be relatively innocuous, but if regular, then the ear appears to assign strong colouration to them, even if it can be proven that the level of the delayed resonances is otherwise insignificant. The location of a speaker system in a room can invoke such a series of resonances, which are subjectively as obvious as those which can occur in the system itself. A further investigation concerned the audibility of irregularities with peaks and dips of magnitude ± 3 dB. If these were more closely spaced than 1 octave, and were not present in the form of a regular series, then they proved inaudible under programme conditions [10].

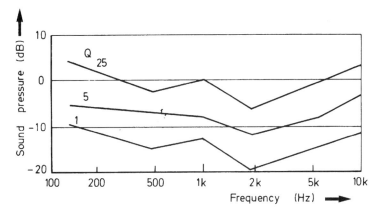

Figure 9.2. Detectability of resonances. 0 dB is a steady-state level. At 2 kHz a Q of 1 may be detected at a level 20 dB below the steady-state response, e.g. classical orchestral programme (after Fryen [8])

*Also Olive and Toole [48].

An isolated peak 4 dB high proves difficult to identify, while any series either of dips or peaks of smaller amplitude, no matter how strange the mathematical relationship, seems to invoke subjective colouration. The ear's response follows a kind of VU meter characteristic where positive peaks are summed but negative ones produce rather less subjective response.

Fortunately it is the broad, 1–2 dB level irregularities which are amenable to adjustment and equalization during system design. Furthermore, while the broad imbalances are easily heard on an A/B comparison basis, it is these for which the ear/ brain processor can make unconscious adjustment after even a short period. Higher Q ringing cannot be accommodated in this way and is equally difficult to deal with by system equalization (see also Reference 48).

It is evident that on a subjective basis, broad deviations in response are more obvious to a listener and that perhaps weighted on octave bands, the speaker designer should be aiming for a deviation of ± 0.25 dB or less in the forward directed response, while according rather less importance to narrow-band deviations of greater amplitude which hitherto may have occupied his attention unduly.

Further confirmation of this 'averaged response' criterion was provided in a series of panel tests which involved the assessment of 200 pairs of commercial loudspeakers [5]. The identities of the speakers were concealed from the panels, and the systems auditioned included both small models with good bass transient performance but necessarily limited bandwidths, and considerably larger systems, with a wider bandwidth but a generally poorer bass transient response.

Interestingly, the panel often voiced a preference for the smaller models, provided that they were not overloaded. In fact, even a miniature loudspeaker of 8 litres volume was found capable of sounding natural in a 'full' and 'spacious' manner — more so in fact than many of the larger systems to which this quality is more likely to be attributed. Subsequent analysis revealed that the favoured speakers were those which possessed very even axial responses over 100 Hz to 10 kHz when measured by third octave and octave averaging.

Non-linear Distortion

Except for gross effects induced by overload or a mechanical failure, there is little evidence to suggest that the distortions measured in electronics such as amplifiers are of the same subjective annoyance as those perceived with loudspeakers [10,11]. While levels of below 0.1% for harmonic and intermodulation distortion are well worth attaining with an amplifier, provided that the transient characteristics are not compromised, higher levels of loudspeaker distortion appear to be of less consequence.

As little correlation appears to exist between subjective sound quality and distortion characteristics in speakers, their interpretation is somewhat arbitrary. Moir compiled a graph of lower limit detectability of harmonic distortion on single tones (Figure 9.3) which shows that below 400 Hz, distortion greater than 1% on second and third harmonics is undetectable. At 60 Hz, over 7.5% of third harmonic content is inaudible, and likewise at 80 Hz, over 40% of second harmonic content lies below the audibility threshold.*

*Many of these earlier tests are for mono stimulus.

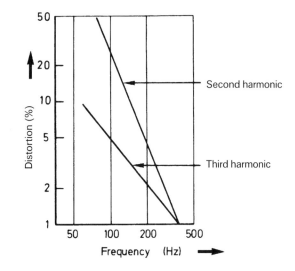

Figure 9.3. 'Just detectable' harmonic distortion on single tones (after Moir [14,15])

At frequencies higher than 400 Hz there is little evidence to indicate that harmonic distortion below 1% is audible. For example, the DIN standard 45 500, which is applied to hi-fi systems, states that for 96 dB s.p.l. at 1 m, the distortion should be less than 3% total harmonic from 250 Hz to 1 kHz, reducing to less than 1% above 2 kHz, this range extending to 5 kHz.

Fryer examined the sensitivity of a small but representative panel to first order (f_1+f_2, f_2-f_1) intermodulation distortion.* The test programme included popular, classical orchestral and solo piano material. Overall, the threshold of detectability of intermodulation distortion lay at the 4%–5% level. Experienced listeners demonstrated a 2%–4% threshold on piano, and unskilled female panellists a similar sensitivity on pop music. On pure tones the results were dependent on the level of the fundamental which was controlled by room resonance effects. Distortion thresholds of about 1% are indicated for pure tone tests [7]. Interestingly enough, neither sound pressure level (within reasonable limits) nor loudspeaker type (from large four-way systems to small two-way bookcase models) had any detectable influence on these results. This finding suggests that the residual intermodulation distortion of the test reproducers were negligible. The conditions for this experiment were related to a domestic situation, with the programme derived from analog disc, which will inevitably contain some intermodulation and harmonic distortion. The music thresholds might have well been closer to the pure tone levels had higher quality programme such as master tapes been employed. (Much of the academic evidence on perception of distortion is either derived using headphones, or a single channel, or both.)

Perception is greatly altered in the presence of two information channels and when speakers are radiating into a normally reverberant room. The author considers that a

---

*Fundamental frequencies (f_1, f_2). The first-order distortion products are the sum and difference frequencies of the fundamentals. reproduced by loudspeakers, this distortion was generated electrically to a high standard of accuracy.

reduction in third harmonic for moving coil designs from a typical 0.3% to 0.1% does result in an audibly more natural timbre, together with gains in clarity and perceived depth for the stereo image.

ATC (UK) have developed a high resistance high permeability alloy for the pole structure which is most effective in controlling third harmonic distortion.

MacKenzie suggests that a maximum of 0.25% harmonic and intermodulation content for the 'mid-range' (200 Hz to 7 kHz) is a desirable limit, and his consumer report on loudspeakers indicates that in the absence of other masking effects, subjective quality degradation is present with systems containing levels of over 1% in the 700 Hz to 12 kHz band. His test method utilized a 300 Hz difference frequency, measuring the upper third product $(2f_1 - f_2)$ from 400 Hz to 20 kHz, with f_1 tracking below [13]. However, the first-order intermodulation product may not represent the entire solution to the problems, and higher orders are well worth investigating. For example, BBC designs have been examined during development for harmonic distortion to the eighth, and intermodulation products to the seventh order (Figures 9.4 and 9.5) [6].

Smoothly changing distortion curves are considered favourable while the presence of sharp discontinuities indicates breakup modes with other serious consequences in terms of irregular directivity and impaired transient performance. If there are any doubts that low distortion is possible from a loudspeaker, the curves for the Yamaha NS1000 should dispel them (Figure 9.6).

This model, a three-way sealed-box system using a 300 mm pulp cone bass, an 88 mm dome mid and 30 mm treble unit, was tested at a 90 dB reference level at 1 m. Above 500 Hz the second and third harmonics were typically below 1%, and the intermodulation product over the same range in general measured 0.15%. To illustrate rather more typical distortion results, the response graph of a contemporary three-way system taken under the same conditions is included for comparison (Figure 9.7).

The large area diaphragm speakers such as the electrostatics should have very low distortion, due to the very small diaphragm excursion required once the low frequency range has been dealt with. Tests on the Quad ESL63 verified this, with harmonic distortion even at 96 dB, 1 m, at around −60 dB for frequencies above 150 Hz. It also showed very good pulse response with regard to compression within its dynamic range.

Frequency-Modulation Distortion

This is often described as 'Doppler distortion', but more correctly it refers to a series of distortion harmonics produced when the frequency of one signal is modulated by another.* This occurs when a drive unit diaphragm or even a local reflecting surface radiating a high frequency is simultaneously radiating another low frequency, where a greater excursion occurs. A typical example would be as a bass driver reproducing low and mid-range signals.

One may speculate about the importance of minimizing FM distortion, since a

*The figures for Doppler distortion are described in two different forms, either as a percentage of frequency shift of the upper frequency, this figure thus constant with respect to the latter, or alternatively as a percentage of the Doppler introduced harmonic sidebands relative to the amplitude of the unshifted upper signal. This second method gives a figure which is proportional to the upper signal frequency and thus varies considerably over the spectrum.

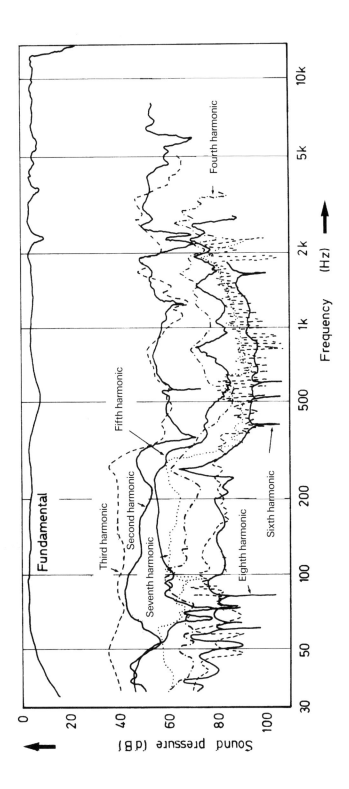

Figure 9.4. Harmonic distortion of LS5/5 loudspeaker measured at 1 N/m^2 at 1.5 m (after Harwood [6])

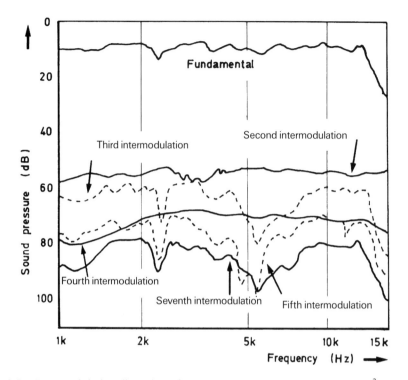

Figure 9.5.   Intermodulation distortion of LS5/5 loudspeaker measured at 1 N/m^2 at 1.5 m (after Harwood [6])

microphone transducing a wide-range programme at high sound intensities will possess FM components in its output. The bass frequencies will cause the diaphragm to move while simultaneously receiving the higher notes. Could a loudspeaker impart some cancellation of this distortion when reproducing such programme? A brief examination of the amplitude of microphone diaphragm motion at typical sound pressure levels shows that such frequency modulation is in fact negligible, and hence such cancellation will not occur.

Several authorities have discussed the problem of FM distortion, and some have suggested that the audibility thresholds are very low; Moir [14,15] has shown that 0.002% is detectable when using a modulating frequency of 20 Hz and a pure tone fundamental. Further work, in this instance using music programme and a revised sensitivity threshold, has suggested that 0.1% might be the level detectable by the experienced listener. A particularly relevant investigation is that conducted by Stott and Axon [16], concerning the high frequency pitch variations classed as flutter, which are also a form of FM distortion. Thresholds of about 1% were determined (Figure 9.8).

In this light, Moir's more recent threshold of 0.1% might appear to be a possible objective to match future equipment and programme standards.

For a given type of diaphragm, however, it would seem that while FM distortion reduces with increasing cone size, and consequently reduced LF excursions, the non-linear amplitude modulation distortions, namely harmonic and intermodulation, remain independent of cone size. With large units, for example of 300 mm

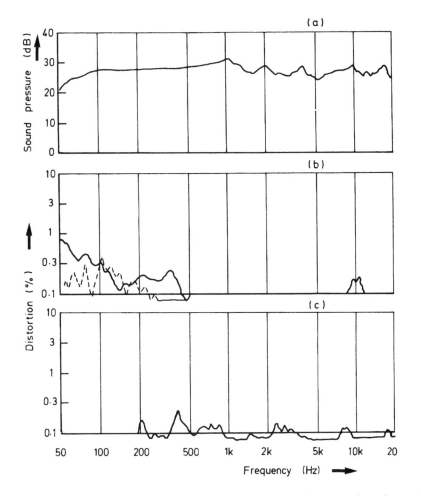

Figure 9.6. A low distortion speaker system by Yamaha, at 90 dB approximately, at 1 m. (a) Axial pressure response, (b) second and third harmonics and (c) intermodulation $(2f_1 - f_2)$ product where $f_2 - f_1 = 300$ Hz (after MacKenzie [13])

diameter, the FM and AM distortions are almost equal at the 1% level (at 85 dB s.p.l.), while for the 100 mm diameter size, the distortion is likely to be dominated by FM components 14 dB greater in level.

However, the ratio of FM to AM components depends strongly on the type of driver. Klipsch [17] was unfortunate enough to measure a rather poor 200 mm mid-range unit (judged from the reproduced frequency response) which produced up to 15% total harmonic and intermodulation distortion within its working band (540 Hz and 440 Hz, f_1, f_2 when producing 100 dB at 0.62 m). Such a performance is not, in my opinion, representative of current high performance examples. He also investigated a full-range compact multiple driver system which was designed for use with an equalizer, and found that at $f_1 = 50$ Hz, $f_2 = 750$ Hz, both set to 95 dB s.p.l. at 0.62 m, the AM content was 14% as compared with the FM content of 3%.

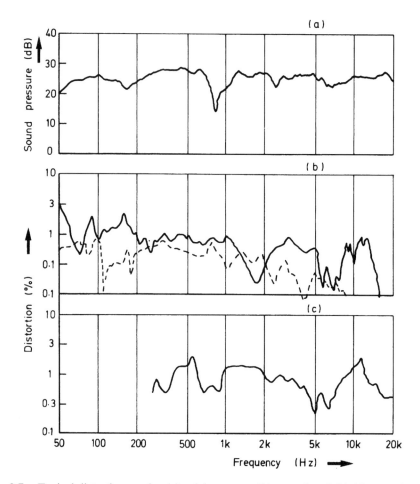

Figure 9.7. Typical distortion results; (a) axial pressure, (b) second and third harmonics and (c) $2f_1 - f_2$, intermodulation product. 90 dB at 1 m, $2f_1 - f_2 = 300$ Hz (after MacKenzie [13])

A relevant point in this context is that significant FM distortion is only produced under high LF excursion which does not occur very often with the average programme.

Judged on subjective grounds, FM distortion in general does not appear to be a major effect. Consumer tests [5] demonstrated that on wide range programme played at a reasonable level (85 dB average in a fair sized room, measured at 2 m), a very high quality 8 litre two-way sealed-box system subjectively outperformed a number of full size three-way systems, of 50–100 litres volume. The small enclosure involved was based on a BBC design and designated LS3/5a. If any system should have failed because of unacceptably high FM content, this would have been the one since at this volume level the harmonic distortion was under 1% and hence could not provide masking.

Published research conducted by Wharfedale [9] indicates the maximum allowable excursion to be 4 mm peak for a wide range driver, before FM distortion becomes audible. Their test method employed electronically simulated Doppler distortion through the use of a variable delay line. The miniature system described above just

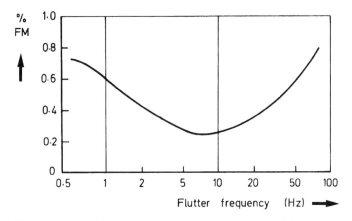

Figure 9.8. Subjective thresholds for piano music, frequency modulated by a pure 'flutter' frequency (after Stott and Axon [16])

meets this criterion at the level used for the consumer test, as in fact do most other quality speakers used within their intended loudness range.

A definitive review of FM distortion by Allison and Villchur [38] clarifies the subject considerably and indicates that at 40 Hz, a 17 mm peak-to-peak excursion lies below audibility. Taking into account subjective flutter weighting this excursion limit remains valid at 20 Hz, this beginning to restrict the maximum allowable output at such a low frequency. However, when the typical spectral content of music programme is considered, this does not present too much of a problem. As noted before, the level of amplitude intermodulation at this level of driver excursion is likely to be much higher.

Transient Response and Colouration

Speech and music are largely asymmetric in waveform structure, and theoretically they demand an accurate pulse or transient response from the loudspeaker. If the standards often applied to the other components in the audio chain were applied to loudspeakers, then the acoustic output of a loudspeaker should decay to negligible levels 0.025 ms after the cessation of a narrow impulse excitation.

Viewed in a pessimistic light, a loudspeaker may be regarded as an assembly of more or less well damped resonances spaced across the audible frequency range, and it thus has little hope of meeting such a requirement. Fortunately the human ear's inherent discrimination of transients is poorer than the above-mentioned standard, and there are indications that a decay rate of 10 dB/ms immediately after the excitation is a realistic level to aim for, with the rate decreasing exponentially after the first 20 or 30 dB decay.

The energy hangover after the impulse has passed is similar to reverberation, and in some speaker designs it has been deliberately encouraged to provide a falsely weighty and spacious subjective effect. Since such a hangover is strongly frequency dependent, 'delayed resonance' is a more appropriate term, and in practice subjective colouration associated with its presence may be ascribed to the frequencies at which it occurs (see Figure 9.33).

Some speaker designers believe that a transient response of the same characteristic as a quality amplifier is necessary from a loudspeaker and have exhaustively pursued aspects such as truthful square-wave reproduction. By definition such a performance requires that the amplitude versus frequency be minimal; and that no differential time delay or frequency unrelated phase shift is present between sections of the audible spectrum. It has been suggested that a speaker with differential time delays, that is, a non-minimum phase characteristic, must sound coloured in this respect, although there is still a matter of dispute.

Colouration remains a major fault with loudspeakers, where the main problem concerns its subjectivity. One listener may prefer a certain system to another solely on the basis of differing colouration, whilst neither may in fact be superior. Colouration can certainly be attributed to delayed resonances whose effects may be imperceptible in terms of an irregularity on the steady-state frequency response, but whose presence is plainly visible on a frequency analysis taken a millisecond or longer after an impulse excitation, and may be equally obvious on subjective grounds (Figure 9.9).*

Delayed resonances are a major cause of colouration and can result from many effects such as unwanted vibration in cabinet walls, drive unit resonances and inadequate damped electrical resonances in the crossover network. Visible broad-band unevenness in the frequency response curve will also result in subjectively perceived colouration (see Table 9.5).

Phase

There has been a resurgence of interest in the general audibility of phase effects, and in particular their relationship to the design of loudspeakers. The measurement of phase was rather difficult until the introduction of modern digital delay-lines, and certain manufacturers, notably Technics, Bang and Olufsen, Bowers and Wilkins, have since produced linear or more correctly 'minimum phase' loudspeaker systems. The Quad Electrostatic speaker belongs to this group, although at the time of its original introduction in 1955, the aspect of minimum phase was not accorded much importance.

Linear phase defines a system with a constant time delay, which gives rise to a linear rate of change of phase versus frequency. Recently a number of papers and articles have been published on this controversial subject [18–27].

It is likely that the imposition of a degree of phase uniformity as an additional design parameter helps to produce better crossover design with improved driver output integration. The latter quality is the smooth transition in output from one driver to its adjacent partner through the crossover frequency range. The misdirected frontal energy lobes in the crossover range are generally easier to handle via a minimum phase approach.

Absolute Phase

Some critical listeners now consider absolute phase as a factor, generally neglected so far. On the basis of the ear's asymmetric response to positive and negative going

*See also audibility of resonances p. 373.

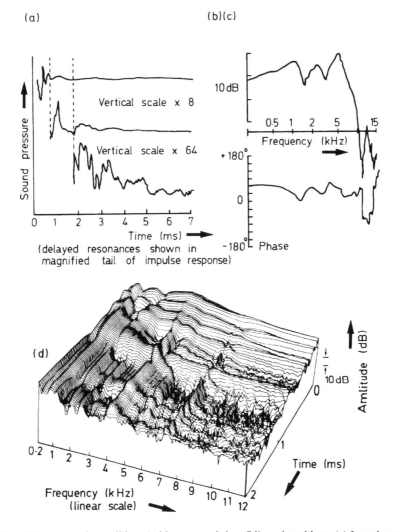

Figure 9.9. 110 mm moving coil bass/mid-range unit in a 7 litre closed box. (a) Impulse response, (b) and (c) frequency response (at $t = 0$, i.e. steady-state equivalent), (d) cumulative delay spectra (showing delayed resonances up to 2 ms after impulse has passed) (courtesy KEF Electronics Ltd)

transients, in ideal conditions it is possible to hear a small difference between a drum beat reproduced with the compression transient leading or vice versa. Subjectively positive or correct phase gives a fuller bodied effect, while phase inversion, with the rarefaction transient leading, gives a slightly 'hollow' sound.

In a listening test an experienced listening panel could reliably distinguish between phase 'normal' (note that the polarity of most recordings is unknown and that many amplifiers phase invert) and 'inverted' on a rock music track with sharp percussive drum sounds. The speaker on which this was heard most clearly was a high dynamic range, high linearity sealed-box design, well damped and possessing particularly good low frequency phase and transient response.

The consumer digital programme source, the compact disc (CD), has been specified as a non-inverting system. In the future, as product and recordings are classified for correct phase, as would be heard by a listener at the performance, absolute phase may become a significant factor. In addition, digital programme bandwidth is far more extended at low frequencies than other types, often down to 2 Hz, −3 dB. While programme level is minimal here, such a wide bandwidth can be important in regard to group delay and phase shift in the audible frequency range. As standards improve, better transient and low frequency response will be demanded from loudspeakers [39].

Except for the historically good phase performance of designs such as the Quad Electrostatic, many designs are now improving to the point where phase is assuming increasing importance. Listening test results for earlier generations of moving coil speaker did not show a significant correlation between phase linearity and sound quality; indeed, it was clear that in many cases attempts by designers to put phase first impaired the overall result. Advances in driver system and enclosure design are leading to the point where good phase linearity is now possible.

Where present, this characteristic allows renewed evaluation of the importance of absolute phase, which may only be carried out with good quality, phase coherent recorded material. In such a case there is a 'familiar' quality to the correctly phased condition, while less realism may be associated with the inverted condition.

The ear is known to have an asymmetric transfer function, with a slight but important difference in response to a positive pressure increase compared with a negative pressure rarefaction. This asymmetry gives rise to inherent and perfectly natural even-order distortion which is part of the fundamental perception of pitch, and why sounds with an odd order* content are felt to be relatively harsh and unnatural.

Subjectively, and this response depends on individuals who also vary considerably in their sensitivity, the 'inverted' sound is perceived as 'thinner', with less space and air around the source, and with a loss in reverberant energy. For some designers, correctly balancing a speaker is judged to be impossible if it or the programme source is phase inverted.

Practical problems arise due to the almost random phasing present on much recorded material, not only from mic to mic, but also track to track and record side to record side. Nevertheless, the AES has recommended that recorded phase be controlled and there are simple devices to help studios get it right. Speaker designers also try to make their contribution to this cause by minimizing phase variations in their designs where possible.

Direct Versus Reverberant Sound

It is not generally appreciated how strong an influence the ratio of direct to reverberant sound can be on subjective quality. Given a room with a reasonably uniform reverberation time with frequency, and taking a normally positioned listener placed 3–5 m from a stereo pair of speakers, most of the sound energy heard is reverberant, not direct. On first inspection one might question the possibility of producing the sensation of a stereo image in the presence of such a dominant diffuse sound field. However the ear can readily distinguish between the arrival times of the 'focused' direct sounds from the speakers and the reflected and diffused reverberant sound arriving later. The

---

*Third harmonic is still relatively 'musical' but higher orders are increasingly harsh sounding.

reverberant sound is still heard, and while a uniform axial frequency response can be perceived under reverberant conditions, the reverberant contribution is influential in determining a proportion of the tonal balance and 'liveliness' in a given room.

While a heavily furnished room will produce a 'dead', rich effect, and a room with largely plastered surfaces, uncovered wooden floors and little furniture will sound hard and bright, it is less well known that the loudspeaker system can impose similar effects, independently of its axial response.

Cabinet size and width, and in addition driver size, control the system's directivity and off-axis energy. The room begins to sum the total output from a speaker and thus the total energy response appears in the resulting equation.

In a room, despite similar axial responses, a large speaker will sound different to a small speaker because the total energy versus frequency response for a small enclosure is quite different to a large one. In free space they could well sound very similar. Assuming for the sake of argument that in a typical (IEC) listening room, a stand mounted 40 litre enclosure sounds natural (with a basically uniform axial frequency response). A 10 litre system will sound 'lighter' and more forward tonally in the mid register, 400–800 Hz. This is due to increased reverberation in this range due to its wider directivity. Conversely, a large 120 litre system will sound 'heavier' in tonal balance with a richer, more recessed mid range due to its narrower directivity in this region.

One manufacturer partially solved this problem with a large system by fitting additional mid and treble drivers on the top surface. Stereo focus is not significantly affected while the extra mid and treble contribution to the reverberant field gives the system a lighter and more 'airy' tonal balance in most rooms.

Another solution consists of separating the bass system from the upper range reproducers, such that a more uniform off-axis response can be maintained over the frequency range. Systems have been vertically tapered in width or successfully fitted with separate 'head' assemblies for mid and treble (see Figure 6.2).

Impedance (As seen by an Amplifier at the Loudspeaker Terminals)

At first sight one might not expect the input impedance characteristic of a loudspeaker to have much influence over the reproduced sound quality, but in practice problems can occur when an amplifier is connected to a speaker system.

With some loudspeakers containing complex crossovers, the designer may have sacrificed the uniformity of impedance to optimize other characteristics, such as axial response. Only the finest amplifiers are able to drive difficult loads without some subjective quality deterioration, and even with these, the maximum output is inevitably reduced. Poor quality amplifiers can thus produce quite disappointing results with loudspeaker systems possessing a demanding reactive impedance characteristic.

A suggested standard which will give good results with the present generation of power amplifiers is for a complex impedance of $Z = 8\,\Omega$, $\pm 20\%$ with a phase angle not exceeding $\pm 30°$ over the range 100 Hz to 10 kHz.*

Non-standard impedances are troublesome on several counts. For example, they make it difficult to specify true sensitivity, because the nominal power delivery of the

*Matching will still be poor with SE/low feedback amplifiers.

matching amplifier is indeterminate. Take the example of a system where the designer aimed to make a speaker 'apparently' more efficient with a standard $8\,\Omega$ based amplifier. Instead of increasing magnet flux, this was achieved by reducing the system impedance to an average of $5.5\,\Omega$, from a nominal $8\,\Omega$. At moderate levels where the amplifier was unstressed, the extra power required was delivered, with a consequent sensitivity improvement, but towards full power the situation was reversed. The low impedance induced premature amplifier clipping of an unpleasant nature, and ultimately the maximum tolerable sound pressure level produced by the combination was less than that for the original $8\,\Omega$ based design.

Dynamic Impedance and Peak Current Demand

It is customary to consider impedance in terms of a swept single frequency steady-state analysis, but an improved assessment [37] of a speaker's dynamic loading on a connected amplifier can be made via an analysis under transient conditions of the complete equivalent circuit including the driver constants.

Once excited or driven by a simulated unipolar transient, a moving coil driver overswings, generating a reverse or negative current. Transients of a type which apply another input pulse during the 'recoil' demand a total current from the amplifier which can peak at double the value predicted by the modulus. In a multi-way system it is possible for further crossover connected drivers to draw additional current. In consequence under certain programme transients of a particular positive and negative timing sequence, it is possible for the effective dynamic impedance to fall to a low value, typically a $3.5\,\Omega$ for a three-way $8\,\Omega$ system, $2.2\,\Omega$ for a $4\,\Omega$ three-way, and as low as $1.7\,\Omega$ for a 'difficult' $4\,\Omega$ load speaker.

From considerations of both amplifier load tolerance and connecting cable requirements a standard '$8\,\Omega$' speaker design is to be encouraged. Conjugate impedance compensation of the complete loudspeaker is also possible (see Figure 6.32(b), (c)).

Power Capacity

The measurement of power capacity is defined in the literature, for example the DIN standard, but in practice can be quite complicated. Essentially drafted to cover single-driver systems, a weighted random noise signal is applied over a defined duty cycle at increasing power until failure occurs. Such destruction testing is worth carrying out at the design stage to pin-point areas of potential weakness but is not helpful in the case of finished multi-way systems. The problem partly lies in the unpredictably transient nature of programme. A system containing a 3 W continuously rated HF driver may well operate successfully in a system rated at 50 W, this applying to the rating of the accompanying amplifier driven on speech and music programme. A specific condition is that the amplifier must not be overloaded. If this were to occur, then the resultant clipping directs increasing energy to the HF unit, resulting in its early demise (see Figure 9.10).

Programme power ratings are commonly applied to loudspeaker systems and these take into account the rating and thermal time constants of the individual drivers and in the bass the additional factor of a damage limit, for peak diaphragm excursion.

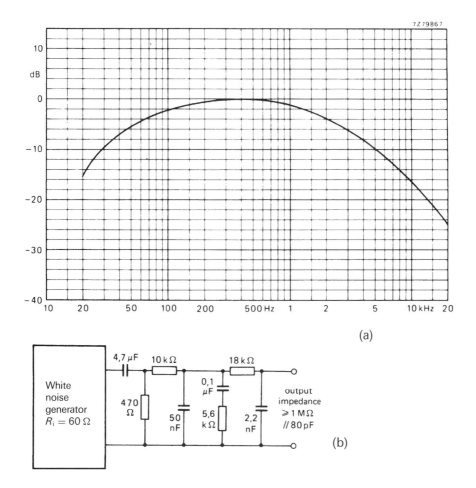

Figure 9.10. (a) Notice weighting curve according to DIN45573 (IEC proposal); (b) the filter necessary to drive this response from a 'white' noise signal (after Hermans)

Amplifier overload is a possibility which cannot be ignored and the speaker designer should aim to allow a reasonable safety margin when specifying the programme power rating. Even so, total freedom from failure cannot be guaranteed unless an electronic protection system is included such as those employed by KEF and B&W in their reference systems. Fuses are often used as a stopgap solution but their sonic behaviour is open to doubt. Certainly, if of low enough value to give complete protection, they behave as non-linear resistors and require two potentially unreliable contacts. The latter, and the atmospheric corrosion of wiper and track in driver level controls, are known causes of audible distortion in systems which have been in service for some years; such contacts are to be avoided wherever possible.

Subjective power testing is possible using broadband music programme where the peak power input is monitored. Using a powerful amplifier (up to 500 W) behaviour may be explored at the dynamic range limits. In one sequence of tests the use of a direct injection bass guitar was found to be very helpful and in addition this instrument's sonic purity proved to be highly revealing of other low frequency performance aspects, for example bass damping, rattles (not so rare as one might expect), bass distortion (reflex port overload, etc.), tune playing ability and box colouration.

All things considered, the quoted power rating remains largely a matter of experience and good judgement.

Compression and Dynamics

Both thermal and magnetic flux factors dictate some degree of compression in moving coil speakers. Sustained high level peaks heat coil wire, increasing resistance and temporarily reducing their sensitivity. Many testers are aware that graphing a loudspeaker response at an input level much higher than 0.5 W will involve some measurable compression. The usual rate of sine generator sweep means that only the HF driver has time to heat up, and by the time 20 kHz is reached, a loss of 1 dB or more is not uncommon in many published graphs. Furthermore, while in theory the varying flux change in the magnet system due to programme should only be a small proportion of the static polarizing flux, under heavy peak drive this may not be so, the peaks then suffering amplitude compression. A high level bass transient may also result in compression of an accompanying lower level mid-band signal, an intermodulation effect.

High sensitivity systems with light diaphragms are generally more linear and suffer less from compression. Ferrofluids can assist in greatly moderating the temperature changes experienced by a motor coil but may suffer from a viscosity variation of their own. Crossover saturation may also play a part; cored crossover inductors can saturate rapidly at peak levels dramatically changing circuit values during the peak.

Reflex systems with a small exit port can also suffer dynamic problems. In one well-known example the low-level bass transients were 'boomy', corresponding to a system Q of about 1.3, well underdamped. Driven to higher levels, 10–30 W peak, the effective port area is smaller and the output is damped via turbulence (5 cm port diameter), with the Q nearer to 1. At still higher power levels severe port turbulence meant that the reflex loading became inoperative, the system changing to sealed-box loading; the result is a compression of high level bass transients relative to low-level ones.

9.2 MEASUREMENT AND EVALUATION: INTRODUCTION

From an engineering viewpoint, loudspeaker assessment might appear fairly straightforward; drive units can be specified to a sufficient degree to guarantee a predictable measured standard. However, in practice objective measurement alone is insufficient to fully describe the sound quality, and subjective evaluation is also essential, with the loudspeaker system preferably judged by reference to live sound.

Subjective appraisal is the final arbiter in the judgement of quality. While engineering theory and mathematics will provide the foundation for a design whose technical accuracy and soundness may be verified by objective measurement, until a valid listening test is undertaken the true merit of the design cannot be assessed.

The total evaluation of a loudspeaker is thus a complex and wide ranging operation whose basic content is outlined in Figure 9.1 under 'loudspeaker appraisal'. The two sections, namely objective and subjective assessment, are dealt with separately, in this section and in Section 9.4.

Objective or Instrument Based Measurements

Figure 9.1 covers the bulk of the useful tests which may concern complete loudspeaker systems, or individual drivers when suitably mounted on a panel or baffle, or representative enclosure.

Test environment

Most loudspeaker measurements utilize a precision microphone to sense the sound pressure from the test loudspeaker. The test environment is of considerable importance, because the readings may be strongly affected by sound reflections under normal reverberant conditions.

For measurements where such interference must be eliminated, the speaker should be taken to a 'free-field' or open-air location and elevated clear of the ground. Cooke [28] indicates that an 8 m elevation is sufficient for a 1 m microphone-to-loudspeaker spacing, provided that the speaker is mounted front uppermost with the microphone positioned above it, in order to minimize reflections. If atmospheric conditions are favourable, then this true free-field location gives more accurate results than an artificial echoless environment, usually an anechoic chamber (Figure 9.11(a)).

In such a chamber, if of moderate size, the optimum working range is limited to 200 Hz to 20 kHz, with a typical absorption of 90% of the sound energy, though with careful calibration the low frequency range may be extended to 50 Hz (near field).

The inconvenience of outdoor measurements where ambient noise (passing aircraft and cars, etc.) is a nuisance, means that anechoic chambers are widely employed. Provided that their imperfections are understood and noted, they are convenient for loudspeaker measurement. Typical anechoic chambers consist of an acoustically isolated room of massive brick or concrete construction, lined internally with wedges of polyurethane foam or fibreglass up to a metre in length. Good absorption of sound is offered down to wavelengths comparable with twice the wedge depth, typically 200 Hz. Below this, the low frequency absorption becomes less effective, and the characteristic or the anechoic chamber gradually reverts to a free-field pressure chamber much like an ordinary room of similar dimensions. A further difficulty encountered with free-field measurement is the necessity for the microphone to be in the far field; that is, several wavelengths distant at the lowest frequency in the range covered. Clearly low frequencies will present the most problems. The microphone must also be positioned at a greater distance from the test system than the largest panel dimension, to avoid near-field diffraction. At 30 Hz the required separation of

Figure 9.11.   (a). Anechoic chamber (courtesy Acoustic Research)

microphone and system makes any kind of free-field measurement almost impossible; either the chambers are not large enough or the signal-to-ambient-noise ratio in open air is likely to be inadequate.

At the normal 1 m microphone spacing, with the tests conducted in an average chamber of $80 \, m^3$, the low frequency section of a curve at typically below 150 Hz will begin to approximate to a pressure response, with the microphone located in the low frequency near field. It is fortunate that for domestic speaker applications the dimensions of the listening room are not too dissimilar from those of the anechoic chamber. Hence both possess a similar broad averaged pressure response and the measurements taken into the chamber still make practical sense. This is however only valid for moderate room sizes, and acoustic conditions will be entirely different if a loudspeaker is used in a large hall, where the free-field response radiation will continue down to a correspondingly lower frequency.

Recently developed methods for response measurements using pulse signals,* where the test room reflections and ambient noise may be suppressed by suitable synchronized

*See also MLS equivalents p. 407.

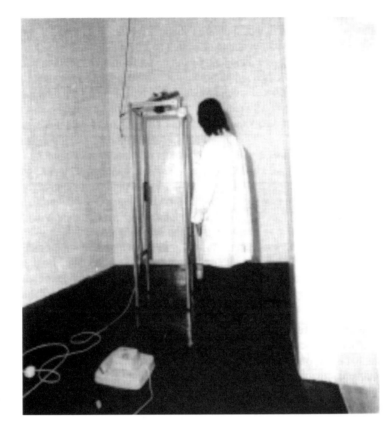

Figure 9.11. (b) Reverberant chamber (courtesy Acoustic Research). Note the hard surfaces, asymmetrical geometry, humidifier (on the floor) and the microphone with motorized motor (on the stand). The test system is flush-mounted in the wall behind the camera

gating, can offer greater freedom, but the theoretical necessity to remove the microphone to the far field still remains.

Test Environment — Anechoic 2π or 4π, Reverberant Chamber or Listening Room

While the author subscribes to general practice, namely measurement in full 4π or anechoic free-field conditions, together with suitable allowance made at low frequencies for intended use and expected local boundaries, some engineers still favour measurement in a 2π or half-space environment. Here an anechoic chamber is used where one wall is solid and reflecting and fitted with interchangeable baffles for flush fitting the test system. Tidy diffractionless graphs free of reflections can be produced and the low frequency response predictably follows the general theory (which assumes 2π working). However 2π use cannot be obtained in practice except where the customer is also prepared to flush mount the system in his room walls, or as may occur with

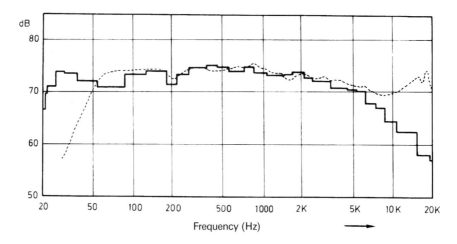

Figure 9.12. Computer-averaged room response in room over listening area (dotted response is anechoic, free field)

miniatures, install them in a bookcase, the books well packed around the systems. In one sense 2π is more accurate than 4π in that one boundary is taken into account at low frequencies, 4π testing taking account of none.

A totally reverberant chamber can give a useful result, not in terms of axial frequency response, but rather in terms of drive unit integration, and overall energy uniformity. An idea of the sound quality of the reverberant field can be obtained such as would be heard in a listening room (see Figure 9.11b).

A real listening room may be viewed as a partially reverberant chamber. While a test reverberant chamber builds a well diffused sound field of long decay time, up to several minutes duration, the listening room R_t* lies in the range from 0.2 s to several seconds. In addition the volume patterns of sound energy are anything but diffuse. However, a well proportioned and acoustically balanced listening room can be used for speaker assessment. It can be worthwhile to take some measure of the room sound field in the region of optimum stereo effect and focus, perhaps represented by a sofa at the listening position. This can be obtained by employing optimum speaker positioning in the room, and by computing an average of the response at a minimum of eight combinations of position and height in the listening space. The averaging covers both left and right speakers, individually excited.

Using third octave pink noise analysis, the results, in the absence of serious loudspeaker colourations, can show considerable agreement with the subjective assessments of frequency response, tonal balance and low frequency extension as heard in the room. Spectral balance between low bass, low–mid bass and mid-range bands is clearly shown, this often hidden in the free field laboratory response (see Figure 9.12).

Poor speaker and subject positioning and/or a difficult room will introduce sufficient aberrations to make this assessment of reduced value.

*R_t = reverberant time.

One cannot expect an 'averaged room response' to remain uniform to the highest frequencies since it consists of a mix of direct and reverberant sound. With the usual narrowing system directivity with increasing frequency, the reverberant contribution naturally and smoothly falls off at high frequencies, imparting a gentle rolloff in the last octave or two, depending on the HF unit diameter. Energy discontinuities, for example at crossover frequencies, are also revealed in this measurement, more or less as they are heard, despite the fact that in many cases their existence may not be obvious from the axial response.

Finally, where accurate low frequency measurements are necessary in the absence of an anechoic chamber this can be carried out by simulating a giant 'chamber' (Gander [43]). This is done by employing a large area of level ground such as an open car park or perhaps a warehouse floor as an acoustic mirror to double half space to full space. Provided that the microphone is within a few millimetres of the ground and the speaker system is placed on its side on the ground, accurate measurements are possible up to frequencies where driver path length to the ground is comparable to wave length. The pressure zone microphone (PZM) music recording method is similar. Apparent sensitivities are doubled due to the reflection; 6 dB must be subtracted from microphone readings.

With gating techniques [29]* and time delayed spectrometry (TDS), room or environmental reflections are eliminated from the measurement and all that is required for wide range speaker measurement is a large enough, temperature controlled volume in which to work. The size is dictated by the need for the lowest desired frequency to propagate sufficiently away from the speaker for its returning reflection to be 'gated' out of the measurement.

9.3   OBJECTIVE MEASUREMENTS

Amplitude/Frequency Responses (4π or Full Anechoic)

Sine excitation

A number of tests may be performed with sinewave excitation energising a power amplifier fed by a suitable automatic sweep-oscillator system. The usual input level is a nominal 1 W referred to 8 Ω (2.83 V r.m.s.) with the microphone (generally a 12.5 mm capsule) placed at 1 or 2 m from the mid axis or the manufacturers' defined listening axis.

Figure 9.13 is virtually self-explanatory and shows the test arrangement using traditional equipment. The trace obtained (Figure 9.14) is a widely used specification for loudspeakers, and with certain reservations is probably the most important. The 4π frequency response at LF may be accurately quantified via a close proximity (< 50 mm) microphone position, valid in the piston range [32]. (See Section 9.4, *et seq.* on subjective assessment.)

Near-field measurement may be complicated by double radiator or reflexed systems and special precautions are necessary here. Each radiating element may be measured and the sum obtained with due account of the radiating areas, including phase where necessary.

*See MLS systems.

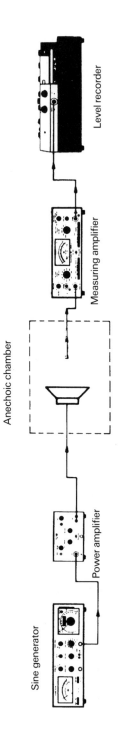

Figure 9.13. Set-up for acoustical frequency response measurements (courtesy Bruel and Kjaer [31])

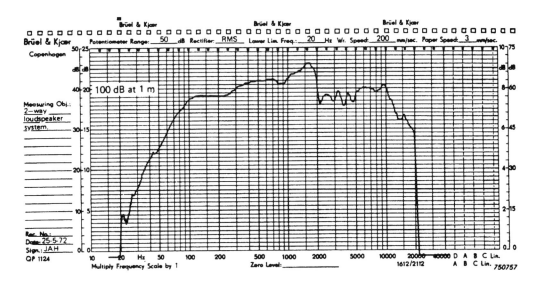

Figure 9.14. Sine response curve of a medium quality two-way loudspeaker system (courtesy Bruel and Kjaer [31])

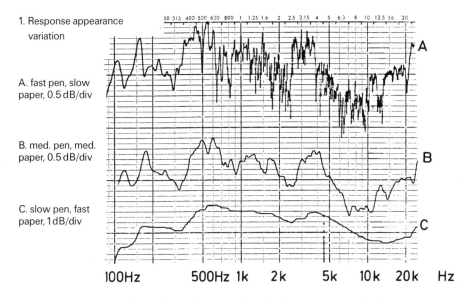

Figure 9.15. The effect of recorder settings on the visual appearance of a frequency response trace. Some marketing departments have exploited this to flatter their products

Clearly only direct radiating systems with forward facing drivers on a plane or nearly plane driver panel can be accurately quantified by the near-field method.

The appearance of a sine wave response trace may be influenced by test conditions; for example, the pen speed on the chart recorder. An exceptionally slow pen speed will tend to smooth sharp resonances or dips, thus giving a false impression of the speaker (see Figure 9.15). Paper speed and scaling are also influential.

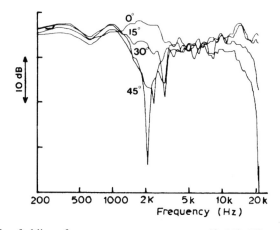

Figure 9.16. Family of oblique frequency response curves at 0°, 15°, 30° and 45°, 1 m horizontal plane (courtesy KEF Electronics Ltd)

With multi-unit systems, interference dips may exist at specific microphone positions, and a lateral or vertical displacement of 10 cm or so can result in one dip disappearing, or another appearing. For these reasons, great care is required in the interpretation of sine wave responses and an initial curve may need confirmation by further tests taken at slightly different microphone positions.

The curve illustrated in Figure 9.14 has the instrument settings recorded on it for future reference, those shown being typical of modern measurement techniques.

Off-axis responses

These are typically ±10° or ±15° in the vertical plane and ±20°, ±30° and ±45° in the lateral plane. A speaker with lateral symmetry need only be plotted in one direction (60° off-axis may also be helpful).

Both sine and noise excitation are common. In addition to the axial response measurement, the loudspeaker may be angled or rotated to allow polar plots at single frequencies off-axis. These plots often reveal irregularities not shown on-axis, for example if a unit's dispersion narrows, or the energy at the crossover point between two drivers is not integrating properly due to phase differences. Such a dip might not be apparent on a single axial response measurement as a result of fortunate microphone placement (see Figures 7.10 and 9.16).

Random Noise Excitation

Whilst a sine test signal is potentially the most accurate and precise, random noise is also valuable for loudspeaker measurement. There is strong evidence that the ear tends to average short-term irregularities and is rather more sensitive to broader trends in energy over third-octave or octave bandwidths.

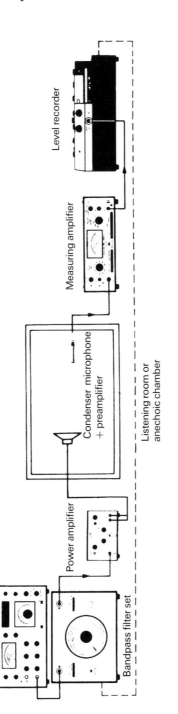

Figure 9.17. Set-up for response measurements using narrow band noise (courtesy Bruel and Kjaer)

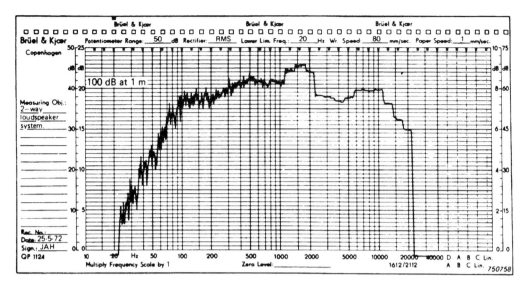

Figure 9.18. Some techniques vary the writing speed during the sweep in order to give the analysis records a more uniform appearance, usually from 10 mm/s at 20 Hz to 160 mm/s at 2 kHz and above. This trace shows one-third octave analysis without such adjustment. (Courtesy Bruel and Kjaer) (same speaker as Figure 9.14)*

Third-octave or octave bandwidth analysed noise provides a convenient method for such averaging. From a philosophical point of view, noise more closely resembles music programme in transient content than sine wave, and hence could be considered more relevant. It also readily permits the measurement of the 'A' weighted or subjective based loudness of the system. If broadband noise is applied to the loudspeaker, then the filtered analysis of the total output provides some ambient noise suppression due to the narrow detector bandwidth. Alternatively the drive to the speaker may be pre-filtered and the microphone amplifier left in the wide-band condition (Figures 9.17 and 9.18).

While an axial sine wave response is a valuable reference, the one-third octave result gives a better idea of the sound quality. Tighter amplitude deviations on the averaged response are worthwhile. If a family of off-axis responses, also in one-third octave, is added to the axial result, a good indication of the forward energy trend, driver integration and general uniformity of response may be gained (see Figure 9.19).

Averaging multiple microphone and loudspeaker position responses in a listening room which includes both room loading and the forward energy produces a curve which correlates well with the perceived response (See Figure 9.21).

The total impedance may also be measured via broadband noise excitation measured as the V/I ratio delivered to the speaker. Such a reading is nearer to the programme value seen by an amplifier than the swept sine impedance graph.

It has been suggested that while one-third octave is satisfactory for the middle octaves, greater resolution, perhaps one-sixth octave would be more valuable at the audible frequency extremes. This is possibly due to the presence of the band edges of

*MLS systems may include octave and 1/3 octave weighting.

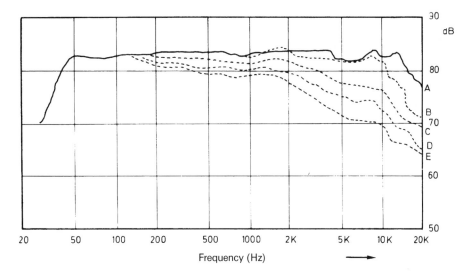

Figure 9.19. Off-axis "forward response family" (ESL 63) A = axial, B = 7.5° vertical, C = 15° lateral, D = 30° lateral, E = 45° lateral (in 1/3 octaves)

most speaker amplitude rolloffs where the slopes require finer analysis resolution for accurate results. Many audio spectrum analysers provide one-third octave analysis as do many of the lower cost computer card test units.

Gated Sine Wave Excitation

Bruel and Kjaer have produced an instrument for gating the output of a signal generator [29] which simultaneously samples and stores the peak amplitude of the received output long enough for a continuous response trace to be recorded (Figure 9.21). Through the elimination of local wall reflections, the need for an anechoic environment is removed, and measurements may be performed at mid and high frequencies in an ordinary reflective room (Figures 9.21 and 9.22). In reasonably open room conditions frequency/amplitude response measurements of fair accuracy can also be made with noise or one-third octave warbled sine wave excitation which suppresses local reflected modes. Depending on the proximity of the nearest surface, measurement from 200 Hz upwards is feasible. This technique has been extended using MLS (maximum length sequence) excitation.

Impulse Excitation

The use of a digital computer equipped with a fast Fourier transform (FFT) processor allows full analysis of the pulse response of a loudspeaker. A pulse-gating technique similar to that described above is employed, providing suppression of room reflections. From the pulse analysis, the steady-state response may be derived automatically and

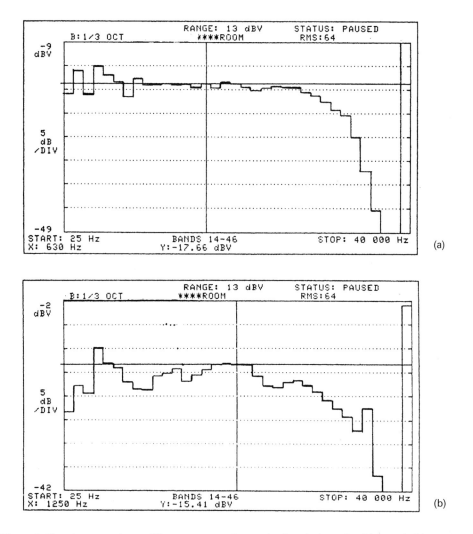

Figure 9.20. A comparison of "room averaged response" technique for (a) a sealed-box system with optimal Q_T and response rolloff, stand mounted in free space; (b) with a similarly sized enclosure with reflex bass loading and an inappropriate alignment. Both possessed visually flat axial anechoic responses while their sound quality related more closely to the room averaged characteristic. (b) Is seen to be less uniform overall while the low frequency peak at 50 Hz is the port resonance magnified by room gain

plotted in the usual way (Figure 9.23). Figures 9.24 and 9.25 show the responses of the same loudspeaker derived by steady-state sine wave, and by pulse analysis [30].

Other systems for gated analysis also allow the investigation of delayed resonances and reflections through the analysis of the output of a system after the initial excitation is over. For example, the arrangement in Figure 9.21, if used with an anechoic chamber to suppress the room reflections, may be adjusted so that the gating control unit reads the output after cessation of the burst excitation.

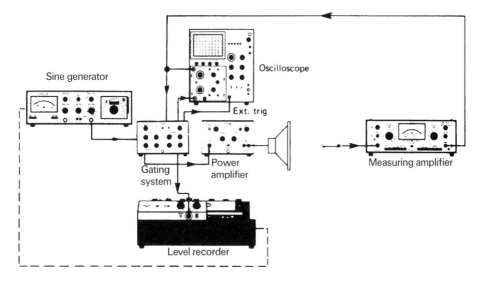

Figure 9.21. Set-up using gating system (courtesy Bruel and Kjaer)

The pulse periodicity is set to exceed the room reverberation R_T. Unlimited signal averaging may then be applied to generate good signal-to-noise ratios. Recent developments have explored full 20 Hz to 50 kHz measurement in a 7.5 m^3 room. Enhancements include the speaker's excitation by alternate $+$ and $-$ impulses whose responses are then subtracted in the computer. D.C. offset and spurious hum are cancelled by this technique allowing some truncation of the measured impulse response to speed-up measurement.

Microprocessor developments have enabled the production of relatively low cost FFT analysers such as the HP3561A which may be set up with a pulse generator to produce an effective impulse test system (Figure 9.26(a)). The required analyser time window may be adjusted to examine the portion of the loudspeaker response of interest via a signal gate, Figure 9.26(b).

An indication of the importance of impulse testing was given by the costly investment carried out by many companies in acquiring the necessary measuring equipment. The Wharfedale research team has also achieved success in delayed resonance analysis, their 'in-house' system developed by using a variation of the gated tone-burst set-up described earlier with burst length proportional to frequency for constant excitation over the frequency range. The output appears in conventional log frequency format from 20 Hz to 20 kHz [34]. The essential features of this method were in fact first established by Shorter at the BBC as early as 1945.

Theoreticians have since examined a host of shaped pulses from raised cosine to controlled tone bursts and have explored a variety of mathematical transformations of the results to aid the visual presentation and analysis.

If a loudspeaker is a minimum phase, as most single moving coil drivers are, then the amplitude versus frequency and phase response versus frequency are uniquely related. One may be obtained from the other via the Hilbert transform.

For assessment as to whether a system is minimum phase, conveniently the phase response is measured, then computed via the Hilbert transform to an amplitude-

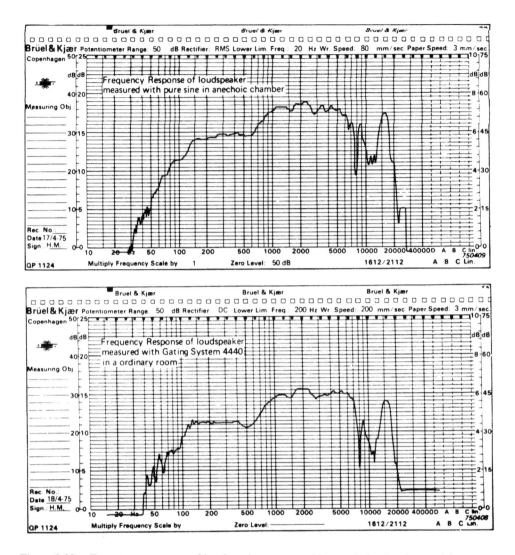

Figure 9.22. Frequency response of loudspeaker measured in anechoic chamber and frequency response of same loudspeaker measured with gating system (courtesy Bruel and Kjaer)

frequency response, which may then be compared with the direct measurement. With computer-based systems Fourier pulse synthesizers can produce almost any required stimulus.

Cumulative spectral decay plotting using apodized tone bursts can provide clearer results than simple cumulative decay representations [41] (see Figure 9.27). Other useful representations have included the Wigner distribution [42] (see Figure 9.28). This method is valuable for multi-way systems which may be examined without the kind of overlapping confusion shown in the basic decay representation. The energy in the Wigner stimulus is properly distributed, allowing a better appraisal by weighting, averaging and transformation.

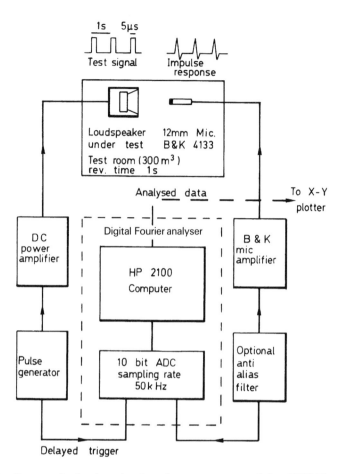

Figure 9.23. Lay-out for loudspeaker impulse measurements (after KEF Electronics [33])

Time Delay Spectrometry (TDS)

This technique, originally explored for audio use by Heyser, has now been made accessible via suitable equipment produced by B & K. In essence, a sine wave is made to sweep the required band pass at a specific rate. At any moment, partially defined by the bandwidth and response time of the tracking filter, the measurement of the moving sine wave does not relate to the single centre frequency, but rather a distribution of frequencies determined by the sweep rate. The slower the sweep, the nearer the side bands and the purer the tone. Given this basic relationship for system resolution, the technique has the ability to exclude boundary reflections and also to explore delayed resonances. As the frequency sweep emanates from the loudspeaker it suffers a time delay due to the path length to the measuring microphone. When the equipment is synchronized, a start frequency of 20 kHz at the speaker will not arrive at the microphone until the generator reaches a lower frequency, depending on the negative or downwards sweep rate and the microphone spacing. The selective analyser is offset to

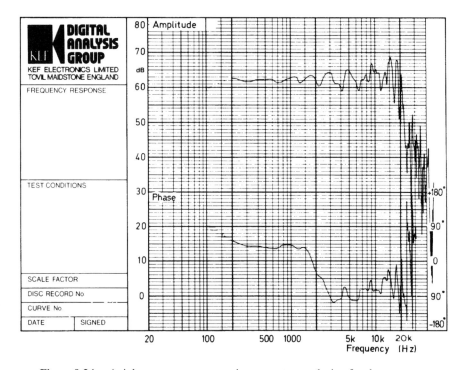

Figure 9.24. Axial pressure response via computer analysis of pulse response

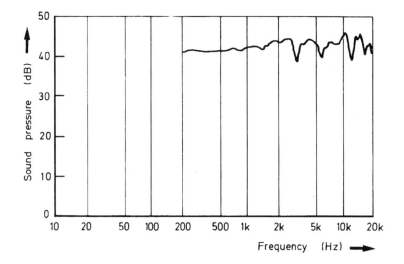

Figure 9.25. System response by sine excitation: anechoic conditions (same model as in Figure 9.24)

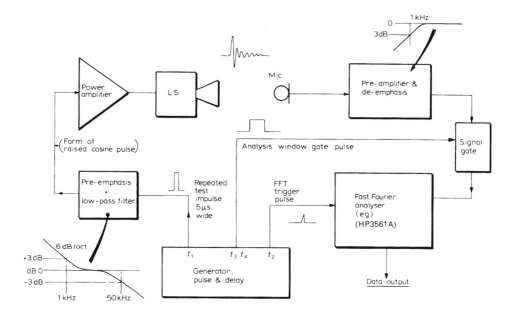

Figure 9.26. (a) The 5 μs pulse is preconditioned before the power amplifier to allow increased energy input to the system. Noise is predominantly low frequency, thus improved via LF pre- and de-emphasis

the lower frequency to maintain the synchronization and offers the facility for selecting the desired time window for measurement. Set to normal, it captures the direct response, and by tracking in synchronism with the source it ignores the reverberant ambient reflections. These occur after the analysis frequency window has passed.

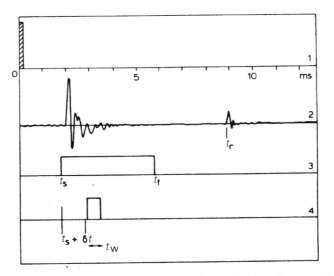

Figure 9.26. (b) Impulse response and capture. (1) Applied impulse typically 5 μs wide

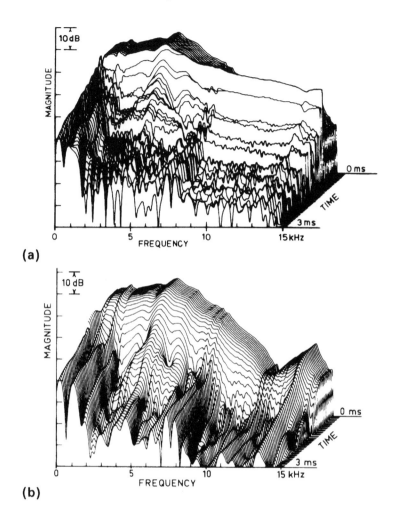

Figure 9.27. (a) Cumulative decay spectra for a later model 110 mm bass/mid-range loudspeaker; (b) 0.45 ms cumulative spectral decay plot for the later model 110 mm bass/mid-range speaker. (Note improved clarity of spectral decay with apodization) (after Bunton and Small [41])

Conversely offsetting the analyser to read slightly after the main swept stimulus energy lobe has past the microphone allows the system to record the delayed resonance output of the test reproducer, the time scale limited by the system dynamic range and by the encroaching reverberant sound. Fourier analysers can be applied with advantage to the measurement system (see Figure 9.30).

The energy/time decay can be obtained via the FFT, a useful general graph allowing quick comparison of the stored energy of different speakers (see Figure 9.31) (Techron also produce TDS analysis systems).

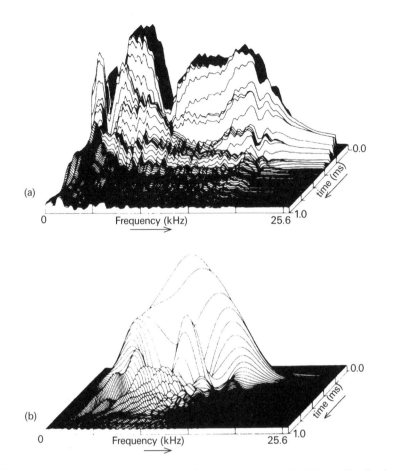

Figure 9.28. (a) Cumulative decay spectrum of a dome tweeter; (b) Wigner distribution of the dome tweeter (after Janse and Kaizer [42])

MLSSA — (Mellissa)

Based on a set of software and a plug-in interface for a personal computer, a loudspeaker and acoustics test instrument has been developed called MLSSA, named after the type of source, namely Maximum Length Sequence System Analyser. This is a special kind of correlated pseudo-random noise which stimulates the speaker under test at a satisfactory power over the full frequency range and over a range of useful time intervals. From a single capture of the output data subsequent or immediate analysis can give the following results; amplitude-frequency response, decay waterfalls, energy-time curve, gated semi-anechoic responses, to whole octave smoothed response, phase and group delay, the Wigner distribution, impedance, Nyquist and Bode plots, both minimum and excess phase, and finally harmonic and intermodulation distortion (the latter in conjunction with an external oscillator) (see Figure 9.32).

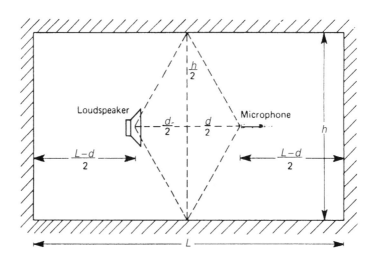

Figure 9.29. Travel distances for the first reflections when the loudspeaker and microphone are centred along all three axes of the room (after B & K)

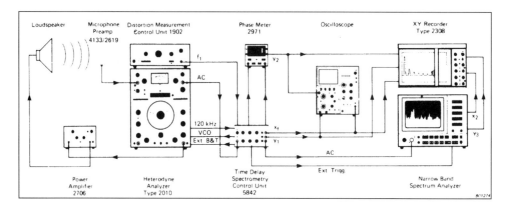

Figure 9.30. TDS system used for loudspeaker development. The same arrangement is used for other types of measurements by exchanging the microphone and loudspeaker systems for the appropriate transducers and amplifiers (after Bierning and Pedersen [47])

Hardware characteristics include 12 bit, 72 dB dynamic range (wideband) and an operating bandwidth up to 50 kHz. Typically the hardware board would be installed in a good quality AT portable to form a complete menu driven instrument.

Waterfall Presentation and Excess Phase

When analysing waterfall-style decay graphs do not be taken in by their seductive appearance. The look of these graphs is strongly dependent on a number of factors,

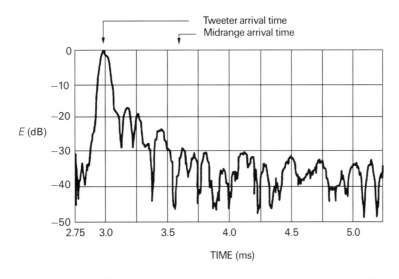

Figure 9.31. Curve of energy vs. time taken at 1 m on axis (after Heyser)

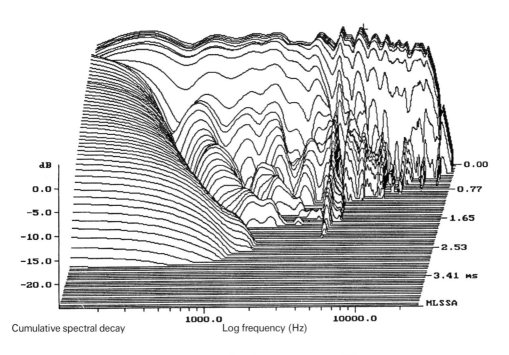

Figure 9.32. Cumulative spectral decay from anechoic simulated impulse response using MLSSA stimulus (exploits correlated noise function for optimum dynamic range). Spendor SP2-2

independent of the device under test. If such graphs are to be compared, it is essential that the test settings be identical, otherwise confusion will reign. Factors which change the appearance so much that false conclusions may be drawn concerning the decay rate of the loudspeaker include:

(a) the number of lines or line density;

(b) the selected filter risetime for the chosen window function;

(c) the vertical scaling, i.e. dB per division;

(d) the length of the time window gate;

(e) the absence of any local reflections within that gate period; these are more obvious than on the axial response transforms;

(f) whether the loudspeaker has a severe peak such as a dome resonance high in the treble and which causes the vertical scale to 'down range' to accommodate the peak and thus give the appearance of a faster decay in the rest of the range;

(g) whether there is significant excess phase in the speaker transfer function, which there usually is, which greatly distorts the waterfall presentation of decay due to the varying delays across the spectrum.

A single waterfall graph does not have the power to properly resolve both the decay rate and the frequency of decay.

It is worthwhile generating at least two graphs, one directed to assessing the decay speed over the frequency range, the other to better resolve the frequency of the decay resonances. For the former, a 5 dB per division scaling with around 50 lines is useful; Blackman Harris-weighting and with a fast 0.1 ms filter risetime. For the latter, a 10 dB per division vertical scale working with a 0.2 ms filter gives enough time for good frequency resolution; the decay ridges in the waterfall field following the initial $t = 0$ 'steady-state response' at the back of the graph.

The excess phase question is more serious. If the drivers for a speaker system are analysed separately, then the decay waterfalls can be surprisingly good and are truthful. Operate the system complete, and the phase content blurs the required information. Only if the speaker is linear phase, or close to it, is the decay response a fair representation of the intent of the test.

The solution is available, if awkward to implement (thanks to M. O. Hawksford for seeking an answer to this problem). Using a suitable math program the numerical data for the impulse response is imported, with the header removed if from MLSSA. It is viewed and windowed, possibly with some sensible corrections at the extremes to avoid the computation of awkward or non-causal numbers. Transformed to the frequency domain using an appropriate filter window the response is also tidied at the band edges, assuming normal rates of rolloff for the speaker technology. By computation of the data array, the excess phase is removed from the frequency data, separating real and imaginary data using a weighting function and summing to deliver a causal impulse where peak energy begins at $t = 0$ and the result is a minimum phase windowed response. This new impulse carries the correct steady-state frequency response but has the phase content removed; it is obvious that the result is like that for a low-phase shift system. When this impulse response is subject to the waterfall display the true decay rate can be seen for the first time. Note that the newly computed result is not linear phase; the latter defining an acausal response which exhibits a symmetric spread of energy on each side of the $t = 0$ position. There are, however, close parallels. An approximation, to be used with caution, may be more simply achieved (e.g. with the MLSSA package) by calling up a stored impulse in the Operations menu and cross-correlating the impulse with itself. Although lacking the proper windowing, the result implied a multiplying out of the phase content. Displaying this impulse on the decay spectrum allows for a clearer estimate of the decay rate for a complex speaker system (note that the sequence is: Library–Operations–Time files–xcorrellation–Data(a)–corr-Data(a) to data(ax).tim]. This pseudo-acausal impulse must be windowed from x (axis) $= 0.0$ over the required reflection free-time span. The transformed frequency response is unchanged (Figure 9.33(a)–(e)).

Audio Precision Test Set

Audio Precision are responsible for the design and manufacture of a powerful audio and acoustics test package in the form of their SYS 1,2 units whose facilities are extraordinarily comprehensive and both measurement accuracy and limit thresholds are particularly good. Designed to be controlled by a desktop PC, it lends itself to a variety of routines, including automatic testing. A recent option consists of a digital processing module which allows for stimulus and analysis in all possible modes, D/A, A/D, both in

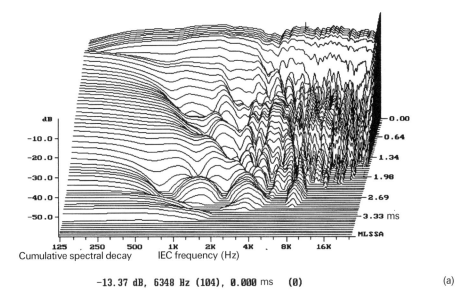

-13.37 dB, 6348 Hz (104), 0.000 ms   (0)   (a)

Figure 9.34. (a) MLS waterfall display of cumulative resonance decay 10 dB per division per 0.2 ms. Filter B/H. Two-way speaker: note metal dome resonance at 23 kHz (gated time window 5 ms)

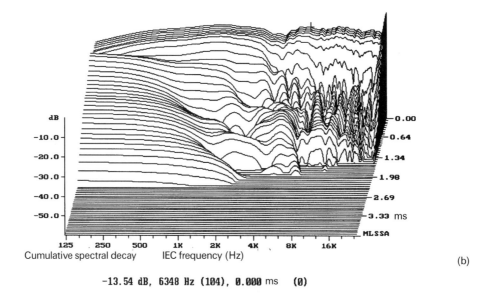

-13.54 dB, 6348 Hz (104), 0.000 ms   (0)

Figure 9.33. (b) As for Figure 9.34(a) but with the gate window truncated at 3 ms. Note the apparent improvement in the longer term decay, even at 2.0 ms

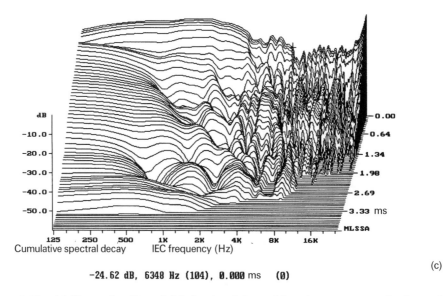

-24.62 dB, 6348 Hz (104), 0.000 ms   (0)

Figure 9.33. (c) Shows the effect of deleting just 0.1 ms of impulse response at the front of the 5 ms window. Much of the 'steady-state' response has been stripped off

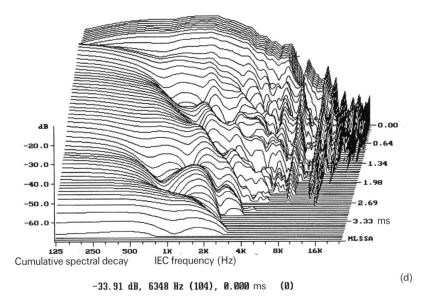

-33.91 dB, 6348 Hz (104), 0.000 ms   (0)

(d)

Figure 9.33. (d) For this two-way speaker the low-pass response is plotted alone. Compare with (a) and note how some interesting metal cone resonances are now revealed in the 4–16 kHz range.

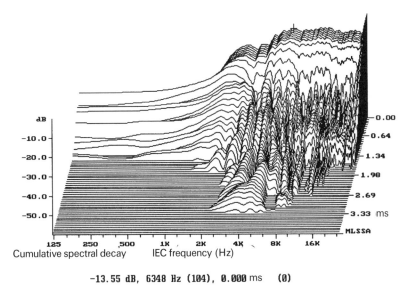

-13.55 dB, 6348 Hz (104), 0.000 ms   (0)

Figure 9.33. (e) The high-pass section of the two-way design. It can now be seen how much this section dominates the range above 8 kHz. Resonance series can now be separated more clearly in the range 2–8 kHz

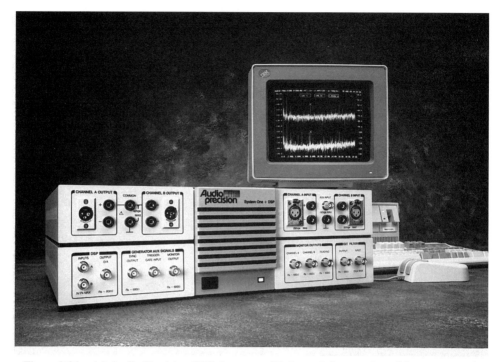

Figure 9.34. (a) Audio Precision SYS-1 test set; (b) (*opposite*) examples of data generated

the analogue and digital domains. Processing software for loudspeakers is to be available including FFT, impulse and related techniques while the digital interfacing versatility is directly applicable to digital filter design and the synthesis of digitally interfaced loudspeaker systems (see Figure 9.34(a) and (b)).

Reciprocity Method for Measurement at Low Frequencies

Measurement below 100 Hz is rendered variously inaccurate due to boundary reflections present even in fair-sized anechoic chambers. Merhaut [49] describes an application of reciprocity for accurate low frequency measurement. Essentially the action of a loudspeaker as a convertor of electrical power to acoustic energy may be reversed or inverted by using the test loudspeaker as a microphone. For testing moving coil loudspeakers the motor coil is loaded by an external resistance much less than that of the coil, e.g. for an 8 Ω model an 0.5 Ω resistor is suitable.

A separate non-critical loudspeaker provides the acoustic stimulus; the sound pressure kept constant at the surface of the test loudspeaker via a precision microphone used as a monitor coupled back into the familiar compressor loop of the generator (B & K or similar). Under these conditions, and when resistively loaded, the output of the test loudspeaker at low frequencies corresponds simply to the first integration of sound pressure and a simple differentiator (6 dB/octave high-pass) network will provide a true voltage output describing sound pressure at low frequencies where the driver is still non-directive, e.g. below 150 Hz (see Figure 9.35).

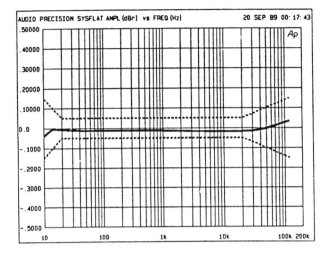

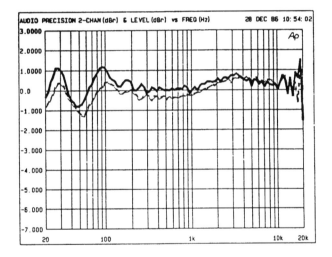

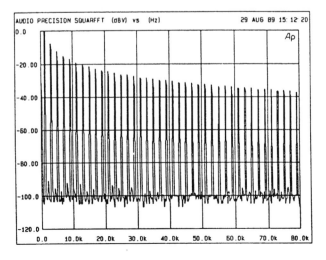

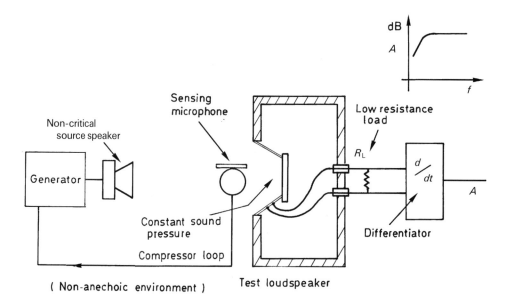

Figure 9.35. A reciprocity method for low frequency measurement in non-anechoic conditions. Driven acoustically by a constant pressure, the electrical output of the test loudspeaker is the first integral of sound pressure when terminated by a low resistance or a virtual earth input and carries the response of the device under test. Simple differentiation results in the desired frequency response

Merhaut also shows how a number of other parameters may be obtained such as η or efficiency, k_a or Bl factor, and the various Q factors.

Harmonic and Intermodulation Distortion

Distortion can vary dramatically over quite a small frequency interval, and for this reason continuous distortion versus frequency sweeps are usually taken. The basic set-up for distortion measurement is similar to the amplitude/frequency response arrangement (Figure 9.36), with the addition of a suitable selective analyser or tracking filter interposed between recorder and microphone. The filter may be offset or displaced by a suitable harmonic interval so that any order of harmonic may be recorded. The use of a two-tone generator with a tracking harmonic multiplier also allows swept intermodulation traces to be recorded.

Figures 9.6 and 9.7 show two such curves. The $2f_2 - f_1$ second-order intermodulation product was traced with $f_1 - f_2 = 300$ Hz, and the combined tone level was set at 90 dB at 1 m. These swept intermodulation measurements may well prove more revealing than the usual single harmonic readings.

One self-evident point is the relationship between sound level and distortion, the latter increasing with the former. It is thus sensible to choose a standard level for comparative purposes, e.g. 90 dB. The DIN standard specified 96 dB, which is on the high side for most low colouration HF units used on continuous tone.* The suggested

---

*Suggests an input power up to 5 Watts.

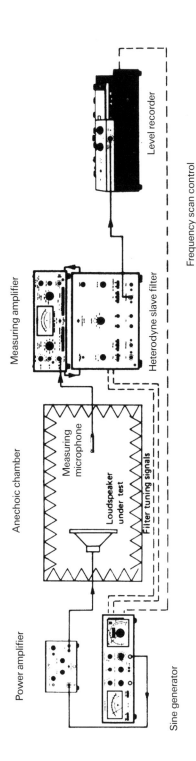

Figure 9.36. Total harmonic distortion measurement (courtesy Bruel and Kjaer)

lower level applies to domestic and low-level monitor systems, and would not be adequate for high-level monitor or large audience broadcast arrays, where 100–120 dB at 1 m would be a more relevant test sound pressure.

Even 90 dB is quite a high level in subjective terms for a distortion measurement, an area reached only during the loudest passages, with the remaining 10–15 dB above occupied by true programme peaks. With the availability of good amplifiers and low distortion digital sources it is worth examining linearity at lower levels to see what improvement is possible. In recent evaluations the author [44] has published distortion graphs taken over a range of levels, 96, 86 and 76 dB at 1 m with a resolution to −70 dB. In some cases distortion fell below −70 dB (0.03%) at the lowest pressure level noted. Computer card based analysers, such as CLIO, are now available, which can measure distortion.

Dynamic or Pulsed Distortion Testing

While 90 or perhaps 96 dB s.p.l. is the highest level which may be safely applied for continuous swept distortion measurement a second approach is required for assessing linearity at peak levels. A 88 dB/W speaker fed 100 W programme will generate peak levels of 108 dB.

Compression may be analysed over the frequency range by means of short tone bursts, around 10 cycles long, with a 10 : 1 mark space ratio so as not to overheat the system. Compression may be measured directly from storage of the reproduced tone bursts, viewing a 'gated' stable group within the tone burst. This area may be captured by an FFT analyser and direct readings of distortion harmonics can also be made. Experience has shown that an otherwise well-behaved system may run into problems towards peak level; in one example the crossover inductor* saturated above 50 W burst. At 400 Hz, 100 W burst, the signal was expanded by 1.5 dB as the inductor saturated. Third harmonic exceeded 30% at this point. In another example a shunt indicator reached overload at 4 kHz and compressed the signal by 2 dB, at 20% of distortion. High power, low frequency tone bursts are also useful in exploring linearity in this higher power range [35].

Doppler Distortion

The measurement of Doppler distortion is not easy as the FM components must be separated from the other accompanying distortions and this tends to restrict the range of measurements (Figure 9.37) (Allison and Villchur [38]).

However, a simple technique exists for separating amplitude components from the total distortion group. This consists of aiming the microphone at 90° to the driver axis. The FM components are essentially axially directed and at 90° the source is no longer moving with respect to the microphone. If a storage analyser were available it would be a simple operation to subtract one from the other, thus leaving the FM contribution.

*Series element in the bass section.

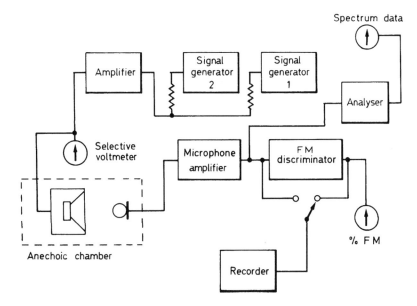

Figure 9.37. Doppler distortion measurement set-up (after Moir [14])

Noise Intermodulation Distortion

Intermodulation distortion may be measured using a random noise signal. The same arrangement as that employed for third-octave analysis is followed (Figure 9.17), but with the addition of another, tracking third-octave filter. One operates on the noise source and the other filters the microphone output signal. With the latter shifted a harmonic interval above the input or fundamental third octave, various combined harmonic/intermodulation curves may be obtained over third-octave bandwidths.

MLSSA techniques can assess distortion as a failure in correlation, while complex digitally derived multi-tone signals may be useful (Andio Precision).

Phase Response

The high quality delay line (e.g. Bruel and Kjaer 6206) facilitates the steady-state measurement of the phase response versus frequency of a loudspeaker (Figure 9.39).

The phase characteristic may also be derived via FFT analysis of the impulse response and by related methods (see Figure 9.39).

Minimum Phase

Several manufacturers have introduced 'linear phase' loudspeakers. In fact, the correct term to describe such systems is 'minimum phase', which implies that a well-defined linear relationship exists between the amplitude and phase response of the speaker; that is, the phase varies linearly with frequency (note that the phase does not have to be

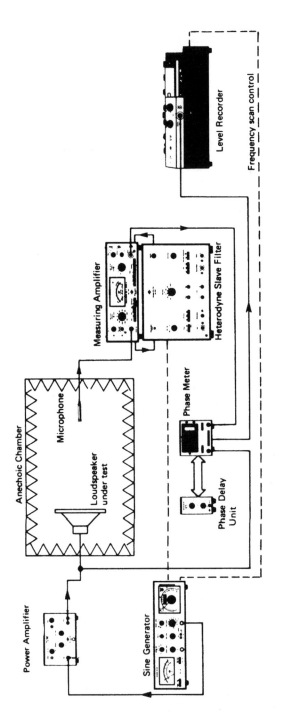

Figure 9.38. Set-up for loudspeaker phase measurements

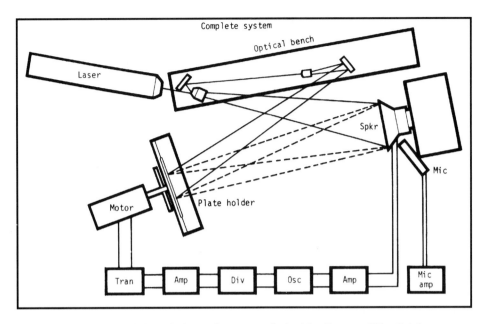

Figure 9.39. A laser holography set-up, devised by Fryer at Wharfedale

constant* with frequency). In such a multi-unit system, the drivers are usually mounted on a flat baffle or front panel, with their effective radiation planes displaced by varying degrees depending on chassis geometry, diaphragm type and panel depth, etc. In such a case a relative time delay will exist between the outputs of the different drive units measured at the listening position, this producing a non-minimum phase characteristic. Using the phase measuring apparatus, these time delays may be quantified and the units may then be physically and electrically aligned so that this differential delay is substantially minimized, thus allowing minimum phase system design, assuming a suitable crossover is utilized. Driver displacements may also be read from the time displaced peaks on an energy/time decay curve.

Square Wave and Impulse Response

A loudspeaker must have a perfect amplitude and phase response to successfully reproduce a square wave with its complex related harmonic structure. However, only minimum phase loudspeakers are potentially able to do this with any degree of accuracy, and other designs will give practically meaningless results on this test.

The impulse response is another matter, as the aim of this measurement is to evaluate delayed resonances in the crossover, the driver diaphragms and enclosure, via detailed analysis of the pulse response and its after effects. Some idea of these delayed resonances may be obtained without measurement, but the use of a FFT program with computer analysis has proved most revealing and has considerably refined the

*This is implicit in the term 'linear phase'.

evaluation of speaker transient performance. The computer system offers a refined averaging system which gives an excellent signal-to-noise ratio. An anechoic chamber is not necessary if the receiving microphone output is suitably gated before analysis, eliminating local wall reflections as described in the section on gating.

Tone-burst Response

Tone-bursts have traditionally been used to examine speaker transient response via analysis of the continued ringing at the test frequency following cessation of the input. This is a less revealing method of examining the pulse, or more strictly the transient response of a speaker, employing gated sections of a steady state sine frequency. By judicious selection of gating period and tone frequency, delayed resonances may be investigated. A key point concerning the toneburst is the requirement for the envelope to consist of a whole number of cycles starting and stopping at the zero crossing point, so as to produce minimum asymmetric pulse disturbance. In addition, the measuring environment must be highly anechoic, or spurious reflections will interfere with the analysis. The recent introduction of gated pulse techniques has made simple toneburst testing largely redundant.

Windowed and specially shaped pulses including tone-bursts fall naturally into the impulse testing category with computed FFT processing and sophisticated post analysis.

Sensitivity, Efficiency and Sound Power Output

Strictly speaking, sensitivity is an alternative way of quantifying loudness* for a given input level. One standard, for example, is the acoustic output in dB ('A' weighted or unweighted) at 1 m on axis for 1 W electrical input. However, most sensitivity ratings are based on a nominal input of 2.83 V (1 W/8 Ω), regardless of the actual value of the usually complex loudspeaker impedance, and are in reality 'voltage sensitivity' ratings.

For an assessment of the true efficiency the total acoustic output should be measured by employing an integrating microphone arrangement—either a multiplex array or a rotating boom-mounted microphone, the latter used in conjunction with a reverberant chamber.

It nevertheless remains difficult to assess the real power input to a complex loudspeaker. A workable method involves the integration of the voltage and current of a random noise signal applied to the loudspeaker, the calculation accomplished by a suitable wide-range RMS measuring instrument, while the s.p.l. is simultaneously measured.

When estimating a linear or unweighted sensitivity figure, a single frequency reading will not suffice due to the usual frequency irregularities of most systems. Good consistency may be achieved by visually integrating a 'best fit' line from 200 Hz to 5 kHz on the anechoic response, taking this level as the sensitivity.

Comparative sound power outputs can be measured in a listening room using noise,

*Strictly, sound pressure level, s.p.l.

or well balanced music programme, in conjunction with a slow weighted sound level meter. In an average $80 \, \text{m}^3$ volume room a stereo pair will deliver at the listening position an s.p.l., 'A' weighted reading around 5 dB less than the mono axial test result at 1 m. For example a 90 dB/W, 100 W rated system will produce maximum room levels of $(110-5)$ dBA or 105 dBA, close to concert hall realism, and subjectively considered very loud in such a room. Modern rock enthusiasts may demand 110 dB in room, attainable with 200 W/ch and 92 dB/W sensitivity.

Laser Measurements

Lasers are finding increasing use in the development of loudspeakers. Using the Doppler interferometric technique the previous need for zero vibration, super rigid test arrangements is markedly reduced. Operating diaphragms and indeed loudspeaker enclosure panels are within reach of the scanned, modulated low power laser beam. Its photo-detector or multiplier output provides a signal which, after processing, provides the displacement and velocity of the areas scanned. The data may be presented as 3D graphs in relief form or stored in a digital memory and fed to a display monitor. With appropriate image processing slow motion animated displays can be produced, greatly aiding the visual assessment of dynamics of acoustic structures (see Figures 9.39 and 9.40).

Electrical Impedance

The DIN standard for impedance is a sensible one and states that the modulus Z should not deviate more than $\pm 20\%$ from its nominal value of 4, 8 or 16 Ω, over the working frequency range.

The measurement is straightforward, involving the use of a current generator, simulated by a normal voltage sweep which is fed to the loudspeaker via a high resistance, for example, 2 kΩ. The variation of voltage with frequency at the loudspeaker terminals reflects the variation of impedance. It is usual to substitute a known value resistor in place of the loudspeaker to confirm the scaling and, strictly speaking, a linear recording scale should be used for clarity (see Figure 9.39).

Most recorders are fitted with logarithmic potentiometers, but a linear amplitude conversion chart will provide the remaining scale corrections required. For current practice, 8 Ω is the recommended speaker standard, although increasingly 6 Ω is becoming prevalent to give better sensitivity specification.

The addition of a phase meter allows the recording of the phase component of impedance — an important factor where amplifier matching is concerned (Figure 9.40). Loudspeakers with large reactive components will often cause premature limiting and a consequent deterioration in sound quality when used with certain otherwise workable power amplifiers. A 'compressor' generator may also be arranged to generate a constant current for accurate impedance measurement (Figures 9.41, 9.42 and 9.43).

At present no straightforward method for analysing dynamic impedance is available. Given recent findings concerning input current requirements of loudspeaker systems

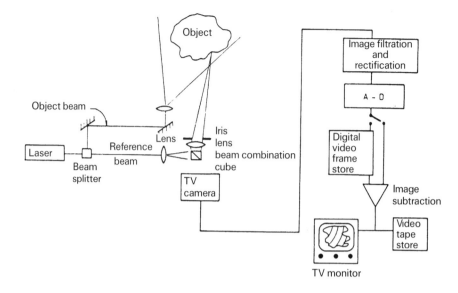

Figure 9.40. (a) Configuration, optical and electronic for ESPI analysis (electronic speckle pattern interferometry)

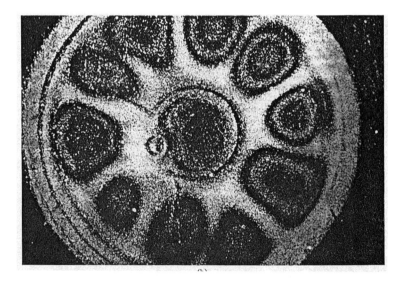

Figure 9.40. (b) Noise reduced image showing modes at 4.2 kHz in a moving coil diaphragm derived using the EPSI technique (after Tyrer [51])

[37], the conventional impedance curve needs to be regarded as only one aspect of the potential loading on amplifiers. In particular, amplifier designers cannot rely on a 8 Ω specified loudspeaker offering a 6.4 Ω minimum impedance modulus under heavy drive, complex programme conditions.

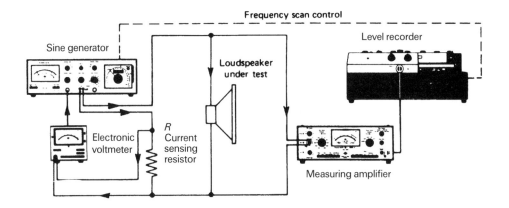

Figure 9.41. Use of beat frequency oscillator as constant current generator (b) Typical loudspeaker impedance curve (courtesy Bruel and Kjaer)

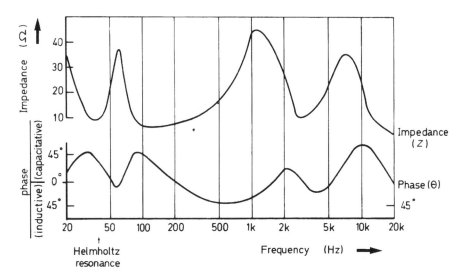

Figure 9.42. Impedance, modulus and phase for a three-way reflex loaded system (crossovers at 3 kHz and 13 kHz, 18 dB/octave equalized)

Computer Controlled Testing

One cartridge manufacturer produces a small microprocessor directed loudspeaker test unit, mainly for production test work.* However, the availability of small computer controllers and an increasing variety of intelligent computer interfaced audio instrumentation from Hewlett–Packard, B & K, etc. means that automated and semi-automated systems are possible for development and production (Colloms [45]). The

*ORTOFON.

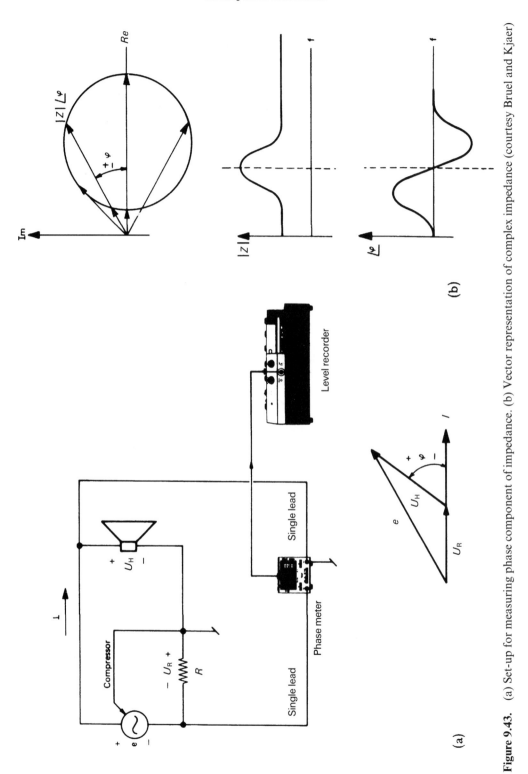

Figure 9.43. (a) Set-up for measuring phase component of impedance. (b) Vector representation of complex impedance (courtesy Bruel and Kjaer)

attached small computer offers great processing flexibility for data captured by such a system, particularly from the FFT analyser. Set-up and calibration routines can be programmed. Spot frequency measurement of distortion is very straightforward and impulse analysis is an obvious part of such a system (Figure 9.44).

In the past few years there has been extensive development in electronic audio measurement systems based on audio 'sound' cards, which are accessories units or mainframe plug-in to PCs. Their operating systems are held on software and subject to frequent and often worthwhile revisions and improvements. At very moderate cost compared with established specialist acoustic measurement products, these PC-based systems offer surprising versatility, often include calibrated microphone options, and are relatively easy to use. DRA Labs MLSSA system is one of the earlier and best known types, is used extensively by this author, and conveniently installed in a Toshiba portable. However, few of these remain which can take a sound card; the alternative is the usual desk-top machine.*

Economical examples include CLIO, by Audiomatica SRL, Italy, which among its many facilities includes distortion and MLS plus a third octave RTA (optional microphone). A similar US design is the Liberty Audiosuite (LAUD) by Liberty Instruments which has some interesting features, including one-sixth octave analysis plus eye-catching displays. LMS is another card-based system designed by Linear X. It is highly accurate and has excellent analysis facilities using wide range sine wave generation. It does not gate or anechoically window responses.

Driver Parameters

While manufacturers who design their own drive units hopefully retain complete information on their research, the system designer who works with drivers from several independent sources needs certain information which may not be readily available. A valuable source is the impedance curve, taken both in free air and with the driver mounted in a suitable air-tight box. The frequency/amplitude characteristic may also be obtained as the unequalized voltage output from a low mass accelerometer temporarily fixed to the cone.

A difficulty both with the practical realization of theoretical analyses and with parameter measurement, is the non-linear variation of relevant parameters with level and/or excursion. Discrepancies of 5%–10% are not uncommon, for example in Bl and suspension compliance. Thus a given Q and f_0 at 0.1 W may be different from the results taken at 1 and 10 W.

Suspension compliance, C_{MS} (N/m)

This may be found by placing a known weight M on the diaphragm (axis vertical) and noting the resulting displacement d; for example, by using a travelling microscope. Now

$$C_{MS} = \frac{1}{s} = \frac{d}{Mg}$$

where g = gravitational acceleration and s = stiffness.

*Or a portable with the 'card' installed in the 'docking' unit.

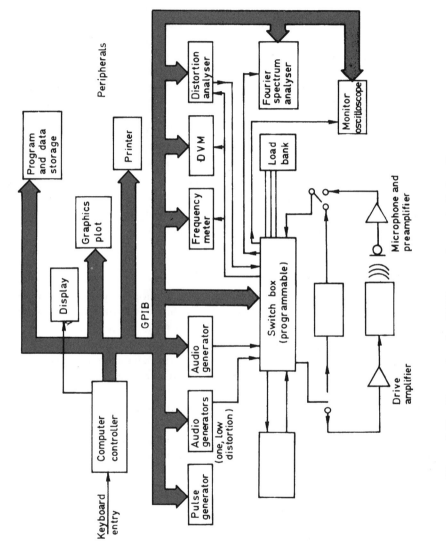

Figure 9.44.   Comprehensive GPIB computer controlled audio test system

An alternative method of obtaining the compliance consists of noting the free air resonance f_0, and the in-box resonance f_c given with the driver mounted in a sealed unlined box of known volume, V_B. The air volume of the box has a compliance C_{AB}. The driver acoustic compliance C_{AS} is related to C_{MS} by S_D^2, the effective piston area, i.e.

$$C_{MS} = \frac{C_{AS}}{S_D^2} \qquad \frac{C_{AS}}{C_{AB}} = 1.15\left(\frac{f_c}{f_0}\right)^2 - 1$$

(the factor '1.15' is a general approximation and is affected by the box size)

also,
$$C_{AB} = \frac{V}{1.4 \times 10^5}$$

so that

$$C_{MS} = \frac{V_B}{S_D^2}\left[1.15\left(\frac{f_c}{f_0}\right)^2 - 1\right] \times \frac{10^{-5}}{1.4}$$

The two values of compliance may be compared to check for measurement accuracy. Note that excessive mass loading will produce non-linearity of compliance and an erroneous value — 2–4 mm of displacement is typical for this measurement.

Moving mass, M_D (kg)

This may be calculated from the free-air or fundamental resonant frequency, f_0, which in turn may be taken from the peak in motional impedance. Having estimated the working diaphragm area (i.e. the projected moving area) by measurement, the equivalent radius, a, may be set. Then

$$M_D = \frac{1}{C_{MS}(2\pi f_0)^2} - 3.15a^3$$

Moving mass may also be found by comparing the free air f_0 with f_m, the result with an additional mass m attached:

$$f_0/f_m = \sqrt{\left(\frac{M_t + m}{M_t}\right)}$$

where M_t = driver moving mass including airload, M_a.

$$M_D + M_a = \frac{mf_m^2}{f_0^2 - f_m^2} \qquad *$$

D.C. resistance (motor coil), R_c (ohms)

This may be measured by using a d.c. ohmmeter or, alternatively note that the minimum impedance value above fundamental resonance falls close to the d.c.

*Audio card/software often generates the T-S or 'SPO' parameters, e.g. CLIO, LINER-X, MLSSA.

resistance. At a suitable low frequency, clamping the diaphragm will also provide d.c. resistance with the impedance measurement method.

Coil inductance, f_c (henrys)

An a.c. impedance bridge may be used at 1 kHz or above, with the driver diaphragm clamped to prevent coil motion. The coil inductance can also be calculated from the impedance curve without clamping if the unit is a low ($<2\%$) efficiency type. The slope of the impedance curve at higher frequencies is dominated by coil inductance.

Mechanical resistance R_{MS} (of suspension components)

The free air resonance peak is controlled by R_{MS}. Hence from the half-power or $-3\,\mathrm{dB}$ voltage point on the impedance curve of f_0, and neglecting the small air load resistive component,

$$R_{MS} = \frac{2\pi f_0}{Q_M}(M_D + M_A)$$

where M_A = air load mass, and

$$Q_M = \frac{f_0}{(f_1 - f_2)}\,(-3\,\mathrm{dB\ points})$$

to a good approximation (if R_E dominates R_{ES} and the impedance test is a good equivalent of current drive).

   Generally Q_{ES} dominates Q_{TS} for higher Q_M values. The impedance curve alone can thus provide much significant data on driver parameters and is often used in quality control sections for speaker production.

Electrical 'Q' factor, Q_E (assuming zero generator impedance)

Q_E is given by

$$Q_E = \frac{2\pi f_0 M_D R_C}{(Bl)^2}$$

Total 'Q' factor, Q_T

This is given by

$$\frac{1}{Q_T} = \frac{1}{Q_M} + \frac{1}{Q_E}$$

Flux density, B

This may be measured by using a flux probe in the magnet gap, although it requires the unit to be disassembled. Alternatively measurement of Bl and prior knowledge of l provides B.

Cone area, S_D *(square metres)*

This is derived via measurement and will include a percentage, usually 50%, of the surround area (i.e. the projected area, not the actual cone surface area).

Force Factor, *Bl* (N/A)

This may be found by applying a known current I and measuring the excursion, d, ensuring that the moving axis is horizontal to exclude gravitational effects. Alternatively the driver may be placed horizontally, its rest position noted and a known mass M applied. A d.c. current is then fed to the coil and adjusted until the extra mass is just balanced and the rest position attained. Then

$$Bl = \frac{9.8M}{I} \quad \text{Wb/m (N/A)}$$

where M is in kg and I is in amperes.

Figure 9.45 is a reproduction of an informative specification published by a manufacturer of drive units. Figure 9.46 shows the IEC test baffle for driver response measurements.

9.4 SUBJECTIVE EVALUATION

Perception of Loudness

Arguing from the aural sensitivity curves, and including the results of long observation the performance of many speaker designs, the subjective judgement of loudness is not easy. When assessing the frequency response of a speaker the usual variations make a definition of the sensitivity quite difficult. Do you try and draw a horizontal line through the 'centre' of the frequency response, and what modification do you make if the response has a slope or is irregular? Some evaluators favour broad band pink noise as a stimulus together with weighting e.g. 'C' or 'A' filters for the sound level meter.

The matter is complicated by the ear's response to variations in frequency distribution. A speaker with a response which is the inverse of the hearing sensitivity curve will by definition 'sound' the loudest. Manufacturers have sometimes exploited this as loudness is usually cheap at this end of the spectrum. The result is generally a hard and fatiguing sound quality. When making comparative judgements of loudspeakers such anomalies make it difficult to calibrate matched levels for the speaker assessments.

Other factors which may affect subjective loudness include a general 'roughness' in the frequency responses [apparently louder], colouration and/or resonances [in practice conferring increased loudness], and finally, significant levels of distortion. Distortion is associated with a shift in power from lower frequency fundamentals to higher pitched harmonics, which also corresponds to the natural shift in tonality and timbre when most musical instruments are played loudly. Thus the harmonic character of non linear distortion is aurally associated with greater perceived loudness, in addition to the rougher, sharper effect of audible distortion.

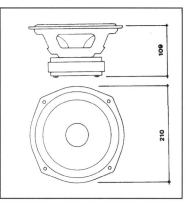

Model B200
Specification Number SP1039

Low/mid range unit with visco-elastic damped
Bextrene diaphragm and high temperature coil
assembly, suitable for use where low distortion and
high power handling are required.

Net weight: 3.0 kg (6.6 lb)

Nominal impedance: 8Ω

Nominal frequency range: 25-3,500 Hz

Typical enclosure volumes:
Totally enclosed box 20-25 litres ($\frac{3}{4}$-1 cu ft)
Reflex 30-40 litres (1-1$\frac{1}{2}$ cu ft)

Power handling:
Continuous sine wave 25 V RMS (see note 1)
Programme 80 W (see note 2)

Magnet:
Flux density 1.2 T (12,000 gauss)
Total flux 1.17 x 10^{-3} Wb (117,000 Maxwells)

Sensitivity: Pink noise input for 96 dB SPL at
1 metre on axis 9 V RMS

Voice coil:
Diameter 32.6 mm (1$\frac{1}{4}$ in)
Inductance 0.25 mH
Max continuous service temperature (30 min) 250°C
Max intermittent temperature (5 sec) 340°C
Thermal time constant 4.5 seconds
Thermal resistivity (temp rise per applied Watt) 3°C/W
Nominal DC Resistance, R_{DC} 7.0Ω (tolerance ±5%)
Typical production spread 6.9±0.1Ω (see note 3)
Minimum impedance (in nominal frequency range)
7.1Ω at 190 Hz

Diaphragm:
Effective area, S_D 232 cm² (36 sq in)
Effective moving mass, M_D 20.2 gm (0.71 oz)
Max linear excursion, X_D 5 mm peak-peak (0.2 in)
Max damage limited excursion 20 mm peak-peak (0.8 in)

Free air resonance frequency, f_s:
Nominal 25 Hz (tolerance ±5 Hz)
Typical production spread 24±2.2 Hz (see note 3)

Total mechanical resistance of suspension, R_{MS}:
0.7 mech Ω

Suspension compliance, C_{MS}: 2.2 x 10^{-3} m/N
(2.2 x 10^{-6} cm/dyne)

Equivalent volume of compliance, V_{AS}: 130 litres
(7,930 cu in)

Force factor, Bl: 7.2 N/A

Damping:
Mechanical Q_M 4.4
Electrical Q_E 0.5
Total Q_T 0.45 (see note 4)

Notes
1 Continuous Power Rating (Pc).
$$Pc = \frac{V^2}{R}$$
V is the RMS voltage which can be applied to the
unit continuously without thermal overload of
the voice coil. At low frequencies the continuous
power rating of the speaker may be reduced
because of limitations imposed on diaphragm
excursion by the acoustic loading.
2 The programme rating of a unit is equal to the
maximum programme rating of any system with
which the unit may be safely used in conjunction
with the recommended dividing network and
enclosure.
The programme rating of any system is the
undistorted power output of an amplifier with
which the system may be satisfactorily operated
on normal programme over an extended period
of time.
3 "Typical production spread" is derived from
statistical analysis of a large number of units, and
is calculated to include 95% of all units.
4 $Q_M = \dfrac{2\pi f_s M_D}{R_{MS}}$ $Q_E = \dfrac{2\pi f_s M_D}{(Bl)^2/R_{DC}}$ $\dfrac{1}{Q_T} = \dfrac{1}{Q_M} + \dfrac{1}{Q_E}$

Figure 9.45.   An example of a useful data sheet for a drive unit (courtesy KEF Electronics Ltd)

During subjective assessment it is often found that the speaker system which sounds quieter than its metered measured sensitivity would indicate, is one of superior quality, possessing low colouration, low stored energy, a smooth frequency response and low distortion; a particular example of such a speaker is a good electrostatic design.

The comparison of live sounds with those recorded and reproduced by a loudspeaker is obviously rewarding, but can be very difficult to organise. There are two major problems, namely that of obtaining a consistent and repeatable live sound, and

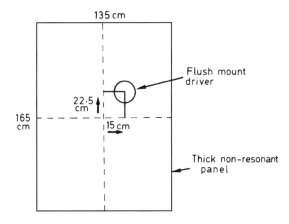

Figure 9.46. IEC Standard baffle, an open panel for driver measurement. The AES also specifies a range of baffles

secondly, assembling a reliable and objective listening panel. It is virtually impossible for a single listener, in particular the designer of the system under test, to make an unbiased assessment of its sound quality.

However, worthwhile methods for subjective sound evaluation are beginning to emerge. A number of factors need to be taken into consideration, some of the most important of which are outlined in Table 9.1 (largely drawn from the IEC draft proposals referring to domestic loudspeakers).

As high fidelity has continued to improve, it has become necessary to introduce further terms to help characterize perceived sound quality. In particular, stereo images can now be seen to offer a respectable dimensional quality where good system transparency is a necessary factor, and where the impression of sound stage depth can be substantial. Furthermore, recent design work has shown that the impression of depth is not necessarily restricted to the mid range, traditionally an area of transducer excellence, but is also a property of both bass and treble frequency ranges.

Other areas worthy of coverage include the perception of ambience and space, of timbre or tonal quality, and of perceived musical pace and rhythm. The psychoacoustic influence of low frequency extension below 40 Hz is also worth discussing.

Stereo Image Depth

In a traditional context, a stereo image is described as an auditory illusion which is intended to create in the mind of the listeners the effect of a performing stage where the angular location of the performer may be approximately recognized and where the acoustic of the recording venue may be heard in proper proportion. Depth is a more subtle aspect where the apparent placement of the performers in the depth plane is a desirable but rarely encountered property. While the practice of stereo recording remains poorly understood and stereo production in the home is just an approximation to the clear transmission of two independent signal channels to the listener, it is permissible to ask whether a more critical discussion of stereo imaging is worthwhile.

Table 9.1.   Electroacoustic standards

Below is a list of the most relevant standards applied to electroacoustics. Many standards in the border area of this subject have been omitted and thus the list cannot be considered as being complete. Furthermore it is known that several Standards Organizations have relevant subjects under consideration which may outdate the present survey in a relatively short period. The Standards are of basic coverage and do not offer much help in the design and specification of high performance systems (see Figure 9.47) (See AES, p 449)

Organization	Number	Date	Status*	Short description
IEC	200	1966	S	Loudspeakers, frequency response, polar plot, resonant frequency, power handling, test conditions, standard baffle
ANSI	S.1.5	1963	S	Loudspeaker, impedance, frequency response, polar plot, distortion, efficiency, power handling, recommended
BS	1927	1953	S	Loudspeaker, physical dimensions, impedance and resonant frequency
BS	2498	1945	S	Loudspeakers, frequency response, distortion, efficiency polar plot, impedance, transient response conditions
IEC	268–5	1972		Loudspeakers** (Second edn. 1987)
DIN	45 500 sheet 1–3	Feb 1971	S	Loudspeakers, frequency response, power handling, distortion, music power
DIN	45 573 sheet 1	July 1962	S	Loudspeakers, test conditions and methods for type test
DIN	45 573 sheet 2	Jan 1969	S	Loudspeakers, power handling and life test
DIN	45 575	May 1962	S	Loudspeakers, standard baffle for measurements

*S Standard
**BS 5428 Pt II, 1977 (the power spectral curve is under amendment)
BS 5492 1980 ($\equiv$ IEC 581–1) Minimum requirements audio systems
BS 5428 1977 ($\equiv$ IEC 268–567) Loudspeakers
IEC 268–5a 1980 Loudspeaker supplement (noise signals)

We are considering the reproduction of a series of audible clues and cues, which given appropriate experience, can greatly help in the mental visualization of a recorded musical event in four dimensions, three in space and one in time.

In the last decade advances in amplifier and speaker technology have opened a window to a more realistic and satisfying level of stereo reproduction. We can now see that a lack of clarity in amplifiers, and poor transient response in loudspeakers has imposed a monochromatic mask over the perceived stereo image, generating a flattened, poster-like effect. Yet the sensation of depth is not unique to stereo and a good mono recording containing sufficient ambience can be reproduced on a single speaker. While the aural impression is something like listening down a short corridor, the sympathetic ear can soon establish perspective relationships between the sound sources, and between the acoustic of the recording hall.

The perception of depth and the related qualities of perspective and space are, in turn, strongly allied to the overall fidelity of reproduction and to the ratio of direct to reverberant or ambient energy in the recordings. At a given location a single recording microphone placed adjacent to the conductor will 'hear' the violins first and with a near-field acoustic quality. The woodwind is generally placed several ranks back, with a recording ratio tipped more in favour of the hall acoustic, the direct sound contribution now considered far field and carrying a matching spectral signature. Finally we move further back to the brass and percussion instruments, whose recorded signal is totally dominated by reverberant acoustic, and if there is a rear concert hall wall, by the encroaching reflections coming directly from this boundary.

A high resolution system is capable of reproducing all this information with sufficient accuracy for the ear and the brain's aural analyser to establish those spatial relationships, thereby creating the subjective impression of depth. A reproducing system of limited transparency may give a vague impression of depth and hall acoustic. A system of high transparency can give the impression of a sound stage extending clearly to the back wall of the concert hall and extending behind the performers.

A proper exposition of the ratio of direct and ambient sounds is essential for the impression of depth and relies on the ability of the loudspeaker to reproduce the complex music signal without adding significant masking due to colouration or overhang. Cues as to ambience and space decay exponentially following the music transient, and speakers with significant diaphragm or cabinet resonance colouration tend to overlay these vital, low level signals.

The impression of space and depth is enhanced if the transient quality and transparency is maintained over the whole frequency range. For example, a system with mild treble emphasis may sound well focused and tidy for pan-potted, front stage orchestration but when asked to reproduce a distant flute will falsify its position preventing the system as a whole from reproducing depth properly. The mid range may well be correctly balanced but the harmonic 'blowing' sounds will be falsely located in the plane of the speakers, too close to the listener. Perspective and focus are thereby disturbed.

Ambience and Scale

Hall ambience and a natural sense of scale are largely the provenance of the low frequency register. Here a slow, coloured bass response can dominate the sound, obscuring the atmosphere and dynamic 'bounce' of a live acoustic, otherwise properly recorded. Our understanding in this area has been considerably advanced thanks to the extended low colouration performance of the larger panel dipole loudspeakers.

Extended Low Frequency Response

Low frequency extension below 40 Hz is assuming increasing importance as digital audio becomes commonplace with its potential bandwidth extending almost to DC. Much as the formant acoustic power in the frequency range 100–300 Hz determines the accuracy of the tonal balance of mid-range sounds; likewise the range 25–50 Hz can be seen to underpin the quality of bass sounds in the 80–160 Hz range. Due to the logarithmic nature of the frequency scale commonly used, it is likely that many designers pay insufficient attention to the low frequency range. Such a careless attitude may be represented by the view that 30–60 Hz is only 30 Hz out of the whole frequency range produced by the speaker. Yet that 30 Hz represents a whole octave of 8 tones, 16 semi-tones, etc. Out of a range 60 Hz to 16 kHz, a range of 8 octaves, it represents 12%; equivalent to the whole octave span from 8 kHz to 16 kHz. Musically the 30–60 Hz range is arguably more important than this high range since only harmonics are present in music above 8 kHz while the 30–60 Hz low frequency range represents fundamentals. This argument is used here to show that much more care is needed in low frequency system design and that every 10 Hz of clean bass extension gained by good design is worthwhile.

Paradoxically, well controlled, low Q bass extension also improves the speed and neutrality in the upper bass range. Sharp low frequency cutoffs in the 50–100 Hz range are often audible as a 'boom' or 'hangover', while the sense of scale on large orchestral works is also diminished by a premature bass rolloff.

Judgement of resonances and their audible effects is extremely complex [48], and depends for example on whether the resonances are present with the formant tone, as in a loudspeaker cone, or delayed and re-radiated from an adjacent structure. The assessment of time decay spectra is complicated by factors such as directivity, local acoustics and the potential audibility of various resonance effects with different kinds of sounds. A wealth of fascinating data on this subject are provided in the above reference [48] which is strongly recommended for further reading.

Tonal Balance, Timbre

Though poorly understood by many designers, critics and loudspeaker purchasers, an accurate tonal balance, i.e. the ability to reproduce timbre or 'tone' correctly, is paramount in loudspeaker design. Given good ingredients, enclosure, drive units, effective design tools, the objective is to secure a performance which sounds 'natural' — lifelike — in the environment where the speaker is intended to be used. Thus a studio acoustic may well differ from a domestic layout and some balance adjustment may be needed.

Systems lacking an accurate perceived tonal balance, by definition, cannot define a good stereo image. The perception of depth, and with it an impression of perspective in reproduced sound, requires that the subtle clues aiding that impression are not confused. The natural timbres of real sounds are familiar; moreover the tonal quality presented gives information about their apparent distance from the listener. Thus a bright, sharp sound appears close, loud, and strongly played, while a warmer sound appears softer and more distant. Such spatial distortion can apply to a single instrument or sound source, or cause further problems.

Take the case of a tonally unbalanced speaker with a response rising 4 dB from 100 Hz to 1 kHz. When reproducing a sound stage it is often found that the depth position of the double bass and similar sounds acts as an anchor, while the reproduction of woodwind, e.g. piccolo, from such a system will be unnaturally thinned and brightened, giving the impression that this source has moved in front of the bass. Thus this element of the orchestral perspective is flattened and distorted.

Consider a speaker with a mildly excessive treble reproducing a wide range sound such a solo violin or a flute. The main formant sounds of these instruments may appear to be correctly placed in the sound stage but the harmonic edges related to bowing for the violin or breath sounds with the flute are exaggerated by the treble imbalance stretching out the image of the instrument in the depth direction. Such confusion may be aurally fatiguing to the listener and seriously weakens the stereo illusion.

Timbre, Frequency Response and Directivity

It has been generally assumed that if the fundamental parameter, namely axial frequency response, was to a good standard and it included specifications for moderate colouration, etc. then a natural tonal quality would inevitably result from a good speaker design. This is not the case. Perceived timbre is based on a complex aural summation of direct and reverberant reflected sound energy as heard in the listening area. The subject is a complex one since it includes the characteristic of the listening room and the disposition of both the loudspeakers and listeners within it. This fact alone largely accounts for many differences in opinion between critics, or between the sound heard at a retail demonstration suite and then obtained at home.

Only by defining some sensible standard for listening rooms and their use can this discussion continue, and the IEC recommendations are a good place to begin. Using a typical well proportioned medium sized room, with a balance of furnishing and acoustic absorption, we will assume that the speakers are sensibly and symmetrically located near one end of the room, spaced away from the boundaries and typically stand mounted. If the three distances from the radiating centre to the nearest boundaries are staggered, e.g. 0.55 m to the floor, 0.75 m to the rear wall and 0.95 m to the side wall, and the floor beneath is carpeted, then reasonably consistent and uniform acoustic energy can be launched towards the listener. The listener hears a mix of direct speaker output and the contribution of immediate local boundaries at frequencies below 200 Hz. Above 200 Hz or so a speaker system generally becomes more directive, while the local coupling is reduced and more direct sound is heard, supplemented by the dominant side wall glancing reflections. These can be significant right through the mid and treble ranges, their strength dependent on the reflective characteristics of the room walls, furnishings, bookcases and the like, and the off-axis response of the speaker. A wide fronted, forward directive speaker will be less affected by the side wall glancing contribution, while a miniature, wide directivity model can suffer considerably.

This explains why two speakers, one large and the other small, but with comparable technical performances with regard to a uniform axial amplitude frequency response and a low delayed resonance behaviour, can and do sound very different in a normal room. It also helps to explain why a single loudspeaker used centrally in a room has a different tonal quality to the same model used as a pair in a more widely separated stereo fashion. This points to a minor weakness in live-versus-recorded listening tests which are generally performed with a single speaker adjacent to a live sound source.

Even assuming a well-ordered design with a good level of integration and uniformity achieved between the drivers, it is clear that the axial frequency response is insufficient to define timbre. At the present state of knowledge, a flat response is a good approximation for medium sized or compact speaker enclosures such as the very common 20–40 litre two-way models with a 170–220 mm bass mid driver and a 20–30 mm treble unit. True miniatures in the 6–12 litre range, even with a reasonably extended bass response, have an in-room tonal quality which is thinner and brighter than the axial or forward responses would suggest. This is because a wider directivity is maintained to higher frequencies, increasing the proportion of mid and treble energy in the reflected and reverberant sound. Conversely a large enclosure, one in the 50–100 litre class, sounds richer and duller than its axial response indication, due to a correspondingly narrower directivity with rising frequency.

Thus the effective acoustic size of a loudspeaker system is an important parameter and timbre is closely associated with the forward energy response, the directivity and the classic measurement of axial amplitude/frequency response. Where unusual speaker geometries are employed, the in-room tonal quality and perceived frequency balance varies again, but is still predictable. For example, some designs have employed a narrow fronted enclosure with two bass mid units, one above and one below the HF unit. Given the variables involved in crossover alignment, the tendency is to run all three in parallel at the crossover region, the combined source approximating to a vertical line. Acoustically this has a narrow vertical distribution and a wide horizontal distribution, increasing the proportion of first reflected side wall energy relative to that in the vertical direction. If the axial response is uniform for such a system, then it is often found that the in-room sound is subjectively dominant in the crossover region with an imperfect timbre, one commonly associated with 'hard' and 'glaring' sounds in the 3 kHz region. To help maintain more constant directivity with frequency, some designers use complex structures; for example, smaller sub-enclosures for the mid range and high frequencies.

Environment

A fairly well specified domestic listening room is suggested for test purposes; most studio control rooms in fact approximate sufficiently to average domestic conditions for this purpose. (The IEC proposals (see Tables 9.2 and 9.3) concern domestic reproducers, and other classes of system such as stage monitors will require an alternative arrangement.)

The room volume should be between 50 and $110 \, m^3$, nominally $80 \, m^3$, with a height of $2.75 \pm 0.25 \, m$, the preferred ratio of dimensions being $L:B:H = 2.4:1.6:1$. This gives the most even distribution of resonances, minimizing room colouration effects. Within the range of heights suggested, the ratio gives the following results for breadth, length and volume.

For reverberation in the 250 Hz to 4 kHz range, the deviation should not exceed $\pm 25\%$ of the average R_T of 0.5 s. Below 250 Hz the R_T should remain below 0.85 s.

The proposal also suggests reflecting walls behind the loudspeakers with the opposing wall behind the listener absorptive. The ceiling should be untreated and the floor immediately adjacent to the speakers is uncarpeted.* Together with this formula, the reverberation time over the range 100–800 kHz should fit the limits specified in Table 9.3.

Climatic conditions are also relevant with the optimal situation given as 20°C, $\pm 2°C$, relative humidity 65%, $\pm 5\%$, and atmospheric pressure 860–1060 millibars.

The IEC suggestions provide conditions which tend to integrate loudspeaker energy output in the vicinity of the speakers, encouraging local boundary reflection.

Investigations into monitoring room acoustics for stereo, particularly those using 'LEDE' life-end-dead-end' methods, indicate that the boundaries behind the listener

_____
*In practice the speaker area is usually carpeted.

Table 9.2.   (a) Recommended room proportions (IEC)

Height (m)	Length (m)	Breadth (m)	Area (m²)	Volume (m³)	
2.5	6.0	4.0	24.0	60.0	(min)
2.8	6.7	4.2	29.0	80.0	(mean)
3.0	7.2	4.8	34.0	101.0	(max)

Table 9.2.   (b) Ideal room ratios for different sized rooms (from Olsen)

Room size	Length	Breadth	Height
Small	1.6	1.25	1
Medium	2.5	1.6	1
Large	3.2	1.25	1

Table 9.2   (c) The first 24 modal frequencies in three typical listening rooms [4]

Room = 6.75×3×2.6 m		Room = 7×3.75×2.7 m		Room = 9×5×3.6 m	
Frequency (Hz)	Mode	Frequency (Hz)	Mode	Frequency (Hz)	Mode
25.19	1.0.0	24.29	1.0.0	18.90	1.0.0
50.40	2.0.0	45.33	0.1.0	34.00	0.1.0
56.67	0.1.0	48.57	2.0.0	37.80	2.0.0
62.00	1.1.0	51.43	1.1.0	38.90	1.1.0
65.39	0.0.1	63.0	0.0.1	47.22	0.0.1
70.10	1.0.1	66.44	2.1.0	50.83	2.1.0
75.57	3.0.0	67.48	1.0.1	50.86	1.0.1
82.54	2.0.1	77.59	0.1.1	58.20	0.1.1
86.52	0.1.1	79.52	2.0.1	60.50	2.0.1
90.10	1.1.1	81.29	1.1.1	61.20	1.1.1
94.44	3.1.0	85.80	3.1.0	66.10	3.1.0
99.92	3.0.1	90.67	0.2.0	68.00	0.2.0
113.3	0.2.0	93.86	1.2.0	70.60	1.2.0
114.9	3.1.1	96.30	3.0.1	73.76	3.0.1
116.1	1.2.0	102.9	2.2.0	77.79	2.2.0
124.0	2.2.0	106.4	3.1.1	81.22	3.1.1
130.7	0.0.2	110.4	0.2.1	82.79	0.2.1
130.8	0.2.1	116.3	3.2.0	88.52	3.2.0
132.2	1.0.2	120.6	2.2.1	91.00	2.2.1
136.2	2.0.2	183.3	1.0.2	96.32	1.2.2
140.2	2.2.1	132.3	3.2.1	100.3	3.2.1
142.5	0.1.2	133.8	0.1.2	100.4	0.1.2

Table 9.3.   Proposed R_T for listening room

Frequency (Hz)	Room reverberation time (s)
100	0.4 ↔ 1.0
400	0.4 ↔ 0.6
1000	0.4 ↔ 0.6
8000	0.2 ↔ 0.6

may be reflective if relatively nearby, but the boundaries adjacent to the loudspeaker should be absorbent. The floor should thus be carpeted, the rear wall curtained possibly with a small portion of some longer wavelength absorption panels behind. For the best psycho-acoustic effect the speakers themselves should be concealed by a full height acoustically transparent but visually opaque curtain. With high quality systems and ancillaries, used with programme of matching quality, it is subjectively possible to conceal the aural position of the speaker enclosure. The perceived sound field can possess properties of height up to several metres, depth up to 20 m or so, and a stage width encompassing at least the room width. Front stage focus can appear so sharp that a sound image of a solo vocalist may appear no wider than a real person. The objectives for the stereo performance of a good loudspeaker are now additional to high clarity, a satisfactory bandwidth, good response uniformity and low colouration.

Ideal Listening Rooms

An exact specification for an ideal room is impracticable due to the wide variation encountered in the acoustics of domestic listening rooms. National taste plays a part in this as the acoustics are largely dependent upon the type of architecture and inside wall finish plus the quantity and style of the furniture.

Little can be done to cure severe or poorly distributed standing wave modes produced by non-ideal proportions. Architects seem to be largely unaware of the need for rooms to possess a natural and even reverberation; this quality is not only relevant to high quality music reproduction but also for the ubiquitous television as well as for clear intelligible human speech. Given reasonable proportions the room may be assessed in a limited way by a finger click or a handclap. Metallic ringing, flutter echo (closely spaced high frequency cross reflections) seriously degrade the reproduction and must be dealt with by a combination of various areas of wall acoustic absorbent. Wall hangings, curtains, window blinds and bookcases, etc. are very helpful in this respect. One of the nicest sounding 'natural' domestic listening rooms found in one of the larger houses was the library (not one with false book spines!). Carpets are a help, especially in the vicinity of the loudspeakers.

Ishii and Mizatani [40] researched the design of a 'standard' listening room for the subjective development of loudspeakers. By domestic standards it is quite large but it gives an idea of what can be done. An R_T of 0.4 s was achieved within a small deviation from 20 Hz to 20 kHz. This design target for the range below 100 Hz is open to criticism on the basis that no domestic room is likely to be equipped with graded bass absorbers to hold the R_T constant. The author can verify that this room had a sufficiently good sound quality for music recordings but was also clearly suited to anechoically flat speakers. The latter tend to suffer from a degree of bass lift in normal rooms.

Positioning

The systems should be positioned in accordance with the designer's or manufacturer's recommendations but if no such information is available, then a spacing of 1 m from the side walls and 0.5 m from the rear wall represents the minimum requirement. The main loudspeaker axis should be elevated 1.25 m above floor level and face the main listening areas. If a stereo pair is tested the left and right systems should be separated by at least 2.5 m, and if several pairs are simultaneously under test, no two speakers should be closer than 0.5 m.

Adjacent speakers or other obstacles spoil both the polar pattern, stereo image and low frequency transient response of the test system. Where accurate results are required, only one stereo pair system should be in the room for auditioning at one time. This mutual interference between test enclosures represents a fundamental weakness in stereo A/B testing. Done in mono, A/B comparisons must be supplemented by a further test series with the positions of the reference and test systems exchanged.

Listening Panel

The panel makeup depends upon the desired size, scope and application of the results, but for hi-fi equipment, a sensible choice would include both male and female enthusiasts familiar with live performances as well as recording engineers and interested musicians. A maximum number of five panellists is suggested and the tester should be aware of the general hearing ability of each panel member, who should ideally be drawn from the twenty to forty-year-old age bracket. The combined results achieved will provide some averaging of the forward radiating characteristics of the system under test, while factors such as the exact placement of the panellists and the rotation of listening places should also be noted. For the most critical stereo analysis the need for the path lengths from each speaker to a panellist to be equal, that is within a few centimetres, dictates the use of a solo listener.

Programme Material

The interaction between most types of music, its recording method and individual loudspeakers, is usually quite marked, and it is suggested that the test programme itself be regarded as a variable. Fairly short 15–30 s programme sections, recorded with adequate pauses to allow the panel members to note their comments, have proved effective. This is satisfactory for A/B pairing. Single presentation methods require more extended programme, typically 2–5 min each.

The programme quality may need to reflect the likely application. For domestic use, an inexpensive loudspeaker is likely to be used with a modest amplifier of possibly bright character with a disc player of significant distortion content, particularly at high frequencies. A speaker with a bright 'ragged' treble might sound reasonable on master quality programme but would be considered unacceptable paired with its likely source material. Both typical and master quality programme should be used, encompassing as wide a frequency range, dynamic range and music type as possible.

Frequently a speaker may be found to suit a particular recording balance or type of music and dismally fail on another. Auditioning cannot be short-cut. Where the recorded sounds are to be compared with life it is best to choose simple sounds, e.g. drum, voice, cymbal, wood blocks, anechoically recorded with a 12 mm capsule at a 1–2 m distance. It will be found that very high dynamic ranges are required in the recording and analogue noise processors tend to spoil the results. Open reel at 38 cm/s half track is acceptable, 16 bit PCM digital is optimal. For realistic reproduction of transient peaks, up to 250 W will be required with the less sensitive speakers.

Analogue and Digital Programme and its Effect on Listening Tests

With the introduction of digital programme some interesting differences have been observed in listening test results as opposed to those taken with analogue sources.

Given a high quality original master, digital programme has altered the emphasis in the balance of qualities judged preferable in the reproduction. Speakers with lower distortion, lower bass system Q factors, more extended tapered bass response, accurate tonal balance, incisive transient definition and high clarity and transparency are favoured. Moderate colouration or delayed resonance effects are also tolerated more than with analogue sources. It is as if the negligible colouration of the digital source allows for a little more lenience with regard to the speaker.

In contrast, analogue sources have tended to favour very smooth axial responses through mid and treble registers. Colouration needed to be as low as possible while a first-rate low frequency performance is less critical.

Speaker performance will remain to some degree programme dependent, not only on the type, the recording acoustic and quality but also whether the replay domestically or on test is analogue (tuner, disc or cassette) or digital (CD or PCM). A custom programme on CDR is most useful.

Loudness

While the preferred listening level is that set by the average panellist, great care must be taken to ensure that the 'loudness' of two speakers successively compared is carefully matched. If the response dissimilarities are severe, then this may be difficult to achieve and loudness adjustments for each programme section may then be necessary. Due regard for the overload point of both speaker and test amplifier is essential, and successive amplifier gain settings and the sound pressure are worth noting, the latter facilitated by a taped calibration signal and an 'A' weighted sound pressure reading. The quality of the results is largely determined by the skill and judgement of the test operator.

Duration

A single test session should not last longer than one hour, but with frequent short breaks and refreshments, good results may in practice be obtained over much longer periods.

The author has extensively used (Colloms [46]) longer duration sessions for single presentation subjective testing, it proving possible to audition 12 models per day. These are done in four groups of three, the groups interspersed with lunch and refreshment breaks. A programme of six programme items lasting a total of 15–17 min is considered sufficient for each speaker presentation.

Test Procedure

Two procedures are discussed in the IEC draft proposals, namely a 'single stimulus' or isolated judgement method where speakers are individually scored, and the A/B method (where speaker A is compared with speaker B), which although more lengthy is less frustrating for the panel, since it affords a greater sensitivity to fine differences. In order to assess the degree of marking variation, the use of a 'control' or reference is valuable, not only on one occasion but repeated several times during the sessions to establish test continuity and to gauge the consistency of marking.

A/B Paired Comparisons and Live Sounds

Random presentation of the speakers is essential, particularly if a reference is involved, and the order of the programme sections should be similarly shuffled. The A and B systems must be given approximately equal periods of audition, and it is also important that the reversed B/A case is also presented. The human ear quickly becomes accustomed to the A or first sound, to the extent of filling in quite major irregularities. When the B sound is heard the ear superimposes the A compensation resulting in an initially incorrect judgement of B's abilities. Given sufficient time for the true assessment of B to become apparent, transfer back to A will give the opposite result. Usually the second or third reversal will result in a reasonably valid judgement of the relative merits of the two systems.

Where 'A' is a live sound, these reversals, although not quite so important, are still worth including. Ideally the comparison should be undertaken in stereo format but in practice the errors involved in stereo recording necessitate the adoption of monural recording. This also applies to the presentation which is best conducted using a single sample of each loudspeaker adjacent to the live sound.

The choice of sound presents some difficulties, an obvious one being the suggestion to use an orchestra. However, single sounds can be effectively employed and these are easy to reproduce consistently and allow the panellists to make quick decisions. The radiation pattern of the source is an important consideration since typical loudspeakers are designed to radiate primarily in the forward plane particularly at higher frequencies, whereas most instruments are more or less omnidirectional. This problem can be overcome to a large degree be providing panel absorbers on the listening room wall behind the loudspeaker and the comparison live sound source.

Useful sounds include those of a percussive or transient nature such as a wood block xylophone, side drum, etc. A large 250 mm plain cymbal is often most revealing of high frequency problems. Acoustic guitar represents a 'mid' centred sound which accurate speakers can mimic very effectively. Though subject to more variation with time and mood, the human voice, particularly male, is a most revealing source spanning 100 Hz to 15 kHz. Distortions result in reduced intelligibility, exaggerated sibilants and the like. The recording technique is important since the accuracy of the replay is vital to the comparisons. The recording location is preferably a large anechoic chamber with a low colouration, flat response microphone employed at a generous distance from the source, typically 1–2 m.

An effective procedure consists of recording the live sounds in the form of short musical or verbal phases lasting 3–5 s interspersed with pauses of slightly longer duration. On test, the tape is run continuously to provide a constant background noise level and the performer 'fills in' the pauses 'live' as they occur, repeating the preceding reproduced phrase. Thus a number of A/B/A/B . . . comparisons are possible with each sound. A total of three to six comparisons are usually sufficient and each sound presentation need only take a minute or so.

Single presentation

Scorned by many authorities on first inspection, in the author's view the single presentation technique provides the best method of subjective appraisal so far. Its accuracy improves with the experience of the panellists and also the quantity of speakers assessed. Up to 70 models may be judged over 12 working days, including the frequent use of control or reference systems. These are randomly introduced throughout the auditioning to assess marking variations. Typically the marks awarded to the control fall within a $\pm 7\%$ error band of its mean value (Figure 9.47).

Panellists are instructed to assess reproduced sound quality against their own experience and, after a short interval, decide a numerical value for the quality reproduced for each programme excerpt auditioned. Having chosen a figure from 0 to 10 the panellist is the asked to write down on a simple form that number, plus any description or justification he might feel is relevant. These characterizations should preferably follow those given in Tables 9.4–9.7. The panellists must score and comment individually, and are requested not to discuss or make any sign concerning their opinions during the session. Only mild stimulants are allowed during the day.

The test controller has to analyse the data, check deviations, eliminate obvious errors and the like and compile a descriptive character appraisal of the speakers from the panellists' score sheets. The speakers are of course auditioned sight unseen, behind a curtain.

Toole [50] has devised a valuable guide to subjective characterization which is represented below.

Definitions: spatial quality

a) *Definition of the sound images* — Refers to the extent that different sources of sound are spatially separated and positionally defined. Images should not move as the pitch of

RATING OF SOUND QUALITY

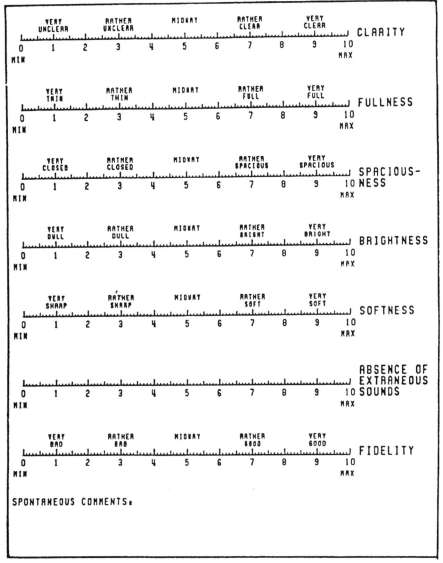

Figure 9.47. Example of a comprehensive test sheet for subjective responses, translated from Swedish. For extraneous sounds include distortion effects, rattles, etc. (after Gabriellson [52])

the music rises and falls. The size of the image should be appropriate to the source of the sound.

b) *Continuity of the sound stage* — Is the display of sound images continuous, left to right, or are there illogical groupings of images, with large gaps in between? Is the reverberation uniformly displayed or is it concentrated in strange places?

Table 9.4. Scale of marking for subjective accuracy

Scale	Characterization
10	Perfect reproduction, completely lifelike
9	Excellent fidelity
8	Very good
7	Good
6	Average Plus
5	Fair or average quality
4	Average minus
3	Acceptable
2	Poor
1	Very poor reproduction
0	No resemblance to or connection with live sounds

Table 9.5. Classification of subjective frequency balance

	Frequency balance		
Range	Region	Deficiency or excess	Degree
Bass	low	+ or −	slight
	mid		moderate
	upper		severe
Mid	low	+ or −	slight
	mid		moderate
	upper		severe
Treble	low	+ or −	slight
	mid		moderate
	extreme		severe

Table 9.6. Colouration

Characterization	Approximate frequency range
Boomy	50–80 Hz
Chesty, plummy	100–150 Hz
Boxy, hollow	150–300 Hz
Tubelike, tunnelly	400-600 Hz
Cuplike, honky	700–1.2 kHz
Nasal, hard	1.8–2.5 kHz
Presence, upper hardness, wiry	2.5–5.0 kHz
Sharp, metallic, sibilant	5.0–8.0 kHz
Fizzy, gritty	10–15 kHz

Table 9.7.

(a) **Stereo assessment**
Image sharpness or focus
Image width
Image depth
Perceived recorded ambience

(b) **Definition and clarity**
Treble clarity
Mid clarity/transparency
Bass clarity or definition

c) *Width of the sound stage* — Refers to the left–right display of sound images. The response scale represents the one in front of you in this room. Mark on it the left and right limits or boundaries of the sounds you hear. Do not include vague reverberant sounds, only those of the orchestra.

d) *Impression of distance of depth* — Should be judged on the basis of a satisfactory impression of instruments at various distances. An unsatisfactory reproduction would have all of the instruments at one distance (two-dimensional), or some of them too close or too far, and so on.

e) *Abnormal effects* — Refer to spatial sensations that do not occur in common experience. For example, it is possible for some sounds to appear to stretch between you and the screen, perhaps even some of the sounds will appear inside your head. Other sounds may appear to have no location, when you know the instrument should be precisely localized.

f) *Perspective* — Refers to your general impressions of the experience. A good reproduction of a good recording with natural room or hall acoustics should suggest that 'you are there' at the performance, complete with a sense of the enveloping ambient sound. A less perfect reproduction could separate you from the performance, giving the impression that you are 'close, but still looking on'. In a still worse reproduction it may seem that you are listening through an opening between the loudspeakers. It is as though you were 'outside looking in' — there is no impression of being within the ambient sound. Other recordings may appear to transport the musicians to the listening room, 'they are here'. The ambience is that of the listening room, and the instruments sound close. Still other recordings are created as abstract special effects, with no attempt to simulate a realistic experience.

Definitions: sound quality

a) *Clarity/definition* — Refers to the ability to hear and distinguish different instruments and voices within complex orchestrations. The individual notes should also be distinguishable, with well-defined attacks, not diffuse or muddled.

b) *Softness* — Refers to the quality of high frequency sounds. These should be smoothly natural, neither overly subdued and mild nor excessively hard, shrill, strident or sharp.

c) *Fullness* — Refers to the quantity of low frequency sounds and their balance with respect to the middle and high frequency sounds. Good sound should be neither too full nor too thin.

d) *Brightness* — Refers to the balance of the high frequency sounds with respect to the middle and low frequency sounds. Good sound should be neither too bright nor too dull.

e) *Pleasantness* — Is an overall rating that concentrates on the pleasantness or lack of aggravations and annoyances in the reproduced sound.

f) *Fidelity* — Is the overall rating that describes how closely the reproduced sound approaches your impression or recollection of the original or 'perfect' sound. This is the one rating that sums up the previous analytical sound-quality ratings. You *must* give a fidelity rating, it is the single-number indication of your opinion. Please report this score as a number (use one decimal if you wish) in the box provided. The number 10 represents perfection. A telephone might score between 0 and 1, and a small portable radio might score 2 or 3.

Scaling

An 11-point scale has given good results, where 0 and 10 represent the respective end points of extreme inaccuracy and excellence, with the numbers in between covering the marking range. The important characterizations can be variously termed as 'accuracy', 'neutrality', 'true to life' or 'true to nature'.

Other relevant characterization of the perceived quality of high performance systems may also be employed, including such terms as 'clarity', 'transparency' (the rendition of detail), 'frequency balance' (describing deficiencies of excesses of level in various parts of the frequency range) and 'colouration', with relevant comments drawn from an approved range of adjectives. As before these may be scaled from 0 to 10, and additional written comments can also be of value.

The Results and their Analysis

The reliability of the conclusions is proportional to the number of results. The IEC document suggests that two or more repetitions of each test combination is essential.

Various approaches may be adopted to the statistical analysis of the data: arithmetic mean, standard deviation and variance, to name but a few. Factors such as the test reliability, dependency on particular system/programme/panellist combinations and fatigue should all be taken into consideration. For example, in the unlikely event of no speaker or programme/speaker dependency, the analysis may be simplified, with averaging becoming the basic technique employed.

Psychological Factors

Before leaving the subject of subjective testing, it is necessary to make a few brief comments concerning its psychological and otological aspects.

The listener or subject is sensitive to many qualities in sound reproduction to a greater or lesser degree, such as non-linear and other distortions, uniformity of frequency response, transient response, colouration and delayed resonances, phase, loudness, reverberation, stereo image and the related depth and position perception. The most favourable or strongest subject reaction is understandably produced by a wide uniform frequency response, a suitable loudness appropriate to the listening environment and the programme, realistic imaging, minimal transient or delayed resonance colourations and a correct reverberation time.

While this may be classed as a basic response, it can be complicated by emotion. A particular fondness for an instrument on the part of the listener, or a certain section of programme, may induce a response out of proportion to that expected on an objective basis, and an error may result in either direction, depending upon the perception of either a pleasing or a disappointing effect. Likewise a subject's response will also depend on such factors as a recent or continuing illness; a cold, for example; his or her mood that day and whether a test occurs at the beginning or end of a session. Listening fatigue is a particularly important factor in this context. When differences are small it may be necessary to account for subject interaction. In one test on amplifier sound quality, such interaction in a listening panel was found to dominate the results [36]. However, despite these many problems, subjective testing, if planned carefully, can be reliably executed and is essential to the advancement of loudspeaker quality. In the author's experience, it is only those models which have undergone successful and exhaustive critical audition prior to release that truly merit the classification 'high performance'.

AES Recommended Practice for Professional Audio — Subjective Evaluation of Speakers AES20 — 1996 (This document is subject to review and represents a consensus of those substantially involved with its scope and provisions)

Much work has gone into this 19 page document with practical and commonsense contributions from major authorities in the field. The coverage is broad and ranges from listening practice to score sheets, evaluation and recommendations for room acoustics, the use of an anchor or control speaker, and speaker placement. Some of the more subtle sound quality aspects of current review practice are absent, but the overall coverage is of great value.

REFERENCES

[1]   Allinson, R. F. and Berkovitz, R., 'The sound field in home listening rooms', *J. Audio Engng Soc.*, **20**, No. 6 (1972)

[2]   Moller, H., *Relevant Loudspeaker Tests*, Bruel and Kjaer Application Notes, 15–067

[3]   Rosenberg, U., *Loudspeaker Measurement And Consumer Information*, statens Provningsanstalt/Rapport (1973)

[4]   Philips Electrical Ltd., *Measuring In The Living Room*, No. 5697E

[5]   Hughes, F. M. (a pseudonym for M. Colloms), 'A group test of thirty pairs of commercial loudspeakers', *Hi Fi For Pleasure*, 4, July, September and October (1976)

[6]   Harwood, H. D., 'New BBC monitoring loudspeaker', *Wireless World*, March, April and May (1968)

[7]   Shorter, D. E. L., 'A survey of performance criteria and design considerations for high-quality monitoring loudspeakers', *Proc. Instn. elect. Engrs.*, 105, p.B., November (1958)

[8]   Fryer, P. A., 'Intermodulation distortion listening tests', *Proc. A.E.S. 50th Convention.* London, March (1975)

[9]   Fryer, P. A., 'Distortions, can we hear them?' *Hi Fi News*, 22, No. 7 (1977)

[10] Harwood, H. D., 'Some aspects of loudspeaker quality', *Wireless World*, May (1976)

[11] Heyser, R. C., 'Geometrical considerations of subjective audio', *J. Audio Engng Soc.*, 22, No. 9 (1974)

[13] MacKenzie, A., *Hi Fi Choice, Loudspeakers*, consumer report, Aquarius Books, London (1976)

[14] Moir, J., 'Doppler distortion in loudspeakers', *Hi Fi News*, p. 817 et seq.

[15] Moir, J., 'Doppler distortion in loudspeakers', *Wireless World*, p. 65 et seq. April (1974)

[16] Stott and Axon, 'The subjective discrimination of pitch and amplitude changes', *Proc Instn. elect. Engrs.*, 102, Pt. B, No. 4 (1949)

[17] Klipsch, P., 'Modulation distortion in loudspeakers', *J. Audio Engng Soc.*, 18, No. 1 (1970)

[18] Bang and Olufsen, Phase Linearity, Publicity note

[19] Bauer, B., 'Audibility of phase distortion', *Wireless World*, 27–28 March (1974)

[20] Hansen, V. and Madson, E. R., 'On aural phase detection', *J. Audio Engng Soc.*, 22, No. 1 (1974)

[21] Harwood, H. D., 'Audibility of phase effects in loudspeakers', *Wireless World*, January (1976)

[22] Heyser, R. C., 'Loudspeaker phase characteristics and time delay distortion', *J. Audio Engng Soc.*, 17, Nos. 1 & 2 (1969)

[23] Ishii, S. and Takahashi, K., 'Design of a linear phase multi-way system', *Proc. A.E.S. 52nd Convention*, October–November (1975)

[24] Klipsch, P., 'Delay effects in loudspeakers', *J. Audio Engng Soc.*, 20, No. 8 (1972)

[25] Moller, H., *Loudspeaker Phase Measurements, Transient Response And Audible Quality*, Bruel and Kjaer Application Notes 15–090

[26] Preis, D., 'Linear Distortion', *J. Audio Engng Soc.*, 24, No. 5 (1976)

[27] Staffedlt, S. H., 'Correlation between subjective and objective data for quality loudspeakers', *J. Audio Engng Soc.*, 22, No. 6 (1974)

[28] Cooke, R. E., 'Misleading measurements', *Hi Fi News*, October (1976)

[29] Moller, H. and Thomson, C., *Electroacoustic Free-field Measurements In Ordinary Rooms, Using Gated Techniques.* (Braul and Kjaer Application Notes, 15–107)

[30] KEF Electronics Ltd., *Loudspeakers Testing Using Digital Techniques*, March (1975)

[31] Moller, H., *Electroacoustic Measurements*, Bruel and Kjaer Application Notes (16–035)

[32] Keele, D. B., 'Low frequency loudspeaker assessment by nearfield sound pressure measurement', *J. Audio Engng Soc.*, 22, No. 3 (1974)

[33] Berman, J. M. and Fincham, L. R., 'The application of digital techniques to the measurement of loudspeakers', *J. Audio Engng Soc.*, 25, No. 6 (1977)

[34] Suzuki, T., Morii, T. and Matsumara, S., 'Three-dimensional display for demonstrating transient characteristics of loudspeakers', *J. Audio Engng Soc.*, 26, Nos. 7 and 8 (1978)

[35] Colloms, M., *Hi Fi Choice Loudspeakers II*, Bunch Books, London (1978)

[36] Colloms, M., 'The panel game', *Hi-fi News*, No. 11 (1978)

[37] Martikainen, I., Varla, A. and Otala, M., 'Input current requirements of high quality loudspeaker systems', AES 73rd conv. P. Print 1987

[38] Allison, R. and Villchur, E., 'On the magnitude and audibility of FM distortion in loudspeakers', *JAES*, 30, No. 10 (1982)

[39] Fincham, L. R., 'The subjective importance of uniform group delay at low frequencies', *A.E.S. 74th convention Pre-print* (1983)

[40] Ishii, S. and Mizatani, T., 'A new type of listening room and its characteristics — a proposal for a standard listening room', *A.E.S. 72nd Convention* (1982)

[41] Bunton, J. D. and Small, R. H., 'Cumulative spectra tonebursts and apodization', *J. Audio Engng Soc.*, **30**, No. 6 (1982)

[42] Janse, C. P. and Kaizer, A., 'Time-frequency distributions of loudspeakers, the application of the Wigner distribution', *J. Audio Engng Soc.*, **31**, No. 4 (1983)

[43] Gander, M. R., 'Ground plane acoustic measurement of loudspeaker systems', *J. Audio Engng Soc.*, Vol. 30, No. 10, Oct '82

[44] Colloms, M., 'Three British loudspeakers', *Hi Fi News*, **28**, No. 6 (1983)

[45] Colloms, M., *Computer Controlled Testing and Instrumentation — An Introduction to the IEEE 488/IEC 625 Bus*, Pentech Press, London (1983)

[46] Colloms, M., *Hi Fi Choice Loudspeakers*, Bunch Books, London (1984)

[47] Bierning, H. and Pedersen, O. Z., 'System analysis and time delay spectometry', Pt. 1 and Pt. 2 B & K Technical Review, No. 5, 1, 2 (1983)

[48] Olive, S. E. and Toole, F. E., 'Modification of timbre by resonances: perception and measurement', *J. Audio Engng Soc.*, **36**, No. 3 (1988)

[49] Merhaut, J., 'Low-frequency measurement of loudspeakers by the reciprocity method', *J. Audio Engng Soc.* (1982). Also, 'Loudspeakers Anthology', Vols. II

[50] Toole, F. E., 'Loudspeaker sound quality and listener performance', *J. Audio Engng Soc.*, **33**, No. 1/2 (1985)

[51] Tyrer, R., 'Use of TV holography (ESPI) for loudspeaker chassis (diaphragm) and cabinet modal analysis', *J. Audio Engng Soc.*, **36**, No. 5 (1988)

[52] Gabriellson, A., 'Planning of listening tests', *Perception of Reproduced Sound*, Aarhus Tech. Eng. College, Denmark (1987)

BIBLIOGRAPHY

AES, *Proc. Int. Conf. The Sound of Audio*, 3–6 May (1990)

AES, Test and Measurement Conference, Portland, 29–31 May (1992)

AES Recommended Practice for Loudspeaker Component Measurement, AES2–1984 (ANSI S4. 26–1984), PP4–9, *J. Audio Engng Soc.*, **32**, No. 10 (1984)

Ashley, R. and Swann, M. D., 'Experimental determination of low frequency loudspeaker parameters', *J. Audio Engng Soc.*, **17**, No. 5 (1969) 525

Baekgaard, E., 'Loudspeakers, the missing link', *Proc A.E.S. 50th Convention*, London, March (1975)

Beranek, L., *Acoustics*, McGraw-Hill, London (1954)

Bücklein, R., 'Audibility of frequency response irregularities, *J. Audio Engng Soc.*, Vol. 29, No. 3, March '81

Celestion, *Celestion's Guide to Living with Loudspeakers*, Celestion International (1983)

Chappelle, P. H., 'The frequency response of loudspeakers on axis or power response', *Proc. A.E.S. 44th Convention*, February (1973)

Christophorau, J., 'Loudspeaker measurement with an accelerometer', *J. Audio Engng Soc.*, **28**, No. 11 (1980)

Corrington, M. S. 'Correlation of transient measurements on loudspeakers with listening tests, *Loudspeaker Anthology*, *J. Audio Engng Soc.*, 1–25 (1979)

Dobbins, P., 'Loudspeaker measurements simplified', *Wireless World*, Vol. 89, No. 1571, August '83

Disler, H., *Psychological Measurement of Acoustic Quality of Sound Reproducing Systems By Means of Factor Analysis*, University of Stockholm, No. 188 (1965)

Fielder, L. D. and Benjamin, E. M., 'Subwoofer performance for accurate reproduction of music', *J. Audio Engng Soc.*, **36**, No. 6 (1988)

Fincham, L. R., 'Refinements in the impulse testing of loudspeakers', *A.E.S. 74th Convention Pre-print* (1983)

Gander, M. R., 'Ground plane acoustic measurement of loudspeaker systems', *J. Audio Engng Soc.*, Vol. 29, No. 3, March '81

Harwood, H. D., 'Loudspeaker distortion associated with low frequency signals', *J. Audio Engng Soc.*, **20**, No. 9 (1972)

Heyser, R. C., 'Determination of loudspeaker signal arrival times', Pt. 1, *J. Audio Engng Soc.*, **19**, No. 9 (1971); Pt. 2, *J. Audio Engng Soc.*, **19**, No. 10 (1971)

IEC Draft Proposals for Listening Tests, Ref. SC29B/WG9; IEC, Technical Committee No. 12, Sub-committee No. 12A, Radio Recording Equipment Draft—Information Guide For Subjective Listening Tests (1975)

Jacobsen, O., 'Measuring loudspeaker constants by a transient method', *J. Audio Engng Soc.*, Vol. 30, No. 3, March '82

Jacabs, J. E. and Lee, C. J., 'Evaluation of acoustic systems utilising correlation techniques', *J. Audio Engng Soc.*, **26**, No. 5 (1978)

Jordan, E. J., *Loudspeakers*, Focal Press, London (1963)

Klipsch, P. W., 'A note on loudspeaker impedance and its effect on amplifier distortion', *J. Audio Engng Soc.*, **26**, Nos. 7 and 8 (1978)

Leclerc, J., 'Time and frequency impulse responses of loudspeakers to windowed tone bursts', *A.E.S. 73rd Convention Pre-print* (1969)

Linkwitz, S., 'Shaped toneburst testing', *J. Audio Engng Soc.*, **28**, No. 22 (1980)

Lipshitz, S. P., Pocock, M. and Vanderkooy, J., 'On the audibility of midrange phase distortion in audio systems', *J. Audio Engng Soc.*, **30**, No. 9 (1982)

Mantel, J., 'Definitions and measurement of fidelity and fidelity index of electroacoustical components and chains', *Proc. A.E.S. 44th Convention*, February (1973)

Olive, S. E. and Toole, F. E., 'The detection of reflections in typical rooms', *J. Audio Engng Soc.*, **37**, No. 7/8 (1989)

Olsen, H. F., *Modern Sound Reproduction*, Van Nostrand, New York (1972)

Preis, D., 'Phase distortion and phase equalisation—a tutorial overview', *J. Audio Engng Soc.*, **30**, No. 11 (1982)

Rife, D. D. and Vanderkooy, J., 'Transfer-function measurement with maximum-length sequences', *J. Audio Engng Soc.*, **37**, No. 6 (1989)

Salmi, J., 'A new psychacoustically more correct way of measuring loudspeaker frequency responses', *A.E.S. 73rd Convention Pre-print*, No. 963 (1983)

Salmi, J. and Weckström, A., 'Listening room influence on loudspeaker sound quality and ways of minimising it', *A.E.S. 71st Convention* (1982) Preprint 1871

Shorter, D. E. L., 'Loudspeaker transient response: Its measurements and graphical representation, BBC *Quarterly Review*, **1**, No. 3 (1946)

Small, R. H., 'Simplified loudspeaker measurement at low frequencies', *J. Audio Engng Soc.*, **20**, No. 1 (1972)

Schroeder, M. R., 'Models of hearing', *Proc. Instn. elect. Engrs.*, **63**, No. 9, p. 1332 et seq. September (1975)

Toole, F. E., 'Listening tests, turning opinion into fact', *J. Audio Engng Soc.*, **30**, No. 6 (1982)

Toole, F. E., 'Loudspeaker measurements and their relationship to listener preferences', *J. Audio Engng Soc.*, Pt. 1, **34**, No. 4 (1986); *J. Audio Engng Soc.*, Part II, **34**, No. 5 (1986)

Villchur, E., 'A method of testing loudspeakers with random noise input', *J. Audio Engng Soc.*, **10**, No. 4 (1962)

West, W. and MacMillan, D., 'The design of a loudspeaker', *Proc. Instn elect. Engrs.* (c. 1937)

Appendix A

CAD Software

1. **AkAbak** [sic.] by J. W. Panzer and Partner, Steinstrasse 15, D-81667, Munich, Germany. US; Bang-Campbell Ass. PO Box 47, Woods Hole, MA 02543-0047. Tel +49(0) 89 177 75 48, Fax +49(0) 89 688 70 76. Akabak is high quality software on two discs and an extensive manual, namely an electroacoustic simulator. While it does not perform automated design optimizations except in respect of low frequency box alignments, it is a comprehensive platform for electroacoustic design, one where the emphasis is on the understanding of the complex interactive processes involved. As such it is a fine educative tool for both engineers and students of the subject. The concept of a design script is introduced, whose language is largely logical, and which constitutes both a valuable discipline and a reference document for the designs. The simulations include driver placement and enclosure diffraction, up to three local room boundaries and a good range of enclosure types with almost unlimited expansion to more complex types. It can use theoretical and imported driver responses, driver impedance and compensation, crossover networks with driver delay, horn loading, piezo simulation, simple electronic/active filters, comprehensive analysis including the predicted power responses, and there is provision for simple models of non linearity, thermal effects, etc. There is a relatively quick, interactive bass simulator, the whole supported by fine Windows friendly graphical and operating interface.

   Multiple user pricing is economical, starting at \$950, down to \$200 for multiple sets, while those engaged in study will find the relevant price worthwhile. The student version (on student accreditation) is heavily discounted. Enquiries welcomed.

2. **Active Filter Design** Viesca, F. G. (budget). Calculates component values for Butterworth filters in four configurations: high and low pass in second and third orders.

3. **Bass Box** Harris Technologies (budget). A modern Windows™ budget program for LF modelling with fine graphics which includes a volume calculator for complex shapes, in-car compensation, and listening room alternatives.

4. **Box Model** Bullock, R. (budget). Significantly improved in the latest version, this high value programme for low frequency modelling includes useful utilities and

handles a variety of conditions well, including vent velocity and variation with power. (There is also a separate disc, TLBOXMODEL, for transmission line enclosures based on the Bradbury model.) Anti-resonant modes are also characterizeable.

5. **Calsod** Waldman, W. Version 1.2 is the budget type tried, 2.0 is a full price professional version. Calsod is well documented, though the 160 page manual needs to be printed from disc first! It can handle up to seven drivers up to four ways and include real or theoretical driver displacements. The whole response may be modelled including the low frequency region. While the directory handling is less friendly it offers good graphics and a simple text processor. The crossover optimization includes the acoustic transfer function and the equivalent electrical circuit of the drivers.

6. **Driver Evaluation** and **Crossover Design** Koonce, G. R. (budget). The first title covers closed vented and passive radiator enclosure design with emphasis on vented boxes. The second adds basic design calculations for first second- and third-order crossovers. The latter are by-the-book results and do not include driver response or motional impedance. The 'quick look' LF software is helpful and quick.

7. **Filter Designer 1** SpeakEasy (middle). A middle priced crossover optimization program with good interfacing for data input and output. Sets of iterated curves may be displayed against variations of a single component value.

8. **Leap** Audio Teknology Inc. (full). A general purpose loudspeaker enclosure analysis program which includes mutual coupling, multiple drivers and ports, losses etc. A comprehensive, professional package, with auto-optimisation.

9. **LMP** and **LMP, SU** Gonzalez, R. (budget). A high value modelling program which includes the crossover, basic driver response characteristics, interdriver delay, etc. The SU improved version is still budget priced and adds good graphics and friendly user interface. Driver impedance is not supported, optimization is user driven.

10. **Loudspeaker Knitted Max** Maximum Effort Software (budget, inc. manual). This package compares with BoxModel and also features basic crossover network design, including schematic output.

11. **Speak** Clark, D. L. (full). Acoustic simulation which matches others on the standard Thiele–Small parameters but which will also extend the analysis to include significant higher order effects such as organ pipe modes in ducted ports. The capability includes double chamber bandpass enclosures with filtering and equalization. While it can handle complex crossovers, off-axis plots and predict distortion, it does not include the primary frequency response data for the drivers, nor does it optimize.

12. **Speaker Designer** Bonney, S. E. (budget). With a friendly operator interface this convenient calculator outputs listings for closed and vented boxes. It includes a manual, present on the disc. While only vent f_p is given it is a convenient LF design guide.

13. **Term-Pro/Term CAD** Wayne Harris Enterprises. Speaker design package for enclosure and crossover design and includes a massive inventory of driver data. The CAD section allows for enclosures to be designed and drawn including production of blueprints, utilization, saving details and materials reporting.

14. **Two-way Active Crossover Design** Galo, G. (budget). A by-the-book filter calculator for two way systems after Bullock. Enter a chosen capacitor value and the rest is determined. Driver responses are not included.

15. **WIN Speakerz** True Image Audio (middle) (also **Mac Speakerze**). A box calculator program which includes basic crossover software plus a facility for box-type generating volume from orthogonal or trapezioid geometries.

16. **XOPT** Schuck, P. L. (full). This powerful iterative optimizer for crossovers supports real driver responses (entered or loaded from an outside file, e.g. MLSSA) and impedances. Offers on- and off-axis responses for systems with up to 5 drivers and inter unit delays. Enclosure and LF factors are not covered.

Old Colony Sound Lab, PO Box 243, Dept. B 90, Peterborough NH 03458, USA.
Note — 'Budget' software is priced up to $100.00; 'full' is in the range $100.00 to $1000.00.

Appendix B

General Design Considerations, Design Tips and Real Examples

Designers work by a set of personal rules, relying on a combination of theory, experience and personal taste. If, by valuing and assessing the comments of others, a designer can recognize known preferences, then he can take account of them in the design process and strive for a more neutral, universally acceptable result. For an older designer it may simply boil down to personal knowledge of some loss of aural sensitivity in the upper treble. If unaccounted for, this could lead such a designer to set higher levels for the tweeter or to miss proper assessment of response irregularities in the last half octave up to 20 kHz. Cross checking using alternative and younger ears is worthwhile, while the designer may also use his experience to take particular technical care with the treble range knowing that his aural judgement may be less satisfactory in this region.

The well-known curves for aural sensitivity and perceived loudness with frequency are widely accepted, but what is less well appreciated is that they represent an average of a wide sample of the population. Variation amongst individuals of good health is large: as much as +5 dB to −10 dB, and each of us has a different natural 'frequency response' to which we are well adjusted. Despite an internal adjustment to what is notionally perceived as a flat response, that intrinsic variation does still affect individual judgements of sound quality. For example, people differ widely on their judgement of loudness—what is realistic and comfortable for one person is painful and unpleasantly loud for another. Likewise, the perceived frequency response of any system, but especially for the loudspeaker, will vary with loudness. It is also the loudspeaker that has rather greater variations than any other component in the listening chain.

We know that if measured with good technique the smoothness of an axial frequency response for a loudspeaker is a fair representation of its behaviour, and is a good indicator of sound quality. But it is still only an indication. Response variations off-axis, the effective energy response and how that interacts with the room, the latter's global acoustic and the effect of immediate boundaries their reflections, as well as many other factors, will all affect the sound quality perceived at the listening position. This is why a skilled designer must use his own aural judgement to determine the tonal balance of a speaker system as it will be used, and not in an arbitrary test location.

457

Even if reasonable tolerances of $\pm 3\,\mathrm{dB}$ are adopted for the overall response variation, for the purposes of specification there remains ample scope for finely balancing the octave-by-octave response to achieve the optimum subjective result.

For example, a $\pm 3\,\mathrm{dB}$ tolerance for the anechoic free space output would even accommodate the difference required for a speaker when wall mounted.

B.1   VOICING A LOUDSPEAKER

Test Sound Level, Check at High and Low Levels

Appreciation of tonal balance or timbre for a speaker system is strongly dependent on sound level owing to the varying frequency response of the average ear at different levels; i.e. the hearing curves. This means that the test sound level is important. When assessing a speaker it is also important not to listen too long to one example, due to the powerful ear/brain interface which will unconsciously adapt itself to the system errors. This adaptation may be so powerful that the otherwise known superiority of a reliable reference speaker may be removed subjectively when switching back to it from the device under test. For example, a speaker with spectral balance flaws may begin to sound reasonable, even plausible, when played loudly enough for the processing in the brain to flatten out the high spots given that the sound level is sufficient to bring the duller depressed regions into aural focus.

It is often very useful to play a speaker at quite low sound levels. It is here that the common flaw of an emphasized upper mid range is exposed, resulting in an unnaturally thin sound, like a small transistor radio. Deep bass will not be heard at low sound levels due to the hearing curve, but accurately recorded speech should sound fairly natural and lifelike.

Lower level listening may also show whether a speaker is lacking in clarity or is weak in reproducing fine detail. A well-balanced, well-integrated system will retain good clarity at low levels, just 50–60 dB at 1 m (very quiet conversation level). Conversely, systems with an uneven join between drive unit ranges and which are poorly balanced may sound surprisingly defective when not driven hard.

Test Location

Ideally the best tonal balance and lowest colouration will be present when the speaker is positioned away from room boundaries, even in free space. In fact some designers swear by free-field testing, listening to domestic speakers in an anechoic chamber or in a real field, up on an open platform.

In practice this makes little sense, since a real room has a profound effect on the sound quality of a given speaker. Wherever possible evaluation should be undertaken in the best possible room conditions, with a reliable representative acoustic.

B.2   DESIGN TIPS

Speaker system design can be a long and tortuous process so here are a few tips which may help speed up the process. These are not a short cut to understanding the process as a whole.

1. Try to associate design changes with sound quality on a constant basis. Freely mix measurement and listening, using music programme of known and familiar quality.

2. Beware of voicing a speaker and/or comparing with a single loudspeaker system (mono). While it is often the starting point in the process, a given design can sound very different when a stereo pair are up and running, properly located with respect to the local room boundaries.

3. It is almost impossible to compare two sets of loudspeakers at a time due to their inevitable placement differences and the acoustic interaction between them. One professional devised a pair of large turntables on which four systems could be placed and successively rotated into position. Loss in quality due to structural weakness and the proximity of the other systems would rule out such a device for critical listening.

4. A CD player equipped with a custom recorded CDR is most useful. This can provide quick access to a number of personally selected representative tracks including one-third octave and pink noise test signals.

5. When designing and trying out crossovers on line with music replay, remote the crossover(s) to the workbench, make sure the cables do not drag them off, and choose a medium power amplifier with fail safe output protection (probably a Japanese type with relay shut down and auto reset). This will save a lot of time with blown amplifiers and or fuses.

6. Have a small test jig handy for the rapid and frequent checking of load impedance. It is all too easy to follow a path to subjective and measured response flatness and find that the input impedance of the system has fallen below tolerable levels.

7. Many designers use double-ended hermaphroditic 4 mm plugs for crossover work using a 4 mm equipped patch panel. The trial components are soldered to the plugs and may be stacked in almost any combination.

8. While primary crossover design may be done on a remote cable connected board, the final balancing and tuning has to be done with the crossover built and placed exactly as it would be for system manufacture. Factors of local acoustic and vibration environment, cables, connectors, specific components and crossover layout can all affect the sound quality, even if little change appears on the measured axial response.

9. Do not be fooled into thinking that speaker system design is not a critical process just because the measuring microphone reads significant variations in response, much larger than, for example, seen in audio electronics. When voicing a speaker 0.5 dB or 5% variation matters, particularly if it is present over a wide range — an octave or more. Quite small changes in an attenuating resistor may not show on a curve, but are easily heard.

10. When assessing frequency response, check the speaker's output from many angles. If there are variations present, are they restricted to a local region and do they balance around the 'zero' or normal sensitivity level on the frequency response? If

not, they may need attention. Variations within a one-sixth octave band are subjectively less important than those one-third octave and wider.

It is helpful to use instrument or visual weighting to average the measured responses on one-third octave and full octave bands, these presented with higher amplitude resolution. For example, a 50 dB vertical scale might suffice for high resolution, narrow-band measurement, one-twelfth octave or better, while one-third octave weighting suggests 25 dB, and whole octave 12 or 10 dB for the overall vertical scale. The purpose of these scalings is to show errors in broadband output more clearly over the frequency range. Given that 0.5 dB of broadband variation in tweeter sensitivity is clearly audible, you will need a 10 dB vertical scale to see it clearly on octave analysis.

11. While a smooth response is not essential for a good sound it is a great design asset since it allows the intrinsic amplitude/frequency response to be seen more clearly. Well-behaved drive units working in low diffraction enclosures have more consistent responses on- and off-axis, facilitating greater design precision.

12. When assessing frequency response it is valuable to keep a constant track of the individual responses of each driver and crossover section. Rolloff slopes which look right should sum correctly if the crossover phase/delay is right in the crossover region. If a simple inversion of the phase of a driver (usually the tweeter) does not correct matters, then fudging the crossover by increasing the overlap between the drivers will not help matters either. Response errors are likely to occur elsewhere, while an unsuspected lobe will be increasing in power somewhere off-axis, to the detriment of the overall sound quality.

13. On the rare occasion where a crossover design is intractable it is usually better to leave it with the correct individual driver responses than to squeeze the system design to try to produce a more aesthetic axial measured response.

14. In a given configuration control of phase/delay at crossover is available via several electrical methods. Reversal of 180° is given by polarity inversion of one driver, steps of 90° can be achieved by changing the order of the crossover, while more subtle shifts are possible by moving the crossover point for either or both sections. For example, the broad acoustic responses of a pair of drivers might permit a choice of upper crossover anywhere in the range from 2.3 kHz to 4 kHz. At 3 kHz a wavelength is 11.3 cm, while the average tweeter in a typical system is physically delayed by 3 cm on a simple flat baffle. This is about 90° at this wavelength (this ignores phase shift due to the mid unit and its crossover). Shifting the precise crossover point provides a sensitive control of phase thanks to this delay.

15. Beware of measurement errors, particularly in the case of taller systems, the current 'speaker fashion'. It is true that at the usual and convenient 1 m mic position much of the measured response is reasonably far field for normal sized speakers. For a tall speaker there is a problem with the differing measurement distances and relative angles and delays to the mic from the more widely separated drive units.

At 1 m a vertical driver displacement of 0.5 m may account for a differential angle of 25° and a path difference of 11 cm, more than 10% of the mic distance.

Remember the inverse square law, i.e. that relative distance error accounts for a 20% loss of sound pressure alone, some 2 dB at the mic position. The 25° worse case angle will account for a further loss due to directivity. It may be better to measure the complex pressure response (e.g. MLSSA, real and imaginary data points) at exactly 1 m for each driver individually, on their axes, and then mathematically sum the final response for an imaginary far-field listening position several metres away.

16. In view of point 15, if a 2 m measuring distance can be used, measurement accuracy for multi-way systems is significantly improved. Where gated measurements are involved, the greater microphone spacing results in a shorter path to the first boundary reflection, unfortunately reducing the length of the anechoic measurement segment.

17. When measuring a tall loudspeaker in a finite room consider placing it horizontally on a suitable stand (taking care first to balance it well!) to bring all the drivers to a uniform position from the boundary. Check that the proximity of the stand does not unduly perturb the measurement. Off-axis measurements are necessarily more awkward and require creative placement according to the frequency range covered.

18. When developing a speaker crossover using the convenient remote cable technique do not use low grade, higher resistance speaker cable. Over the path length from the design position to the speaker itself, losses may be sufficient to modify the results.

19. Where it is essential to A/B test the axial sound quality of single examples of a loudspeaker system use free space placement, well away from boundaries.

20. Check for proximity of installed crossover inductors with local metal work, magnets etc.

In one example prototype which I had to trouble-shoot, the designer had placed the crossover in close proximity to the magnetic shielding steelwork on the back of a driver. Nearby inductor values shifted more than 15% while further unwanted coupling was occurring between several other inductors and between the steelwork itself. The measured responses were almost within the defined production tolerance but the sound quality had shifted unacceptably from the prototype.

B.3   EXAMPLE OF A TWO-WAY CROSSOVER AT 2.2 kHz THE EFFECT OF HIGH PASS CROSSOVER ALIGNMENT ON SOUND QUALITY

Using CAD the system response was very flat and well proportioned with a $7\,\mu F/$ 0.33 mH two element second-order network to a $5\,\Omega$ HF unit.

However, the system sound quality showed some 'glare' or excess in the 800 Hz to 1.2 kHz range. Looking at the graph, the tweeter rolloff was fine, a clean 12 dB/octave. Trying alternatives, an equally valid 'flat' alignment for the overall system response was found with the 0.33 mH treble shunt inductor reduced to 0.23 mH. The visible effect on the response trace was almost undetectable, but the subjective effect was one of greater ease and aural comfort in the upper mid range, while the 'glare' which had hitherto been unfairly ascribed to the mid-range driver had been removed from the sound. Viewed technically, the crossover point has been moved slightly apart; however, the axial responses continue to sum as desired, while the excess energy input at crossover has been satisfactorily controlled.

B.4   MULTIPLE DRIVER COMBINATIONS, SENSITIVITY AND IMPEDANCE

(a) Take a reference driver of $8\,\Omega$ impedance, $90\,dB$ sensitivity.

(b) Wind it to $4\,\Omega$ with no change in build, *Bl* held constant. $2.83\,V$ delivers $2\,W$ power and $93\,dB$ s.p.l. output.

(c) Take two $8\,\Omega$ drivers, and run in parallel for $4\,\Omega$; for $2\,W$ total input you now get $96\,dB$ s.p.l. (that extra factor of $3\,dB$ is thanks to the doubled 'shove' from two magnet systems).

(d) Wind the reference driver for $16\,\Omega$ and with $2.83\,V$ it will draw an $0.5\,W$ input. The result is $87\,dB$ s.p.l.

(e) Parallel two $16\,\Omega$ drivers and the load is $8\,\Omega$; $1\,W$, $2.83\,V$ gets you $93\,dB$ s.p.l.

(f) Series connect two $4\,\Omega$ drivers and for $1\,W$ $8\,\Omega$ you get $93\,dB$ s.p.l.

In practice the $16\,\Omega$ case is preferred since it is generally $1-1.5\,dB$ louder than the $4\,\Omega$ case due to the better gap utilization efficiency of the $16\,\Omega$ winding.

(g) Four $8\,\Omega$ drivers in series–parallel, $8\,\Omega$ load, $1\,W$ input will give a $96\,dB$ output, since we now have four magnets worth of efficiency boost ($10 \log 4 = 6\,dB$).

(h) Series connect two $8\,\Omega$ drivers of $90\,dB$ each and input $2.83\,V$, drawing $0.5\,W$, and get $90\,dB$ because the $3\,dB$ efficiency gain compensates for the $3\,dB$ drop in input power.

B.5   SYSTEM DESIGN EXAMPLE

Here is an example of what can be done with a pair of real-world drivers for which a crossover is required in the $3\,kHz$ region. The bass unit is a $5\,in$ ($120\,mm$) unit of $87\,dB$ nominal reference sensitivity, the tweeter a $1\,in$ ($25\,mm$) soft dome of $88\,dB$. Plotted from $200\,Hz$, the '$17\,dB$' level in the graph is equivalent to $87\,dB$ s.p.l., while for the bass unit the reference sensitivity at $200\,Hz$ is seen to be $3\,dB$ less at $84\,dB$.

Note that these drivers are mounted in the required enclosure, a 7 litre miniature which will be used in free space on high stands. While the maker's reference sensitivity is $87\,dB$, based on the usual Thiele–Small calculation, this is typically valid for 2ii space and not for a small box system intended for use some distance away from room boundaries. Thus the practical sensitivity, equivalent to 4ii full space, actually is $3\,dB$

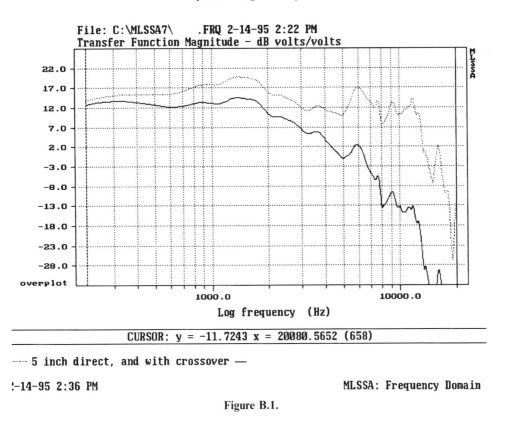

File: C:\MLSSA7\   .FRQ 2-14-95 2:22 PM
Transfer Function Magnitude - dB volts/volts

CURSOR: y = -11.7243 x = 20080.5652 (658)

----- 5 inch direct, and with crossover —

2-14-95 2:36 PM   MLSSA: Frequency Domain

Figure B.1.

less, at 84 dB W.* In Figure B.1 that reference level appears at 200 Hz. The average response trend of the raw bass mid unit rises some 5 dB from 200 Hz to 1.5 kHz, quite typical behaviour (dotted trace). Above 1.5 kHz the response is uneven; ideally it should be maintained to well beyond the crossover point and then decay smoothly. In this example it remains peaky and loud up to 14 kHz; at first sight an unpromising candidate.

From the response for the high frequency unit (Figure B.2) (dotted) the sensitivity is 4 dB too loud over its respectably smooth primary range, 4–15 kHz. Unfortunately, the response peaks by 3–4 dB at 1.5 kHz and a cross check against the impedance curve confirms that this is due to an underdamped fundamental resonance. If the option is available, the system designer could use the addition of Ferrofluid or similar damping to control and suppress this resonance.

Nevertheless, it is still possible to design a fairly straightforward crossover network for this driver pair. To begin, the mid-range rise needs to be equalized to an approximately level response, this mainly achieved with the first inductor of the low-

*Pressure and power are broadly equivalent for low frequency nearfield, approximately below 300 Hz in this case.

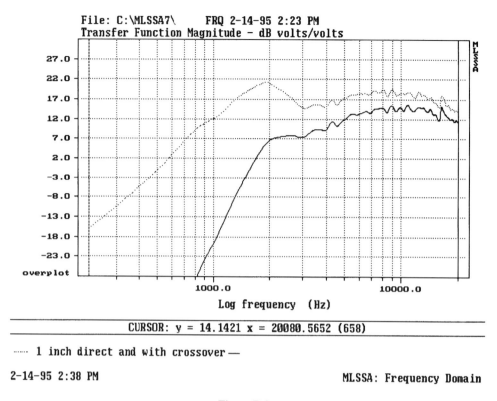

File: C:\MLSSA7\   FRQ 2-14-95 2:23 PM
Transfer Function Magnitude – dB volts/volts

CURSOR: y = 14.1421 x = 20080.5652 (658)

······ 1 inch direct and with crossover —

2-14-95 2:38 PM   MLSSA: Frequency Domain

Figure B.2.

pass crossover filter. In this case a value of 2.1 mH delivers the appropriate correcting slope. Our target is also a −6 dB level at the crossover point of 3 kHz and a maximally flat response can be achieved by resistively damping the shunt capacitor, resulting in a second-order network.

With too low a value of damping the acoustic combination will peak at 1.5 kHz. Thus the values of C and R are balanced to provide the smoothest response, i.e. the required −6 dB point and a good out-of-band rolloff. The high frequency unit response may well be adjusted to provide some reinforcement in the 2 kHz range.

On the treble side, in the absence of driver redesign, first the treble sensitivity must be tamed by 4 dB; a lead resistor is most common, in this case 7 Ω. The final value is arrived at after the crossover is itself complete since the latter can also act as a level control, according to the degree of response equalization required from it.

A second-order network was tried initially but was rejected on two counts.

First, at crossover the phase agreement between the drivers was unsatisfactory leading to an inconsistent response above and below axis. A good check on this is to invert the phase to one unit and see how good a cancellation notch can be achieved at the crossover point.

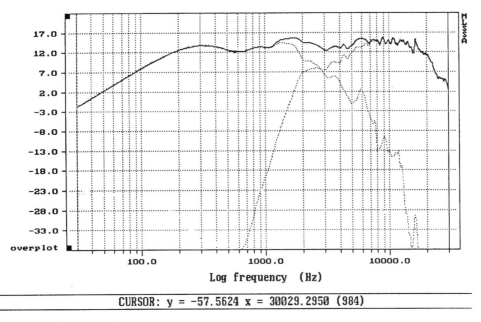

CURSOR: y = -57.5624 x = 30029.2950 (984)

Complete System with crossovers

Figure B.3.

Secondly, going to third order for the high-pass network gives an additional phase rotation or shift of up to 90°, which gives more freedom in integrating the phase responses through the crossover region. The higher rolloff slope of the third-order network also helps to tame the 1.5 kHz peak, although it will not wholly suppress it. Crossed over, the output is nominally −6 dB at 3 kHz, but a proper rolloff slope is not achieved until below 1.5 kHz.

In this design the enclosure is bass reflex loaded, which in this size of box tends to add some extra energy in the low mid range (partly due to port output) in addition to the desired, tuned increase in the bass. Consequently, the system may be run a little brighter in the treble than theory suggests and the final response for the system is seen in Figure B.3. Amplitude limits of ±2.5 dB are met for an 80 Hz to 18 kHz frequency range without the need for smoothing, although it is certainly true that the output is fairly 'strong' from 1.3 kHz to 2.5 kHz; this is a mild but audible feature of the design, lending a touch of 'bite' and 'crispness'.

Good uniformity is shown for the off-axis measurement, confirming the good phase characteristic attained through the crossover range.

This example illustrates how at first sight unpromising drivers can in fact be guided into a satisfactory alignment without great complication. Ferrofluid damping for the tweeter would add that final touch and help to soften output in the 1.5 kHz region.

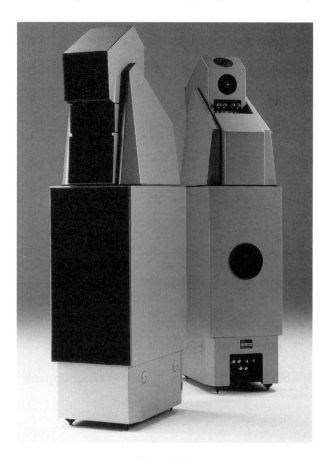

Figure B.4.

Wilson Audio X-1 Grand SLAMM a Seven Driver System

This is an example of the art of high performance loudspeaker system design. While designed mainly for domestic use it stands nearly two metres high, is transported as a set of modules in crates of total weight 1,800 kg. Much of the construction uses heavily cross braced, high density phenolic panels.

Delay-compensated overall, low frequencies are handled by a pair of bass reflexed 380 and 310 mm drivers, offset tuned and sharing a common 140 litre volume tuned to 24 Hz by a high power duct, 155 mm dia. Arranged in a vertical array, two 170 mm polypropylene coned bass-mid drivers flank a 25 mm driver fitted with an inverted titanium dome. Working beyond 7 kHz, two high frequency drivers are located on the rear facets of the enclosure to add diffused, upper octave energy to the ambient field.

Extremely heavy and inert in construction, this design delivers a 20 Hz to 25 kHz ± 3 dB, 500 W power handling, 6 Ω nominal impedance and a high 95 dB/W sensitivity. It stands on massive floor locking steel spikes. Despite obvious logistical difficulties and a very high price, production has been steady at a remarkable 10 pairs per month since its introduction in 1994.

Figure B.5.   Nautilus by B&W Loudspeakers Ltd

Nautilus by B&W, a Low Diffraction, Line Terminated System

An example of loudspeaker engineering as art, this is an advanced four way pure piston system designed for active drive with effective, graded, transmission line loading for each driver. The name derives from the spiral line terminating the 310 mm bass unit, modelled on the Nautilus shell.

B.6   CROSSOVER PHASE UNIFORMITY

A good cross check for crossover performance is to reverse the phase of one of the drivers, measuring on the design axis. The phase cancellation is good when optimally designed and the inversion of one driver should result in a clean, symmetrical notch at the crossover point plus equally good cancellation above and below the target frequency. Notches and bumps in response that appear at unexpected places in the crossover region indicate poor phase control and unwanted overlaps in the operating range of the drivers. If such overlaps are required to shape the overall energy response of the system, then the designer should at least be aware of the consequences for response consistency and uniformity off-axis.

B.7 DO NOT BE TAKEN IN—THE PERILS OF CAD

CAD looks very pretty on paper but does not always work. Hours of interactive CAD can be explored, but when the system is built as a working model there can be a nil result—poor sound and poor correlation with the theory. First of all the modelling must be based on good data, representative of the consistency, character and directivity of the drivers. Baffle size and consequent considerations of diffraction effects, driver spacing and relative driver delays all complicate the issue. It is a help that most well-behaved drivers are, in themselves, minimum phase devices and consequently their phase characteristic may be computed from the frequency response.

Tonal balance is one of the most difficult aspects of speaker design; it is easily heard and yet often remains hidden in the complexity of response measurement and equivalently the theoretical targets indicated for CAD design. For example, the tonal quality, the subjective, perceived frequency response for a bass–mid driver will depend not only on its axial response but also on its directivity function with frequency, how well it integrates with the pass band of the high frequency driver, its inherent colouration, its 'sound' when coupled into the chosen enclosure and, not least, the bass alignment, damping, response shape and extension and any contribution from the port which may arise.

Perhaps the referenced driver responses used for CAD should be an average of the forward 10° solid angle to try to give a more uniform integration envelope for the data. Blips and minor discontinuities are inevitable given the complex interaction of phase, resonances, diffraction and directivity.

B.8 COMPARING PISTONIC AND BENDING WAVE RADIATORS

For a small piston the air load has a reactive characteristic (see Section 2.2, fig. 2.5 for ka less than 2.). In this range power and pressure are equivalent and to generate a flat frequency response this piston requires driving with constant acceleration with frequency.

This range is termed mass-controlled since the current i in the coil results in a force F acting on the moving mass M to produce constant acceleration with frequency. Outside of this range, typically less than 2 octaves e.g. 100 Hz to 300 Hz, the piston radiator will not have a naturally flat response.

By contrast the distributed mode acoustic panel loudspeaker radiates sound from bending waves in the panel, i.e. it is not pistonic. In this case the air load is resistive and is not frequency dependant (see the upper range of Figure 2.5). In addition the summation of all the complex bending waves in the panel defines a resistive mechanical impedance at the driving point. It can be shown that for a low loss bending radiator nearly all the mechanical input power is radiated as sound energy.

The radiation efficiency of the bending wave panel approaches 98% of that for a piston of the same area.

The requirement for a flat power response is for a mechanical power inout which is constant with frequency and this is achieved by a constant velocity drive. The input current i does achieve constant velocity for coil motion due to the resistive mechanical impedance loading provided by the panel for the coil.

For the piston $$F = Ma$$ [Newton's second law]

[Force = Mass × acceleration]

and Force = Bli

[where i is the current, l is the length of coil wire and B is the magnetic flux density].

$$a = Bli/M$$

For a flat frequency response the acceleration must be constant, and is mass-controlled. This provides a piston velocity which falls with frequency to match the 6 dB/octave rising Ω term in the equation for air load reactance.

For the bending wave radiator, the air coupling is in the resistive region, this matching the purely resistive mechanical load seen at the driving point on the panel. Thus the transfer from mechanical to acoustic energy is independent of frequency.

$$F = Rm \ U$$ [Rm is the mechanical resistance of the panel]

[Force = Resistance × Velocity]

and $F = Bli$ for the motor coil

thus $U = Bli/Rm$

In this equivalent circuit a constant current delivers constant velocity and a rising acceleration with frequency; the desired match.

There is no mathematical or acoustic limit to the upper frequency range of a bending wave radiator assuming that constant current is maintained over the desired frequency range.

Index